"Suffering and evil often challenge our understanding of, and love for, the God who created and rules over the world. Focusing on God's work of creation and providence, Highfield is swimming against a stream in evangelical theology that has revised or modified aspects of the traditional understanding of God's nature (simplicity, immutability, eternity and omnipotence). He cogently reaffirms classical theism, and he applies it very helpfully in our present context. Whether or not readers agree with Highfield's criticism of recent departures from the classical tradition, they will benefit, as I have done, from interacting with his superb restatement of the tradition and his own constructive appropriation of it."

Terrance Tiessen, professor emeritus of systematic theology and ethics, Providence Theological Seminary

"In this splendid book, Ron Highfield carefully argues that a proper theological understanding of God's creative action and providential care for all things may afford peace and assurance in this age of anxiety. Solidly grounded in Scripture and masterfully engaging key figures in both classical and contemporary theology, Highfield employs the notions of analogy and secondary causality to preserve God's transcendence and immanence while affirming human freedom and the authentic causality of creatures. The reader is invited to find 'comfort, hope, courage and joy' in the reality of God's providence—in the truth that 'God so orders and directs every event in the history of creation that God's eternal purpose for creation is realized perfectly.'"

Michael J. Dodds, Dominican School of Philosophy and Theology

"*The Faithful Creator* is an extraordinarily rich scholarly treatment of the subjects of divine creation and providence. Highfield situates the doctrine of providence where it belongs—as an outworking of the doctrine of divine creation, which in turn is grounded in the triune nature of God. And the discussion culminates in a wise and extensive treatment of the diverse ways in which the problem of evil threatens to undermine our belief in divine faithfulness. Highfield seems equally adept when discussing biblical, philosophical and historical-theological dimensions of the issue, enabling him to bring into the conversation an impressive range of thinkers—ancient to contemporary. This study demands consideration by those on all sides of the providence debate."

James S. Spiegel, Taylor University

The FAITHFUL CREATOR

AFFIRMING CREATION AND PROVIDENCE IN AN AGE OF ANXIETY

✳

RON HIGHFIELD

IVP Academic

An imprint of InterVarsity Press
Downers Grove, Illinois

InterVarsity Press
P.O. Box 1400, Downers Grove, IL 60515-1426
ivpress.com
email@ivpress.com

InterVarsity Press® is the book-publishing division of InterVarsity Christian Fellowship/USA®, a movement of
students and faculty active on campus at hundreds of universities, colleges and schools of nursing in the United
States of America, and a member movement of the International Fellowship of Evangelical Students. For
information about local and regional activities, visit intervarsity.org.

All Scripture quotations, unless otherwise indicated, are taken from THE HOLY BIBLE, NEW INTERNATIONAL
VERSION®, NIV® Copyright © 1973, 1978, 1984, 2011 by Biblica, Inc.™ Used by permission. All rights reserved
worldwide.

Cover design: Cindy Kiple
Interior design: Beth McGill
Images: Hands of God and Adam, detail from The Creation of Adam, from the Sistine Ceiling (fresco),
    Michelangelo Buonarroti, at Vatican Museums and Galleries, Vatican City/Bridgeman Images
    Cloudy, starry sky: ©imagoRB/iStockphoto

ISBN 978-0-8308-4082-3 (print)
ISBN 978-0-8308-9896-1 (digital)

Printed in the United States of America ∞

Library of Congress Cataloging-in-Publication Data

Highfield, Ron, 1951-
 The faithful creator : affirming creation and providence in an age of anxiety / Ron Highfield.
  pages cm
 Includes bibliographical references and index.
 ISBN 978-0-8308-4082-3 (pbk. : alk. paper)
 1. Creation. 2 Providence and government of God—Christianity. 3. Good and evil—Religious aspects—
Christianity. 4. Trust in God—Christianity. I. Title.
 BT695.H54 2015
 231.7'65--dc23
    2015022989

| P | 23 | 22 | 21 | 20 | 19 | 18 | 17 | 16 | 15 | 14 | 13 | 12 | 11 | 10 | 9 | 8 | 7 | 6 | 5 | 4 | 3 | 2 | 1 |
| Y | 34 | 33 | 32 | 31 | 30 | 29 | 28 | 27 | 26 | 25 | 24 | 23 | 22 | 21 | 20 | 19 | 18 | 17 | 16 | 15 |

CONTENTS

Acknowledgments 9

Abbreviations 11

Introduction 13

PART ONE: CREATION

1 "Let There Be Light": Creation in the Old Testament 25

    The Old Testament Sources 26

    Old Testament Theology of Creation 34

2 "In The Beginning Was the Word": Creation in the
New Testament 40

    New Testament Theology of Creation 40

    Christ and Creation 44

3 "Creator of Heaven and Earth" 56

    One God Beginning and End 57

    The Trinity and Creation 69

    Divine Freedom 72

4 How God Creates the World (Part One):
Probing a Mystery or Solving a Problem? 74

    Creation as Divine Action 75

    The How of Divine Creating 82

    Ancient Theories of Mediation 84

    John Polkinghorne—A Contemporary
Theorist of Mediation 89

5 How God Creates the World (Part Two):
Assessing the Quest 95

    Creation and Causation 95

    Creation and Relationality 99

    The Quest for How—An Assessment 104

    Why God Made the World 108

6 Jesus Christ as Creator   113

    Hellenistic Jewish Background   113

    Two Models for a Theology of Creation   115

    Jesus Christ in Modern Theology of Creation   120

    Christ and Creation   127

7 Creation from Nothing: Creation as an

    Act of Sovereign Generosity   131

    Development of the Doctrine of Creation from Nothing   131

    The Biblical and Theological Meaning

       of *Creatio ex Nihilo*   134

    Divine Ideas and Possible Worlds   135

    Matter—The Third Principle   140

8 Divine Creation and Modern Science   143

    Ancient Science   144

    Early Modern Science   147

    The Competence and Limits of Natural Science   149

    Relating Natural Science and Christian Theology   152

    Big Bang Cosmology and Evolutionary Biology   155

    Gould, Dawkins and Dembski   159

    Why the Controversy?   164

    Augustine or Bellarmine?   168

9 Creation and Time   172

    The Bible and Time   173

    Augustine and Aquinas on Time   174

    Christ and Time   181

PART TWO: DIVINE PROVIDENCE

10 Biblical Theology of Divine Providence   187

    Providence in the Old Testament   187

    Providence in the New Testament   191

    Biblical Theology of Providence   194

    Interpretation of the Bible in the Theology of Providence   196

11 All Things Work Together   209

  Defining Providence   209

  Unfolding the Doctrine of Providence   210

12 Models of Perfect Providence: Foreknowledge   226

  The Foreknowledge Models   228

  The Problem with Molinism (And Other
    Foreknowledge Models)   232

13 Models of Perfect Providence: Omnipotence Models   240

  Thomas Aquinas   241

  Post-Reformation Reformed Orthodox Theologians   242

  Karl Barth   247

14 The Open Theist Model of Providence   257

  Open Theism's Revisions to the Traditional
    Doctrine of Providence   257

  Gregory Boyd on Divine Providence   258

  Critical Observations on the Open Theist Model   265

15 Creation, Providence and Freedom   274

  The Biblical and Theological Idea of Freedom   275

  The Compatibility of Providence and Freedom   294

PART THREE: THE CHALLENGE OF EVIL

16 Creation, Providence and Evil   301

  Dualism—Moral or Metaphysical?   302

  Sin, Error and Corruption   306

  The Fall and Original Sin   308

  How Do We Know Sin and Evil?   316

17 The Philosophical Problems of Evil   322

  Ancient Forms of the Problem   322

  The Modern Problem of Evil   325

  The Logical Problem of Evil   325

  The Evidential Problem of Evil   328

18 The Rhetorical Argument from Evil   332

    Leibniz's Theodicy and Voltaire's Satire   333

    Ivan Karamazov and Protest Atheism   339

    The Karamazov Argument After Dostoevsky   348

    Responses to the Rhetorical Problem of Evil   350

    Responding to the "Rhetorical Argument from Evil"   355

19 "Do Not Be Afraid"   360

    The Faith of Jesus   361

    Faith in Jesus   363

Selected Bibliography   367

Author Index   386

Subject Index   388

Scripture Index   390

ACKNOWLEDGMENTS

I
N WRITING ON A THEME AS BROAD AS creation and providence,
I found myself learning from scores if not hundreds of other writers. I
came to appreciate as never before the communal nature of theological
scholarship. I'd like to take this occasion to thank those who give their lives
to this great cooperative effort, many of whose names appear in the foot-
notes and bibliography of this book. Your labor has not been in vain. I wish
also to thank Pepperdine University for granting me a sabbatical leave for
the spring semester of 2011, which enabled me to write a first draft of this
book. I want to express my profound gratitude to my wife, Martha Ellen
Farrar Highfield, who for thirty-seven years has given unwavering support
for my theological work.

ABBREVIATIONS

ANF   *The Ante-Nicene Fathers*. Edited by Alexander Roberts and James Donaldson. 1885–1887. 10 vols. Repr., Peabody, MA: Hendrickson, 1994.

CD   Karl Barth. *Church Dogmatics*. 4 vols. Translated by Geoffrey W. Bromiley and T. F. Torrance. Edinburgh: T & T Clark, 1936–1969.

NPNF¹   *The Nicene and Post-Nicene Fathers*. Series 1. 14 vols. Edited by Philip Schaff. 1886–1894. Repr., Peabody, MA: Hendrickson, 1994.

NPNF²   *The Nicene and Post-Nicene Fathers*. Series 2. 14 vols. Edited by Philip Schaff. 1886–1894. Repr., Peabody, MA: Hendrickson, 1994.

PRRD   Richard Muller, *Post-Reformation Reformed Dogmatics*. 4 vols. Grand Rapids: Baker Academic, 2003.

SCG   Thomas Aquinas, *Summa Contra Gentiles*.

SumTh   Thomas Aquinas, *Summa Theologica*.

INTRODUCTION

We LIVE IN AN AGE OF ANXIETY, and we always have. Anxiety is built into the structure of human existence. Unlike the sparrow and the lilies of the field, which live only in the moment, we bear the past into the present—nostalgia for lost happiness and regret for missed opportunities. The future spreads out before us like an infinite horizon teeming with possibilities for good or evil. In the present moment, standing always at the crossroads, we have no choice but to act, knowing that we do not know enough to choose wisely. This is the ground of our anxiety: that we can imagine many futures but cannot control which of them comes to be, that we must act even though we cannot fully determine the outcome and that our forced choices may make the difference between ultimate happiness and final perdition.

Jesus also lived in an age of anxiety. To his little band of anxious disciples, he said, "Don't be afraid" (Lk 12:7).[1] Do not worry about your life, whether it will be long or short or how you will sustain it (Mt 6:25, 34). Do not let your hearts be troubled by scorn or threats or rumors of wars. But we are still afraid. And our hearts are often troubled. Anxiety about what may come finds voice in a barrage of questions: Will I find a good job? Will I find love? Will my children turn out all right? Will I have enough money to retire? What if I am injured in an accident? What if I get cancer? If these voices grow quiet still other troubles take their place. We find ourselves disturbed by a misplaced phone, an unexpected repair bill or a lower grade than we expected on an exam. There is no limit to the number of bad things

---

[1] All Scripture quotations are taken from the New International Version.

that can happen, and our imaginations are prodigious at calling them up one after another.

Still, Jesus said, "Don't be afraid." I imagine that some in his audience were thinking to themselves, "Why not be afraid? It's a dangerous world!" To this silent doubt Jesus answers, "You are worth more than many sparrows" and "Your Father has been pleased to give you the kingdom" (Lk 12:7, 32). For Jesus, God's power and faithfulness make fear unnecessary and even unreasonable. In these sayings Jesus draws on a theme deeply rooted in the Old Testament— the faithfulness of God. The Lord is faithful to his covenant with Israel: "Know therefore that the LORD your God is God; he is the faithful God, keeping his covenant of love to a thousand generations of those who love him and keep his commandments" (Deut 7:9). For Paul and the rest of the New Testament, God is faithful to his promises given in Jesus Christ: "He will also keep you firm to the end, so that you will be blameless on the day of our Lord Jesus Christ. God is faithful, who has called you into fellowship with his Son, Jesus Christ our Lord" (1 Cor 1:8-9); and "For no matter how many promises God has made, they are 'Yes' in Christ" (2 Cor 1:20).

To the harassed and persecuted believers in Asia Minor, Peter writes, "So then, those who suffer according to God's will should commit themselves to their faithful Creator and continue to do good" (1 Pet 4:19). This text is the only occasion in the Bible where God is referred to as "faithful Creator." Peter does not draw on the theme of God's covenant faithfulness or even God's faithfulness in raising Jesus Christ from the dead. He invokes God's faithfulness as Creator. The pagan culture demands that these believers compromise or abandon their faith. But Peter urges them to continue to live as disciples of Jesus, give themselves into the Creator's care and place themselves under his protection. Throughout the Bible the act of creating the world is viewed as the quintessential demonstration of God's power. Peter reminds his hearers that the faithful God of the covenant and Father of Jesus Christ is also the Creator of heaven and earth.

I chose Peter's term "faithful Creator" as the title of this book because it combines the ideas of creation and providence in one thought at once profound and encouraging. *The Faithful Creator* can be viewed as a long meditation on 1 Peter 4:19. I ask first about the character of God's act of creating heaven and earth and then seek to understand the biblical faith that the Creator will remain faithful to creation and bring it to its glorious consummation. Even after reflecting on the challenge evil and suffering poses to our confidence in the Creator's power and faithfulness, I find that Peter's advice to place ourselves into the hands of a "faithful Creator" rings as true today as it did when he first counseled it.

This book is also a sustained argument with evangelical open theism, liberal process theism and other models of creation and providence that deny that God accomplishes his good will in all things and, hence, undermine the confidence that Peter commends. This perspective in all its manifestations views God and creation as inhabiting the same causal space, so that where God is active the creature must be passive and where the creature acts God must remain quiescent. It fails to take God's transcendence seriously enough, thinking that God cannot relate to creation unless God possesses certain properties, powers and activities in common with creatures. This commonality of being manifests itself in open and process theism's use of univocal terms to refer to God and creatures, so that the limits of creatures are transferred to the Creator by way of language. In contrast to these perspectives, this book will keep in mind that all our language about God is analogical, that our knowledge of God must be revealed by God and that God's relationship to creation in all its dimensions is always beyond full human comprehension.

However, this book will also attempt to avoid the danger from the other side. Some forms of theological determinism display too much confidence that we know how God accomplishes his perfect will. Certainly any view that denies freedom of will and hence denies human responsibility is impossible to reconcile with biblical faith. Additionally, no particular philosophical or psychological theory of the compatibility of free will and determinism should be granted canonical status. And great care should be shown in speculating on God's use of sin and evil to bring about his will.

Part One: Creation

The book contains three parts. Part one examines the Christian doctrine of divine creation, unfolding it in nine chapters. Chapters one and two survey the Old and New Testament texts and theology of creation, developing five theses on theology of creation in dialogue with contemporary biblical scholarship. The New Testament points to Jesus Christ as the means through whom God created heaven and earth. To explore this theme I examine four

New Testament texts that affirm Christ's role in creation: John 1:1-5, 10; 1 Corinthians 8:4-6; Colossians 1:16; and Hebrews 1:2, 10. Chapter three deals with the identity of the Creator mentioned in Genesis 1:1. Whether one understands God according to the view of process theology or open theism or classical theism makes a huge difference in the way one develops the doctrines of creation and providence. Hence it is necessary to state clearly at the beginning how one views the doctrine of God. This chapter lays out the basic tenets of the traditional or classical view, explaining and defending the traditional interpretation of the central classical divine attributes: simplicity, eternity, immutability, omnipotence and divine freedom. I also include a section on the Triune identity. In this chapter and the book as a whole, I maintain a respectful tone toward those with whom I disagree, for I do not believe that all disagreements in theology vitiate solidarity in the commonly confessed faith.

Chapters four and five take up the issue of the nature of God's act of creating the world. This section clarifies the ordinary meaning of "creating" and carefully differentiates it from divine creating. It contrasts divine creation with Plato's world-formation theory and with John Polkinghorne's theory that God creates and guides the universe in ways understandable in the categories of modern physics. I examine the concepts of causality and relation for their usefulness and limits in describing divine creation, concluding in opposition to the process and open views that the ideas of causality and relation should be used analogously when applied to God's act of creation. Finally, I address the issue of the motivation and purpose of creation. Creation is motivated by divine love and aims to bring creatures into the triune life, bringing glory to God by glorifying creatures.

Chapter six pursues the New Testament teaching that Jesus Christ is the one through whom God creates. This chapter places this teaching into the context of ancient theories of mediation and examines the thought of select church fathers, especially Ireneaus and Origen. The chapter concludes with examinations of the theologies of creation of four contemporary theologians: Karl Barth, Jürgen Moltmann, Wolfhart Pannenberg and Robert Jenson. Each of these theologians roots creation in the trinitarian life of God, while denying that God must create out of natural necessity. Chapter seven examines the meaning, biblical basis and the patristic and modern formulation of the doctrine of creation from nothing. The chapter finds some modern theologies of creation lacking in their understanding of creation from nothing.

Chapter eight explores the relationship between the Christian doctrine of creation and the cosmological theories of modern natural science. It tells the story of ancient cosmology and natural philosophy and the rise of modern science in Galileo and others. With the help of contemporary philosophers of science we get clear on the idea of scientific explanation, which defines what natural science can and cannot do. The chapter then considers and modifies Ian Barbour's model of the proper relationship of science and theology, articulating a hybrid model, which we then apply to big bang cosmology and evolutionary biology. In the final sections the chapter examines proposals from Stephen Jay Gould, Richard Dawkins and William Dembski for the relationship of theology and science. Finally, I contrast Augustine's and Robert Bellarmine's interpretive strategies for relating the Bible to scientific discoveries. Chapter nine addresses the question of the relationship of time to eternity. It examines the theories of Augustine and Thomas Aquinas, and rejects the views of those who argue that God's time is mere everlastingness and argues along with Augustine and Aquinas that time is a mode of being of the creation.

PART TWO: DIVINE PROVIDENCE

Part two develops a doctrine of divine providence that incorporates the view of the God-creature relation developed in part one.[2] The Scriptures express unequivocal confidence that God works in all things to accomplish his will. Nevertheless, theologians differ in their understandings of just what this confidence entails. These differences center on two questions: (1) How can a robust view of divine providence be harmonized with a level of human freedom able to support the height of dignity and the depth of moral responsibility that the Christian faith attributes to human beings? (2) How can a robust view of divine providence be harmonized with the presence of suffering and sin in the world? Much of the doctrine of providence is devoted to these two questions. But a third question also needs to be addressed: How can one win through the natural tendency toward the negative emotions of

---

[2]Charles M. Wood correctly observes that to speak of anything theologically is "to try to grasp it in its God-relatedness" (*The Question of Providence* [Louisville, KY: Westminster John Knox, 2008], p. 15).

anxiety, fear, despair and self-pity to a state consistent with strong faith in divine providence: peace, courage, hope and love?

Chapter ten surveys the direct and indirect biblical teaching about divine providence and draws five conclusions: (1) God the creator of all things continues to sustain his creation in being and to act wisely and powerfully in nature and history to realize his plan. (2) God acts in creation in direct and obvious ways in what are called miracles, and in indirect and subtle ways, such as in natural processes and the free actions of human beings. (3) God acts even in or through evil acts and intentions so that God's plan is realized without God doing evil. (4) God's providence encompasses "all things" (Rom 8:28), so that everything that is and everything that happens is made to serve the divine purpose. (5) For those who love God the preceding affirmations about God's providence impart comfort, hope, courage and joy.

In chapter eleven I lay out my theology of divine providence. I define providence as *that aspect of the God-creation relationship in which God so orders and directs every event in the history of creation that God's eternal purpose for creation is realized perfectly.* This simple definition contains the essential components of the doctrine unfolded in the chapter. (1) Providence is not a totally separate series of divine acts but an aspect of the one God-creature relationship. (2) Providence is God's own personal action, not delegated to angels or left to impersonal causes. (3) God "orders and directs" the history of creation, not leaving creation to chance or fate or misguided freedom. (4) Divine providence covers every event in the history of creation, great and small, good and bad, contingent and necessary. (5) God's eternal purpose guides God's providential work. God does not need to adjust his plan or improvise in response to unexpected events. (6) God realizes his aims perfectly. God cannot fail.

In chapters twelve through fourteen I examine three models of providence. The first two models affirm that God exercises perfect providence in which God controls everything that happens.[3] Chapter twelve considers

---

[3]By "perfect" providence I mean simply that every event in the history of creation happens according to the divine plan. Thomists, Calvinists and Molinists agree here. When I say below that open theists abandon "perfect" providence, I do not mean to denigrate their view. We cannot win arguments with labels. What I call perfect providence they call "meticulous" providence. Meticulous sounds a bit fussy, but it amounts to the same thing. Perhaps open theists would also want to claim the word *perfect*. The reader can judge whether such a claim can be sustained.

foreknowledge models in which God's complete knowledge of the future allows God to plan the course of history down to the last detail. As an excellent example of the foreknowledge model this chapter examines the middle-knowledge model of providence as articulated by William Lane Craig. Chapter thirteen surveys theologies of providence that root God's perfect providence in God's omnipotence. God exercises providence simply by making the world God wishes and guiding it perfectly to its appointed end. As examples of omnipotence models I examine the work of Thomas Aquinas, select post-Reformation Reformed Protestants and Karl Barth. Although I express certain reservations and admit certain unresolved problems, my sympathies lie with this model. It seems to me that it does greater justice to the deity and praiseworthiness of God.

Chapter fourteen deals with the open theist model of divine providence. Unlike the two previous models, this one abandons perfect providence. As my theologian who defends open theism I chose Gregory Boyd. According to Boyd and other openness theologians, God cannot foresee the future perfectly and hence cannot determine in advance how world history will turn out. God possesses perfect knowledge of the present moment only and depends on persuasion in dealing with human beings and angels. In this chapter I explain as objectively as I can Boyd's model and then address what I believe are its defects.

Chapter fifteen addresses one of the major objections to omnipotence models of providence, that is, that God cannot both control the course and outcome of world events and allow human beings to exercise free will. I examine the freedom theme in the New Testament and in the history of theology, giving special attention to Bernard of Clairvaux, Thomas Aquinas, Jonathan Edwards and Paul Helm, who in different ways contend that divine determination and human freedom are compatible. Next I deal with thinkers who deny the compatibility of divine control with human freedom. In conclusion, I argue that of the available perspectives Aquinas's view of the compatibility of human free will and divine sovereignty best preserves the biblical picture of God's relationship to human freedom.

PART THREE: THE CHALLENGE OF EVIL

Chapter sixteen addresses the questions of the nature and origin of evil. I deal first with the challenge to the Christian doctrine of creation posed by

metaphysical dualism and the answers given by such early theologians as
Basil of Caesarea and Augustine of Hippo. To make clear the difference
between the good creation and its corruption in the fall, the next section
distinguishes between fallibility, peccability and corruptibility, which are
inherent weakness of creatures, and error, sin and death, which are not inherent in the nature of creatures. The chapter also addresses the biblical
concepts of the fall and original sin, and it ends with the question of how we
gain theological knowledge of sin and evil.

Chapter seventeen deals with the problem that the occurrence of evil
poses to belief in God. The chapter begins with a historical overview of different formulations of the problem of evil and then deals with two modern
forms of the argument: the logical and the evidential problems of evil. The
section on the logical problem lays out the argument formulated by J. L.
Mackie and the response by Alvin Plantinga. The next section summarizes
the evidential problem of evil as set out by William Rowe and the answer to
Rowe given by William Hasker. Chapter eighteen deals with what I believe
is the greatest challenge to Christian faith arising from the experience of evil.
I am calling it the rhetorical argument from evil, though it is sometimes
called "the emotional" problem of evil.[4] The rhetorical argument begins by
rehearsing in exquisite detail excruciating and nauseating accounts of horrendous cruelty and suffering. It does not move to a conclusion by way of
inference but ends with a challenge, explicit or implicit: how dare you diminish the horror of human suffering by making it a means to greater good
or a higher harmony! In this chapter I examine two powerful rhetorical
arguments: Voltaire's ironic critique of optimism in his *Candide* and Ivan
Karamazov in Dostoevsky's *The Brothers Karamazov*. I also look at the
twentieth-century movement "protest atheism," which developed in the
wake of the two world wars and the Holocaust. I examine Dostoevsky's own
answer to Ivan Karamazov in the character Zosima the Elder. I also consider
the answers given by John Roth and Jürgen Moltmann. In the last section I
develop my own approach to the rhetorical argument from evil, which carefully takes into consideration its rhetorical form.

Chapter nineteen addresses the mood and attitude toward life that arises

---

[4]I prefer the expression "the rhetorical argument from evil" rather than "the emotional problem
of evil" because I am focusing on the use to which stories of horrible suffering are put in argumentation.

out of the doctrine of providence defended in this book. It examines Jesus' teaching about God's care of his children. As a summary of Jesus' teaching I use Matthew 6:25-34, which begins "Therefore I tell you, do not worry about your life. . . ." The chapter also examines Paul's teaching on divine providence in Romans 8:28-30, which begins, "And we know that in all things God works for the good of those who love him. . . ." We can summarize Paul's perspective in two brief statements. (1) The glory of our eschatological destiny will infinitely outweigh any suffering we must endure in this life. (2) God will order everything that happens, no matter what it is, to our good. The first statement allows us to contemplate our lives as a whole—or even the history of the world as a whole—from an eschatological perspective. Even before the end, in the unsettled thick of life's battles, we know the final judgment about suffering. It is "*not worth comparing* with the glory that will be revealed in us" (Rom 8:18). The second statement assures us that even in the moment of trial God will not allow suffering's absurdity to stand. It will be forced to serve God and fit into an order that leads to our glory.

Part One

✳

CREATION

1

"Let There Be Light"

Creation in the Old Testament

LEARNING HOW A THING BEGAN tells you much about how it will end and the course of its journey. In our experience everything begins from nothing and returns to nothing. From dust to dust, sunrise to sunset, in the end everything returns to its beginning. And if our origin really is nothing, our end will be nothing as well and our story a meaningless tale. But the Bible's story does not begin with nothing, and it does not end with nothing. It begins and ends with God. And because God is our beginning and end, our journey will not be meaningless, for God surrounds and enfolds our time in his eternity. God alone is our origin and our creature-relationship to God defines our essence, and this makes the study of divine creation supremely relevant to our existence.

In this chapter I want to introduce the biblical theme of creation in order to provide a foundation for systematic reflection. I don't claim to develop a thorough biblical theology of creation or enter into exegetical debates best dealt with by specialists. Nor will I enter into the critical debates about the tradition and literary history and history of interpretation of the canonical texts. Rather, I will survey the central texts of the Old and New Testaments dealing with creation and place them in the broad context of the history of Israel and the Christian church. This will enable us to summarize with some confidence the basic principles of the biblical view of creation. These principles will serve as guides and limits on our theological critique and construction.

THE OLD TESTAMENT SOURCES

Though references or allusions to God's creation of the world can be found in many parts of the Old Testament, they are concentrated in Genesis 1-2:4a; Genesis 2:4b-25; Job 38–41; Psalms 8; 19; 24; 33; 74; 89; 90; 93; 102; 104; 148; Proverbs 8:22-31; Isaiah 40-45; and Jeremiah 10:12-16.; 27:5. Gerhard von Rad classifies the texts into two classes, the "theologically didactic" and the "hymnic." Only Genesis 1 and Genesis 2 fall into the didactic category, for only these chapters provide "large complexes" of "direct theological statements about the Creation."[1] So it's not surprising that throughout Christian history Genesis 1–2 has been the focus of theological reflection on the theme of creation.

> In the beginning God created the heavens and the earth. Now the earth was formless and empty, darkness was over the surface of the deep, and the Spirit of God was hovering over the waters.

> And God said, "Let there be light," and there was light. God saw that the light was good, and he separated the light from the darkness. God called the light "day," and the darkness he called "night." And there was evening, and there was morning—the first day.

> And God said, "Let there be a vault between the waters to separate water from water." So God made the vault and separated the water under the vault from the water above it. And it was so. God called the vault "sky." And there was evening, and there was morning—the second day.

> And God said, "Let the water under the sky be gathered to one place, and let dry ground appear." And it was so. God called the dry ground "land," and the gathered waters he called "seas." And God saw that it was good.

> Then God said, "Let the land produce vegetation: seed-bearing plants and trees on the land that bear fruit with seed in it, according to their various kinds." And it was so. The land produced vegetation: plants bearing seed according to their kinds and trees bearing fruit with seed in it according to their kinds. And

---

[1]Gerhard von Rad, *Old Testament Theology*, trans. D. M. G. Stalker (New York: Harper & Row, 1962), 1:139-40.

God saw that it was good. And there was evening, and there was morning—the third day.

And God said, "Let there be lights in the vault of the sky to separate the day from the night, and let them serve as signs to mark sacred times, and days and years, and let them be lights in the vault of the sky to give light on the earth." And it was so. God made two great lights—the greater light to govern the day and the lesser light to govern the night. He also made the stars. God set them in the vault of the sky to give light on the earth, to govern the day and the night, and to separate light from darkness. And God saw that it was good. And there was evening, and there was morning—the fourth day.

And God said, "Let the water teem with living creatures, and let birds fly above the earth across the vault of the sky." So God created the great creatures of the sea and every living thing with which the water teems and that moves about in it, according to their kinds, and every winged bird according to its kind. And God saw that it was good. God blessed them and said, "Be fruitful and increase in number and fill the water in the seas, and let the birds increase on the earth." And there was evening, and there was morning—the fifth day.

And God said, "Let the land produce living creatures according to their kinds: the livestock, the creatures that move along the ground, and the wild animals, each according to its kind." And it was so. God made the wild animals according to their kinds, the livestock according to their kinds, and all the creatures that move along the ground according to their kinds. And God saw that it was good.

Then God said, "Let us make mankind in our image, in our likeness, so that they may rule over the fish in the sea and the birds in the sky, over the livestock and all the wild animals, and over all the creatures that move along the ground."

So God created mankind in his own image,
in the image of God he created them;
male and female he created them.

God blessed them and said to them, "Be fruitful and increase in number; fill the earth and subdue it. Rule over the fish in the sea and the birds in the sky and over every living creature that moves on the ground."

Then God said, "I give you every seed-bearing plant on the face of the whole earth and every tree that has fruit with seed in it. They will be yours for food. And to all the beasts of the earth and all the birds in the sky and all the creatures that move along the ground—everything that has the breath of life in it—I give every green plant for food." And it was so.

God saw all that he had made, and it was very good. And there was evening, and there was morning—the sixth day.

Thus the heavens and the earth were completed in all their vast array.

By the seventh day God had finished the work he had been doing; so on the seventh day he rested from all his work. Then God blessed the seventh day and made it holy, because on it he rested from all the work of creating that he had done. (Gen 1:1–2:3)

This classic text begins with God and presents no genealogy of the gods.[2] It refers all things to God as their origin. God created "the heavens and earth," an expression that means "absolutely everything," the whole universe.[3] There is no struggle among the gods, no battle with a chaos monster. Heaven and earth are not made from the body of a god; they are "created." They do not spontaneously flow out of God. Nor does God build them out of preexisting material. Creation is a free and personal act unconstrained by any external law. The narrative structures God's acts of creation in a six-day series with each day seeing the separation or development of a new sphere of creation, ascending toward the crown of creation, human beings. Each day begins with "And God said" and after the second day ends with "and God saw that it was good." God created all things by his word, by the act of speaking. No physical exertion is required nor time expended. For God, creation is as easy as speaking; indeed, creation for God is even easier than speaking is for us. God's declaration of creation's goodness indicates that God accomplishes exactly what God wanted to accomplish. Creation really exists before God.

---

[2]See Nahum M. Sarna, *Understanding Genesis: The Heritage of Biblical Israel* (New York: Schocken, 1970), pp. 9-12.

[3]Gerhard von Rad, *Genesis: A Commentary*, trans. John H. Marks (London: SCM, 1961), p. 47. Hebrew does not have a word, like the Greek word *kosmos*, that means the organized whole. Bernhard W. Anderson, "Creation," in *The Interpreters Dictionary of the Bible*, ed. George A. Buttrick (Nashville: Abingdon, 1962), 1:725-32. Nor does it contain the modern concept of the universe. The ancient Greek concept of *kosmos* also differs from the modern concept of universe.

On the first day God creates light, an indispensable condition for life and knowledge. God separates light and darkness and named the first "day" and the second "night." The second day God finds a world flooded with water uninhabitable by any creature. So God's first act is to create a barrier (the sky) to separate the water above from the water below. You can see already the trajectory of this creation account. God is preparing a place for his creatures, especially his human creatures.[4]

On the third day God separates the land from the water and for the first time involves creation itself in the further production of other creatures. Almost any non-Israelite in the ancient world would have been shocked at what happened on the fourth day. God created the sun, moon and stars in the sky to give light and mark the seasons. Israel's neighbors viewed the sun, moon, and stars as divine beings that rule the earth and humankind. But here they have been deprived of their divine status and have been given the role of servants to human beings. The fifth day sees the creation of fish and birds to populate the seas and the sky. For the first time God blesses creatures and bids them reproduce. The power of reproduction is thus declared a divine gift to creatures rather than a divine power to be worshiped.[5]

The sixth day brings the creation of land creatures and human beings. Only the creation of human beings is preceded by divine deliberation and declaration of intent.[6] Human beings are made in the image and likeness of God and given rule over the rest of creation. They are God's "crowning achievement."[7] Then God rests on the seventh day.

Genesis 2:4b-25 provides a second approach to creation, focusing not on the construction of a grand cosmic cathedral with humanity as its keeper, priest and king, but on humanity and human community.[8] Having made the heavens and the earth, God looks out on a barren and parched land. What is wrong? No rain and no farmer. So God takes some of the dust, makes a human body and breathes life into him. Human beings are taken from the earth and remain connected to the earth. God places the

[4]Von Rad, *Genesis*, p. 54.
[5]See Walter Elwell, ed., *Baker Theological Dictionary of the Bible* (Grand Rapids: Baker Academic, 2001), s.v. "Create, Creation."
[6]See William P. Brown, *The Seven Pillars of Creation: The Bible, Science and the Ecology of Wonder* (New York: Oxford University Press, 2009), pp. 41-44, for a description of the creation of human beings on the sixth day.
[7]Elwell, *Baker Theological Dictionary*, s.v. "Create, Creation."
[8]Claus Westermann, *Creation* (Philadelphia: Fortress, 1974), pp. 66-88.

man in the garden God has planted and puts him in charge. But human beings are not meant to live alone. Plants provide no company, and animals cannot enter into meaningful fellowship. So God makes a woman, a counterpart who is his equal in every quality that makes him human. William Brown points out that the narrative contains a series of lacks or "not-goods made good" by God.[9] Two of the lacks are made good by the creation of man and woman: the lack of a farmer to tend the land and the lack human partner for the man. This creation story highlights the central role humanity plays in relation to the rest of creation, for good or ill. The creation story of Genesis 2:4b-25 flows seamlessly into the story of the fall (Gen 3:1-24).[10] The narrowing of focus to the creation of human beings and their consequent fall provides a fitting prelude to the chapters that follow. These chapters describe the moral and spiritual decline of humanity to the point that God's only alternative is to undo creation in the great flood and begin again (Gen 6:1–8:22).

Job 38–41 falls into von Rad's second category of creation materials, hymnic. At this point in the poem Job has made his accusations of injustice against God and has been rebuked by his friends. In Job 38, God begins to speak: "Who is this that obscures my plans with words without knowledge? . . . Where were you when I laid the earth's foundation? . . . Who shut up the sea behind doors?" (Job 38:2-8). Job is quizzed about the mysterious forces of nature and the habits of a series of animals. The poem devotes a whole chapter (Job 41) to the monstrous Leviathan. He terrorizes the earth and incarnates all that is wild and sublime in nature.[11] These references to creation and all the others in these four chapters are designed to put Job in his place, to show his weakness and ignorance in comparison to the Lord's power and wisdom.[12] There is no intention to teach about the beginning or about the workings of nature as such. Nevertheless, the creative activity of

---

[9]Brown, *Seven Pillars of Creation*, p. 80.

[10]In light of this seamless transition Brevard Childs argues that "it is very unlikely that the J creation account ever had an independent existence apart from its role as an introduction to Chapter 3" (*Biblical Theology of the Old and New Testaments: Theological Reflection on the Christian Bible* [Minneapolis: Fortress, 1993], p. 113).

[11]See Brown, *The Seven Pillars of Creation*, pp. 115-40, for his interpretation of why nature is portrayed as wild and beyond human control.

[12]N. H. Tur-Sinai, *The Book of Job: A New Commentary* (Jerusalem: Kiryath Sepher, 1967), p. 522, and Marvin Pope, *Job: A New Translation with Introduction and Commentary*, Anchor Bible 15 (Garden City, NY: Doubleday, 1973), p. 291.

God is assumed, and God's status as Creator and ruler of nature's forces is taken as instructive of his attributes and identity.

In the Psalms we discover other functions of creation language. Richard J. Clifford finds creation references concentrated in two types of psalm: communal laments and hymns.[13] Psalm 74, for example, laments God's anger that has given Israel over to its enemies. The supplicant urges God to do what he did in past times:

> It was you who split open the sea by your power;
>> you broke the heads of the monster in the waters
> . . . crushed the heads of Leviathan. . . .
> The day is yours, and yours also the night;
>> you established the sun and moon. (Ps 74:13-16; cf. Ps 44; 77; 89)

Creation here is not an independent theme; rather God's powerful victories at the beginning ground the psalmist's appeal for intervention in the present. The thought is that if God can create heaven and earth in the beginning, crushing and taming the great forces that resisted him, how much more can God help and deliver today! As a hymn, Psalm 136 gives thanks for the wonderful works of the Lord:

> Give thanks to the LORD for he is good.
>> *His love endures forever.*
> Give thanks to the God of gods . . .
> who by his understanding made the heavens . . .
> who spread out the earth upon the waters . . .
> who made the great lights . . .
> the sun to govern the day . . .
> the moon and stars to govern the night. (Ps. 136:1-9; cf. 93; 96; 114; 148)

Concerning Psalm 136 and creation, Brevard Childs says, "Ps. 136 has joined praises to God as creator with a full liturgical recitation of God's redemption of Israel in history without any sign of inner friction."[14] These psalms urge the worshiper to praise and thank the Lord for his great deeds of mercy, power and knowledge. Again, the goal of these texts is not to teach about creation but to take what is already well known into the language of praise.

---

[13]Richard J. Clifford, "Creation in the Psalms," in *Creation Accounts in the Ancient Near East and in the Bible*, ed. Richard J. Clifford and John J. Collins, Catholic Biblical Quarterly Monograph Series (Washington DC: Catholic Biblical Association of America, 1994), pp. 57-69.
[14]Childs, *Biblical Theology*, pp. 113-14.

The first chapters of the book of Proverbs extol the blessings wisdom bestows. Those who reject wisdom

> will . . . be filled with the fruit of their schemes.
> For the waywardness of the simple will kill them,
>   and the complacency of fools will destroy them. (Prov 1:31-32)

But

> Discretion will protect you,
>   and understanding will guard you.
> Wisdom will save you from the ways of wicked men. (Prov. 2:11-12)

Wisdom is more precious than gold or silver:

> For those who find me find life
>   and receive favor from the LORD.
> But those who fail to find me harm themselves;
>   all who hate me love death. (Prov 8:35-36)

In Proverbs 8:22-31, wisdom explains the reason she is of such great value:

> The LORD brought me forth as the first of his works,
>   before his deeds of old;
> I was formed long ages ago,
>   at the very beginning, when the world came to be.
> When there were no watery depths, I was given birth,
>   when there were no springs overflowing with water;
> before the mountains were settled in place,
>   before the hills, I was given birth,
> before he made the world or its fields
>   or any of the dust of the earth.
> I was there when he set the heavens in place,
>   when he marked out the horizon on the face of the deep,
> when he established the clouds above
>   and fixed securely the fountains of the deep,
> when he gave the sea its boundary
>   so the waters would not overstep his command,
> and when he marked out the foundations of the earth.
>   Then I was constantly at his side.
> I was filled with delight day after day,
>   rejoicing always in his presence,

rejoicing in his whole world
    and delighting in mankind.

Wisdom does not discuss creation for the sake of instruction but to show her superiority to everything else in creation. According to R. B. Y. Scott, "What is here being affirmed is not that Yahweh is Creator, but that Yahweh's attributes of wisdom 'existed' prior to its expression in his acts of creation."[15] She was "brought forth" or "formed" or "given birth" before creation. Wisdom was present at the beginning, standing next to the Creator and delighting in his presence. Hence when one receives wisdom and follows her lead one comes into contact with the primeval mystery of how things came to be and how they work. One participates in an attribute of the Creator.[16] Wisdom literature witnesses clearly to the divinely given order that structures creation.[17]

The prophetic references to the work of creation present us with another case of the use of creation language. In the second part of Isaiah, where the prophet addresses the defeated and exiled people of Jerusalem, the despondent people are reminded of God's great works of creation and are encouraged to expect future deliverance. The messenger comes with "good news" for Zion (Is 40:9):

The sovereign LORD comes with power,
    and he rules with a mighty arm. (Is 40:10)

Throughout these chapters God's original creation is seamlessly blended with his present control of the forces of nature, which then flows smoothly into his determination to act in history for the salvation of his people.[18] The event of creation extends into the present and encompasses Israel's salvation

---

[15]R. B. Y. Scott, *Proverbs, Ecclesiastes*, Anchor Bible 18 (Garden City, NY: Doubleday, 1965), pp. 71-72.

[16]Scott prefers to take wisdom as a divine attribute despite the mythological references and wisdom's personification (ibid., p. 72).

[17]Childs, *Biblical Theology*, p. 115. Steven Parrish echoes many other scholars when he argues that the wisdom emphasis on the order inherent in creation necessitates modification of the thesis that for the OT creation is merely a prelude to salvation history. "The observation that a divinely established order inheres in the cosmos cautions against subordinating creation to salvation history and calls for a balanced estimation of this relationship" ("Creation," in *Mercer Dictionary of the Bible*, ed. Watson E. Mills [Macon, GA: Mercer University Press, 1991], pp. 181-82).

[18]Claus Westermann gives this section of Is 40 the title "The Creator Is the Savior" (*Isaiah 40–66: A Commentary*, trans. David M. G. Stalker (Philadelphia: Westminster, 1969), p. 46.

"into God's continuing work of creation."[19] Call to mind, the prophet urges, that God "measured the waters in the hollow of his hand" and "with the breadth of his hand marked off the heavens" (Is 40:12). The writer asks his hearers to

> Lift up your eyes and look to the heavens:
>> Who created all these?

and answers,

> He who brings out the starry hosts one by one
>> and calls forth each of them by name.
> Because of his great power and mighty strength,
>> not one of them is missing. (Is 40:26; cf. 42:5; 45:12)

If the exiles were wondering whether God still had power enough to help, this affirmation is a strong assertion of God's power over all other powers, including gods of his people's captors. For, as Claus Westermann points out, the Babylonian gods were astral; Marduk himself was the sun god.[20] Jeremiah uses creation language much to the same end as does Isaiah. In Jeremiah 10, the prophet mocks the gods of the nations for being "worthless" idols made by craftsman. In contrast,

> God made the earth by his power;
>> he founded the world by his wisdom
>> and stretched out the heavens by his understanding. . . .
> He who is the Portion of Jacob is not like these,
>> for he is the Maker of all things. (Jer 10:12-16; cf. 27:5)

OLD TESTAMENT THEOLOGY OF CREATION

From these representative texts and others we can draw several theological theses that articulate the basic aspects of the Old Testament theology of creation and will provide guidelines for further systematic reflection.[21] These

[19]Brown, *Seven Pillars*, p. 198.
[20]Westermann, *Isaiah 40–66*, pp. 57-58.
[21]For a picture of the changing landscape in OT theology of creation from the early to the late twentieth century, see Walter Brueggemann, "The Loss and Recovery of Creation in Old Testament Theology," *Theology Today* 53 (1996): 177-90. Gerhard von Rad's seminal essay of 1936, "The Theological Problem of the Old Testament Doctrine of Creation," set the agenda for studies in the OT theology of creation for the next thirty years (reprinted in *The Problem of the Hexateuch and Other Essays* [New York: McGraw-Hill, 1966], pp. 1-78). Von Rad minimized

theses are theological and hence do not contain everything the Old Testament says about creation. They represent a final, canonical statement of faith rather than a developmental or pluralist picture. For a biblical assertion about creation to qualify as *theological* it must speak about God as Creator or God's relationship to the creature. Statements in which the Bible speaks of creatures in ways subject to empirical observation or introspection must be excluded from the theological category. Job 39:13-18 notes that the ostrich lays its egg on the ground unmindful of the traffic that may crush it and observes that "God did not endow her with wisdom" (Job 39:17).[22] Proverbs 6:6-8 tells the lazy person to take notice of the ant:

> It has no commander,
>> no overseer or ruler,
> yet it stores its provisions in summer
>> and gathers its food at harvest. (Prov 6:7-8)

Psalm 32:9 warns,

> Do not be like the horse or the mule,
>> which have no understanding
> but must be controlled by bit and bridle
>> or they will not come to you.

Even if these statements are true and inspired, they are based on observation and are comparisons meant to teach a moral. They do not intend

---

Israel's interest in creation in view of its focus on God's acts of salvation and judgment in history. Von Rad gained many followers, including the influential American OT scholar G. Ernest Wright. Claus Westermann's 1971 essay "Creation and History" (in *The Gospel and Human Destiny*, ed. Vilmos Vajta [Minneapolis: Augsburg, 1971], pp. 11-38) signaled the end of the dominance of the von Rad thesis. Westermann argues that both creation and history were important in Israel's theology and life. Frank Moore Cross in *Canaanite Myth and Hebrew Epic* (Cambridge, MA: Harvard University Press, 1973) argued that there is no huge distinction between Israelite and Canaanite understandings of the divine activity in nature. Other scholars, such as Hans Heinrich Schmid, Walther Zimmerli, the older Gerhard von Rad, Bernhard Anderson, Walter Harrelson, Samuel Terrien, Rolf P. Knierim, Jon D. Levenson and Terence E. Fretheim, continued the move away from the early von Rad thesis. Indeed, we find in Levenson and Fretheim a complete reversal of the early von Rad theology of creation with history being replaced by creation as the "universal horizon" of Israel's thought (Brueggemann, "Loss and Recovery," p. 185).

[22]Other qualifications may be needed. The statement that God did not give the ostrich wisdom, despite its form, is not a theological statement any more than the statement that God did not have Ron Highfield born in 1900 or God did not make it snow in Malibu, California, on July 4, 2011, are theological. Just being true does not make it theological.

to teach about the habits and characteristics of creatures.[23]

Thesis one: *The one God is the absolute origin and sovereign ruler over all that is not God.* The Old Testament does not offer a genealogy of God. God is already there at the beginning, and every other thing in existence originated in God's creative will. In Walther Eichrodt's words, "By their unqualified rejection of theogony the Old Testament affirmations establish the unconditional dependence of the world on God."[24] Because God is the absolute origin of all creatures, of all that is not God, he is their sovereign ruler by right and by power.[25] All other rulers and powers derive their right and their power from God. No other power commands a sphere of independent action.[26] The manifold world of creatures finds its most basic unity in "the completely personal will of Jahweh their creator."[27]

Thesis two: *The one God freely established the Creator-creature relation, which is characterized by generosity, freedom and power on the Creator's side and dependence and debt on the creature's side.* God creates freely without inner compulsion or external coercion or motivation. God's act of creation is "effortless," deriving only from the divine word and will.[28]

---

[23]We will deal at greater length with the relationship between science and theology in a later section.

[24]Walther Eichrodt, *Theology of the Old Testament*, trans. J. A. Baker (Philadelphia: Westminster, 1967), 2:99. And Pieter Smulders asserts that according to the Old Testament "the whole world owes its being entirely to the free, sovereign action of God" ("Creation," in *Encyclopedia of Theology: The Concise Sacramentum Mundi*, ed. Karl Rahner [New York: Crossroad, 1975], p. 314).

[25]Werner Foerster, "κτίζω [to create]," in *Theological Dictionary of the New Testament*, ed. Gerhard Kittel and Gerhard Friedrich, trans. Geoffrey W. Bromiley (Grand Rapids: Eerdmans, 1964s), 3:1010-11.

[26]See Terence Fretheim, *God and World in the Old Testament: A Relational Theology of Creation* (Nashville: Abingdon, 2005) for a different view. In direct critique of Bernhard Anderson (*From Creation to New Creation: Old Testament Perspectives* [Minneapolis: Fortress, 1994]), Fretheim argues that though creation is "dependent" or even "deeply dependent" on God, by creating the world God has chosen to put himself into an "interdependent" or even a "dependent" relationship with creatures (pp. 270-73). In this book and others Fretheim attempts to develop a "relational" view of God and creation. Fretheim uses a process-philosophy-inspired concept of "relation" to place God and creatures in an interdependent relationship. He mistakenly applies the concept to the God-creature relation in the same sense as it applies to the creature-creature relation. I will address this blunder in a later section.

[27]Von Rad, *Theology of the Old Testament*, 1:141. See also Anderson, *From Creation to New Creation*, p. 8: "The creation depends totally on the will of the transcendent God."

[28]Bernhard Anderson, "Creation," in *The Interpreter's Dictionary of the Bible*, ed. George A. Buttrick (New York: Abingdon, 1962), 1:728. See also Sarna, *Understanding Genesis*, p. 11. Speaking of the use of *bārāʾ* ("to create") in Gen 1:1, Karl-Heinz Bernhardt observes: "This verb does not denote an act that somehow can be described, but simply states that, unconditionally, without further intervention, through God's command something comes into being that had not existed before" ("בָּרָא," in *Theological Dictionary of the Old Testament*, ed. Johannas Botterweck and Helmer Ringgren, trans. John T. Willis et al. [Grand Rapids: Eerdmans, 1975], 2:247).

Creatures contribute nothing to their own coming to be; their origination relies wholly on the generosity and power of the Creator. If we speak at all of a motivation for creation—and we must be careful in using this concept—we must speak of divine love or grace or benevolence. God creates out of fullness and joy and not from emptiness and need. Even if some creatures, especially human beings, participate in their own growth toward perfection and the coming to be of other creatures, the Creator-creation relation retains its characteristic asymmetry. The real independence of creatures is relative, ever retaining its relation of "absolute dependence" to the Creator.[29] Speculating further about the nature of the creature's dependent independence lies outside biblical theology and falls to the systematic theologian.

Terence Fretheim challenges the traditional idea that the Creator-creature relationship is completely asymmetric, arguing rather that the Old Testament views the relationship as interdependent. Creatures cannot possess authentic existence distinct from God unless God gives up some power to creatures. In Fretheim's words, "God moves over and makes room for the Other."[30] Fretheim here uses spatial terminology univocally between God and creatures. He treats the sphere of God's freedom and action as if it were physical space and applies rules that govern relations between physical objects—that two things cannot occupy the same space at the same time—to the God-creature relation. As we will see in the course of this study, Fretheim's mistake plagues many modern theologies of creation and providence.

Thesis three: The creation really exists before God and stands before him as good; that is, as the result of God's act of creation, the creature really is what God intended it to be and can be used for the purpose God intended. God treats creation as a genuine other, giving it space and a role to play and making it responsible to God. At the conclusion of the six days of creation, God pronounces the work of the six days "very good" (Gen 1:31). To say that the world of creatures is good is to say, "In every part it corresponds to the intentions of its Creator; no hostile power was able to frustrate his design."[31] Creation possesses its own being and its own powers that can be turned to God's use in achieving his grand design as the history of creation moves

---

[29]Anderson, "Creation," 1:728.
[30]Fretheim, *God and World*, p. 272.
[31]Eichrodt, *Theology of the Old Testament*, p. 108.

toward its goal.[32] This divine judgment opens creation for human use and enjoyment.[33] The nonhuman creation gives human beings an opportunity to praise God for the "ordered, reliable, life-giving character of the world as gift of God."[34]

Thesis four: The Creator-creature relation established at the beginning, with its characteristic qualities, endures for all time. God's act of creation is not left in the dim past while its results continue to unfold. The act of creation applies to the whole creation from beginning to end. God continues to be the Creator whether in reference to present action in history or nature or in view of the hope for a new creation.[35] Creation continues in the relation of dependence, responsibility and debt throughout its duration even as it acquires other types of relationship: disobedience, reconciliation and redemption.

Thesis five: Human beings possess a unique relationship to the Creator characterized by their image and likeness to God and responsibility to him.[36] God addresses other creatures and even nonexistent things, and these things respond to God's creative word by coming into being, by helping to bring other creatures into being or by reproducing after their kind (Ps 33:9; cf. Rom 4:17).[37] But human beings are unique among the creatures God addresses; they are his "chief work" in creation.[38] They are the "crown of creation."[39] For they have been given power to understand

---

[32]John H. Walton understands the "it was good" judgment to be a statement not about creation's moral state but its "functional" suitability for God's purposes (*The Lost World of Genesis One: Ancient Cosmology and the Origins Debate* [Downers Grove, IL: IVP Academic, 2009], p. 50).

[33]Westermann, *Creation*, pp. 60-65.

[34]Walter Brueggemann, *Reverberations of Faith: A Theological Handbook of Old Testament Themes* (Louisville, KY: Westminster John Knox, 2002), pp. 40-41; see also Fretheim, *God and World*, p. 284.

[35]Anderson, *From Creation to New Creation*, pp. 27, 93-96.

[36]David Fergusson draws four "features" from his study of the creation theme in Scripture: (1) the God-creature relationship is defined, (2) the goodness of the world is asserted, (3) creation anticipates preservation and consummation and (4) the act of creation evokes praise and points to the excellence and goodness of the Creator (*Creation*, Guides to Theology [Grand Rapids: Eerdmans, 2014], p. 9).

[37]See Foerster, "κτίζω [to create]," 3:1010, for discussion of this the "contradiction" of God's speaking to things that do not yet exist.

[38]Von Rad, *Theology of the Old Testament*, 1:140.

[39]Anderson, *From Creation to New Creation*, p. 11. Anderson concludes that Old Testament theology of creation is "anthropocentric" ("Creation," 1:729-30). With his heightened ecological consciousness Fretheim combats the anthropocentric view of creation, arguing rather for "an interdependent mutuality of vocation" between the human and the nonhuman creation. The nonhuman as well as the human possesses a direct relationship to God (*God and World*, pp. 273-81).

God's command and respond freely. Since they have been given power over themselves, they also possess the power to rule over the rest of creation.[40] The nature that enables human beings to exercise these powers is called the "image and likeness" of God. Herein resides the dignity and the vocation of God's human creatures. In contrast to other creation stories in the Ancient Near East, where humans were created to serve the gods' needs, in the Bible creation serves humanity and humanity represents God to the rest of creation. So much so that John Walton can call the Old Testament view of creation "anthropocentric."[41]

---

[40]See Eichrodt, *Theology of the Old Testament*, pp. 118-50, for his survey of the anthropology of the Old Testament. Eichrodt qualifies the human superiority and ruling authority over the nonhuman by setting the human-nonhuman relationship within the creature-Creator relation. All things serve God, and even if the nonhuman creation serves human beings, it serves them only within the universal service to God.

[41]Walton, *Lost World*, p. 68.

2

"IN THE BEGINNING WAS THE WORD"

Creation in the New Testament

T HE NEW TESTAMENT REFERENCES to creation presuppose and echo the Old Testament creation texts and affirm the five theses stated above.[1] On many occasions New Testament authors refer to the created order to emphasize that a principle is universal or that a state of affairs is unparalleled. Jesus rejects divorce because "at the beginning of creation God 'made them male and female'" (Mk 10:6). Or in Mark 13, Jesus tells of a coming time of distress "unequaled from the beginning, when God created the world, until now—and never to be equaled again" (Mk 13:19). For Paul, the creation has been testifying to God's existence and nature "since the creation of the world" (Rom 1:20). Some references to creation speak of a "before" creation, that is, of eternity. Jesus asks his Father to return the glory he had with him "before the world began" (Jn 17:5; cf. Jn 17:24). According to Paul, God chose his elect "before the creation of the world" (Eph 1:4), and Peter speaks of Christ as "chosen before the creation of the world" (1 Pet 1:20).

NEW TESTAMENT THEOLOGY OF CREATION

As a way to organize the New Testament's theology of creation we will first summarize the teaching that falls under the five headings used in the discussion of the Old Testament theology. Second, we will examine the

---

[1]For studies of the New Testament documents' views of the nature, composition and structure of the world, see the essays contained in *Cosmology and New Testament Theology*, ed. Jonathan T. Pennington and Sean M. McDonough (New York: T & T Clark, 2008).

new insights into creation achieved in the New Testament by viewing creation through the lenses of the work of Jesus Christ and the Spirit, that is, the triune pattern of the New Testament understanding of the economy of salvation.

Thesis one: The one God is the absolute origin and sovereign ruler over all that is not God. For the New Testament, God is the sovereign Creator of everything. Jesus praises his Father as "Lord of heaven and earth" (Mt 11:25). As opposed to the idols of Athens, Paul offers an alternative vision of God: "The God who made the world and everything in it is the Lord of heaven and earth and does not live in temples built by human hands. And he is not served by human hands, as if he needed anything. Rather, he himself gives everyone life and breath and everything else" (Acts 17:24-25). Paul praises the Creator in these words: "For from him and through him and for him are all things" (Rom 11:36; cf. 1 Cor 8:6; Heb 2:10). God's universal rule is emphasized dramatically by the angel who speaks in Revelation 10: "And he swore by him who lives for ever and ever, who created the heavens and all that is in them, the earth and all that is in it, and the sea and all that is in it, and said, 'There will be no more delay!'" (Rev 10:6).

Thesis two: The one God freely established the Creator-creature relation, which is characterized by generosity, freedom and power on the Creator's side and dependence and debt on the creature's side. Paul affirms that, with the same generosity in which God called light out of darkness at the beginning, God has now "made his light shine in our hearts to give us the light of the knowledge of God's glory displayed in the face of Christ" (2 Cor 4:6). Peter assures his readers that even in persecution we can trust our "faithful Creator" (1 Pet 4:19). It is only in God that "we live and move and have our being" (Acts 17:28). Paul quotes approvingly Isaiah 29:16, which compares the relationship of the Creator to the creature to the relationship of the potter to the clay (Rom 9:20). The writer of Hebrews reminds us that "by faith we understand that the universe was formed at God's command, so that what is seen was not made out of what was visible" (Heb 11:3).

Thesis three: The creation really exists before God and stands before him as good; that is, as the result of God's act of creation, the creature really is what God intended it to be. In dealing with issue of the permissibility of eating meat sacrificed to idols Paul permits the Corinthians to "eat anything sold in the meat market without raising questions of conscience, for 'The

earth is the Lord's, and everything in it'" (1 Cor 10:25-26, quoting Ps 24:1). In
1 Corinthians 15, Paul argues against his Corinthian opponents for the resurrection of the body. His main proof is that Christ was raised from the dead—
something even his opponents accepted—which presupposes that he possessed a real body (1 Cor 15:1-20). Salvation as Paul sees it is salvation from
sin and death, not from entrapment within the physical world and the slug-
gishness of mind caused by matter. In response to the heretical teachers
plaguing the churches under Timothy's charge, Paul refutes them with
Genesis 1: "They forbid people to marry and order them to abstain from
certain foods, which God created to be received with thanksgiving by those
who believe and who know the truth. For everything God created is good,
and nothing is to be rejected if it is received with thanksgiving, because it is
consecrated by the word of God and prayer" (1 Tim 4:3-5; cf. Mk 7:14-23;
Rom 14:14). James is insistent that the "Father of the heavenly lights" is the
source of "every good and perfect gift" (Jas 1:16-17).

   *Thesis four: The Creator-creature relation established at the beginning,
with its characteristic qualities, endures for all time.* Like the Old Testament, the New Testament acknowledges the present world as created and
does not leave God's act of creation in the misty past. God created everything that now exists; it is just as much a divine creation as the first appearance of the physical world. When New Testament texts say "all things"
were made by God they mean *all things that were, are and will be.* Typical is
Revelation 4:11:

> You are worthy, our Lord and God,
>   to receive glory and honor and power,
> for you created all things,
>   and by your will they were created
>   and have their being. (See also Jn 1:3; 1 Cor 8:6; Eph 3:9; Col 1:16)

God's continuing creation finds expression in Hebrews, where God creates
the universe through the eternal Son of God, who also sustains "all things
by his powerful word" (Heb 1:3). For Paul, creation exists not only *from* God
but also "through him and for him," encompassing all creatures from beginning to end (Rom 11:36; cf. 1 Cor 8:6; Heb 2:10). When Jesus told us to
pray to our Father, asking, "Give us today our daily bread" (Mt 6:11), he
wanted us to believe God really gives bread to us today. When we give

thanks for our food, we should not simply thank him for creating the laws of nature and processes that produce our food; we thank him for creating our food: "For everything God created is good, and nothing is to be rejected if it is received with thanksgiving" (1 Tim 4:4; cf. Mk 7:14-23; Rom 14:14).

   Thesis five: Human beings possess a unique relationship to the Creator characterized by their image and likeness to God and responsibility to him. Jesus refers to Genesis 1:27 and Genesis 2:24 in his teaching on marriage and divorce (Mt 19:4-6), declaring that humans do not have a right to rip apart what God joined together. Perhaps more relevant to the point of this thesis is Jesus' teaching about the value of individual people to God. We need not worry about our lives as if we were at the mercy of chance and fate, for God takes care of the birds of the air. Besides, "are you not much more valuable than they?" (Mt. 6:26). Jesus seeks even *one* lost sheep, *one* lost coin and *one* lost son (Luke 15). For Paul, God's love for believers precedes creation, "for he chose us in him before the creation of the world" (Eph 1:4). James roots the dignity of human beings in creation: "With the tongue we praise our Lord and Father, and with it we curse human beings, who have been made in God's likeness" (Jas 3:9). However, the ground mentioned most often as marking human beings as special is God's love demonstrated in Jesus Christ. The Gospel of John declares that God so loved the world "that he gave his one and only Son" for its salvation (Jn 3:16). This theme is continued in 1 John: "This is how we know what love is: Jesus Christ laid down his life for us. And we ought to lay down our lives for our brothers and sisters" (1 Jn 3:16; cf. 1 Jn 3:1; 4:9-12). Paul is no less eloquent: "But God demonstrates his own love for us in this: While we were still sinners, Christ died for us" (Rom 5:8; 2 Cor 5:14-15). In Romans 5 and 1 Corinthians 15, Paul compares and contrasts Adam and Christ. Adam brought sin and death into the world, but Christ brought righteousness and life (Rom 5:12-21). God made the first man "a living being," but Christ has been made "a life-giving spirit." Just as we bore the image of Adam, "so shall we bear the image of the heavenly man" (1 Cor 15:45-49). In Genesis, God made human beings in his image, but in the New Testament Christ is held to be the image of God in an archetypical sense: "The Son is the image of the invisible God, the firstborn over all creation" (Col 1:15; cf. 1 Cor 15:49). It is only through Christ that human beings can truly realize the image of God in their actual existence (Rom 8:9; 2 Cor 4:4; Col 3:10).

CHRIST AND CREATION

The New Testament theology of creation cannot be understood apart from its proclamation of the identity and work of Jesus Christ. Jesus is the Son of God (Mk 1:1; Gal 2:20; 1 Jn 5:20), Lord (Rom 10:9; Phil 2:11), image of God (2 Cor 4:4; Col 1:15; cf. Heb 1:3), Messiah (Mt 16:16; Acts 2:36; 1 Pet 1:11), Savior (Acts 5:31; Phil 3:20; 2 Tim 1:10), the Word of God (Jn 1:1, 14; 1 Jn 1:1; Rev 19:13), and the wisdom and power of God (1 Cor 1:24).[2] Jesus is the savior of the world (Acts 4:12; Heb 1:3; 1 Jn 4:4), the revealer of the true God (Jn 1:18; 17:6; Col 2:2-3; 1 Jn 5:20) and the eschatological judge (Mt 25:31-46; Jn 5:22; 2 Cor 5:10).

Given the exalted nature of Christ's identity and work, it is not surprising at all that Christ is also understood to be the Creator, the one through whom God created and sustains the world. It is simply incomprehensible that Jesus Christ could have been with God in the beginning as his exact image, Son, wisdom and power; have become in time the savior of the world and revealer of the true God; and be expected to return as the eschatological judge and redeemer of the world, but not be thought to have participated in the creation of the world. As Richard Bauckham points out, the divine work of creating, ordering and bringing all things to completion is "properly indivisible."

> The participation of Christ in the creative work of God is necessary, in Jewish monotheistic terms, to complete the otherwise incomplete inclusion in the unique divine identity. . . . [Including Jesus] precisely in the divine activity of creation is the most unequivocal way of excluding any threat to monotheism . . . while redefining the unique divinity of God in a way that includes Jesus.[3]

Christ is specifically designated Creator in John 1:1-5, 10; 1 Corinthians 8:4-6; Colossians 1:16; and Hebrews 1:2, 10. We could also include the nature miracles recorded in the four Gospels, the healings, resurrections, the feedings of the five and four thousand, walking on water, and the calming of the stormy lake as showing that Jesus was viewed as wielding power only the Creator could exercise.

Four christological texts. I will now briefly examine the four christo-

---

[2]I have chosen only a few representative references from different New Testament authors. There are scores of these references.

[3]Richard Bauckham, *God Crucified: Monotheism and Christology in the New Testament* (Grand Rapids: Eerdmans, 1998), p. 36.

logical texts mentioned above: John 1:1-5, 10-14; 1 Corinthians 8:4-6; Colossians 1:15-17; and Hebrews 1:1-4, 10. My twofold goal is modest: (1) to assure ourselves that these texts in their immediate, proximate and canonical contexts actually teach that Christ participated in the creation of the world, and (2) to draw out some theological implications of this teaching.

John 1:1-5, 10-14.

> In the beginning was the Word, and the Word was with God, and the Word was God. He was with God in the beginning. Through him all things were made; without him nothing was made that has been made. In him was life, and that life was the light of all mankind. The light shines in the darkness, and the darkness has not overcome it. . . .
>
> He was in the world, and though the world was made through him, the world did not recognize him. He came to that which was his own, but his own did not receive him. Yet to all who did receive him, to those who believed in his name, he gave the right to become children of God—children born not of natural descent, nor of human decision or a husband's will, but born of God.
>
> The Word became flesh and made his dwelling among us. We have seen his glory, the glory of the one and only Son, who came from the Father, full of grace and truth.

Like creation texts in the Old Testament, which relate historical actions of the Lord to God's creative action in the beginning, the Gospel of John relates the enlightening and saving work of Jesus to creation. This relation gives cosmic meaning to the salvation Jesus brings and christological meaning to creation. In the body of the Gospel, Jesus confronts people with signs, teaching and claims that force them to ask about his identity. What qualifies Jesus to be savior, judge and revealer? Does he really possess power to save completely? The reader of the Gospel, however, already knows the answers to these questions. The Word who became flesh in Jesus (Jn 1:14) existed eternally with God and was God. Hence, the salvation brought by Jesus derives from the metaphysically ultimate reality and, therefore, is universally imperative. The Word through whom salvation comes is the means, the instrument through whom God created all things and in whom all things live. Every creature bears the stamp of the Logos. Hence the Word made flesh possesses power and authority over all creatures to heal, judge and save.

John 1:1-2 makes four assertions about the Word.[4] (1) Clearly this text
plays off of Genesis 1. "In the beginning" echoes the "In the beginning" mentioned in Genesis 1:1, that is, the beginning of the creation. In Genesis, God
is already there, so there is no account of God's origin. This "omission" is
glaring in light of ancient creation myths that include genealogies of the
gods, and it speaks volumes about Israel's understanding of God. Likewise,
there is no account of the origin of the Word in the Gospel of John. But John
states explicitly about the Word what was understood in Genesis about God:
the Word *was already there*; thus the Gospel calls attention to the eternity of
the Word. (2) The next assertion remains in the sphere of "the beginning"
but adds that the Word was *with* God. So at the beginning of creation the
Word was already there in intimate relation to God. The eternal Word is
coeternal with God. This relationship is mentioned again in John 1:18 (see
also Jn 17:5, 11, 21-24), where the Word is called "the only begotten God" (or
"God the only Son") who is "at the Father's side" (author's translation). The
relation of being "with" indicates a distinction of some kind from God (*ho
theos*). (3) The distinction between the Word and God calls for the third assertion: "the Word was God." So the Word was already there in the beginning
with God *as God*. To maintain harmony among the first three assertions, the
author makes a threefold distinction that is somewhat obscure in translation.
In English, we use the word *God* as a proper noun and as a common noun;
that is, it can mean God or deity. For John, there is the Word (*ho logos*), God
(*ho theos*) and divinity (*theos*). The Word was God (*theos*) with God (*ho theos*)
in the beginning! (4) The fourth assertion combines (1) and (2): "the same
one was in the beginning with God" (author's translation). This element
closes the ontological part of the prologue and by mentioning the "beginning"
again prepares for the cosmological part that follows immediately.

John speaks next of the Word's role in creation: "Everything through him
came to be and without him came to be not one thing" (Jn 1:3).[5] All creation
came to be *through* the Word (*di' autou*). "God" (*ho theos*) is the implied
subject of this sentence, and the Word is the instrument through which God
created all things. The last half of John 1:3 underlines the point made in the

[4]Ronald Cox sees John 1:1 as asserting the Word's threefold "ontological pre-eminence" in the
"temporal, relational and the substantial" (*By That Same Word: Creation and Salvation in Hellenistic Judaism and Early Christianity* [Berlin: de Gruyter, 2007], p. 233).
[5]Translation by Cox, *By That Same Word*, p. 235.

first half of that verse: *Everything* came through the Word; not even *one thing* escapes the creative agency of the Word. Everything that fits into the category of having "come to be" also is subsumed into the category of things that came to be through the Word.[6] Unlike Hebrews 1:3, which speaks of the Son as sustaining all things through his powerful word, or 1 Corinthians 8:6, which speaks of the one Lord "through whom we live," or Colossians 1:17, which says of Jesus Christ that "in him all things hold together," John does not speak directly about the Word's role in sustaining creation. However, John 1:4 speaks of this role indirectly. Unfortunately, there is a translation problem caused by an ambiguity about where the end of the previous sentence falls and the next one begins. It says either, "In him was life, and the life was the light of human beings," or "That which came to be [*ho gegonen*] in him was life and the life was the light of human beings."[7] In either case, however, the life that was in the Word became the "light" in human beings. Without denying that the Word continues to work in the world in general, the role of the Word in the ongoing of creation is here narrowed to the human world. The reference to "light" refers back to the creation of light in Genesis 1:2 and forward to the light brought into the world when the "Word became flesh and made his dwelling among us" (Jn 1:14). Apparently John sees a double meaning in Genesis 1:2. God created physical light on the first day of creation, and "it was good." But light can also mean the intelligible world, mind and intellect; hence, the Word is the light that gives life to intelligent creatures. Life and light came to be in the one who is life and light of himself. The Word is the source who makes these precious gifts available to human beings. In its blending of the two meanings of light, the Gospel of John deftly moves from the subject of creation to the revealing and saving work of the Word, which is the main concern of the Gospel.

1 Corinthians 8:4-6.

So then, about eating food sacrificed to idols: We know that "An idol is nothing at all in the world" and that "There is no God but one." For even if there are so-called gods, whether in heaven or on earth (as indeed there are

---

[6]Emphasis on the universality of the creative agency of Jesus Christ can also be seen in Col 1:16, which will be treated below: "For in him all things were created: things in heaven and on earth, visible and invisible, whether thrones or powers or rulers or authorities; all things have been created through him and for him."

[7]See Raymond Brown, *The Gospel According to John 1–12*, Anchor Bible 29 (Garden City, NY: Doubleday, 1966), pp. 6-7, for discussion of this issue.

many "gods" and many "lords"), yet for us there is but one God, the Father, from whom all things came and for whom we live; and there is but one Lord, Jesus Christ, through whom all things came and through whom we live.

In the larger context of 1 Corinthians we see the conflict between two very different approaches to salvation, anthropology and ethics, which manifest themselves in the practical issues of love among believers and exclusive loyalty to Christ.[8] The one against which Paul argues contends that salvation is attained by acquiring enlightenment through wisdom. This approach divides Christians into the "knowing," "strong" and "perfect" on the one hand and the unenlightened, "weak" and "imperfect" on the other. And this division results in arrogant and unloving behavior in the community and a cavalier attitude toward demonic powers. In contrast, Paul argues that salvation comes through faith in the crucified and risen Christ. All are united in the one body of Christ and are equally dependent on Christ. Following Christ means exemplifying Christ's humility and love and seeking God's blessings in Christ alone. This division plays itself out in many different ways that Paul addresses in the letter: in divisions in the community over teachers (1 Cor 3:1-23), in arrogance toward Paul's "weakness" (1 Cor 4:1-21), in differences in sexual morality (1 Cor 5:1-13; 6:12-20), in the nature of marriage (1 Cor 7:1-40), in the controversy about eating meat sacrificed to idols (1 Cor 8:1-10:33), in relation to worship and Lord's Supper (1 Cor 11:1-34), in relation to spiritual gifts (1 Cor 12:1–14:40) and concerning the resurrection of the dead (1 Cor 15:1-58). In 1 Corinthians 8–10, Paul addresses a tension in the church caused by some who are so assured that idols are nothing and that the gods they represent do not exist that they do not

---

[8]For a study of this text in light of Hellenistic Judaism, see Cox, *By That Same Word*, pp. 141-61. Gordon D. Fee, *Pauline Christology: An Exegetical-Theological Study* (Peabody, MA: Hendrickson, 2007), pp. 88-94, 102-6, argues against interpreting this text as drawing on the personified wisdom tradition as developed in Hellenistic Judaism, specifically that of the Wisdom of Solomon and Philo of Alexandria. Sean M. McDonough in *Christ as Creator: The Origins of a New Testament Doctrine* (New York: Oxford University Press, 2009), while acknowledging the possibility of Hellenistic influence, sees the decisive impetus for concluding that Christ is Creator as the life and fate of Jesus. I owe my understanding of the issues of "love" and "loyalty" in this text to McDonough (ibid., p. 155). Larry Hurtado, *Lord Jesus Christ: Devotion to Jesus in Earliest Christianity* (Grand Rapids: Eerdmans, 2003), pp. 123-26, also dismisses the idea that the Hellenistic notion of preexistent wisdom, rather than the person of Jesus, was the agent of creation. "Instead they [claims to Jesus' creative agency] were prompted by profound religious convictions about the transcendent significance, unique status, and role of Jesus Christ, who was sent forth from God for the redemption of the world" (p. 126).

hesitate to eat food sacrificed to idols in pagan temples. Apparently they think eating such food has no religious significance and may even serve to highlight their spiritual strength. But others are not so sure about the benignity of eating meat sacrificed to idols.

Paul provisionally takes the side of the bolder party by agreeing that an idol is nothing at all. There is only one God. But notice that he does not say of the so-called gods and lords what he says of idols. Of course, there are no *legitimate* gods and that is why he calls them "so-called." As Sean McDonough expresses it, "Paul rather is stating that none of these entities—whether the dumb wood or stone of the idol, or the nonhuman spiritual forces in the world—can lay claim to the status of the one true God or his Messiah, the Lord Jesus Christ."[9] But Paul is not willing to rule out the existence of demonic beings that pretend to be gods and lords and thus qualify as so-called gods and lords. He returns to the subject in 1 Corinthians 10 in the context of the Lord's Supper, the Christian counterpart to sacred meals in pagan temples. Paul there forbids eating meat sacrificed to idols because it has been "offered to demons, not to God" (1 Cor 10:20). In this section (1 Cor 10:1-22) he shows what happened to the Israelites when they participated in idol feasts. They died because of their disloyalty to Christ, who for Paul has always been the one "through whom" God has bestowed blessings. In 1 Corinthians 8, not pursuing the demonic possibility there and delaying an outright prohibition, Paul urges that love may forbid what knowledge permits. Those who have knowledge of the nonexistence of other gods and lords should take care to avoid occasioning others to violate their consciences.[10]

Reading 1 Corinthians 8:4-6 in its larger context makes clear why Paul refers to belief in one God "the Father, from whom all things came and for whom we live." This assertion reminds the Corinthian community of its most basic confession (see Deut 6:4), confirms its belief that idols are nothing and asserts that pagan gods, whatever their existential status, are not truly gods.[11] The one God is the origin and goal of all creation. There is

---

[9]McDonough, *Christ as Creator*, p. 154.

[10]Fee, *Pauline Christology*, argues that Paul, though denying the objective reality of pagan gods and lords, attributes to them a "subjective reality" for the consciences of recent pagan converts (p. 89).

[11]Bauckham argues persuasively that Paul carefully adheres to the wording of the Shema of Deut 6:4, using every word of that formula: "The LORD our God, the LORD is one." However, Paul applies "God" to the Father and "Lord" to Jesus. See *God Crucified*, pp. 38-40.

no need to seek help from other powers or to divide one's loyalty among many lords. But why add the next assertion? "And there is but one Lord, Jesus Christ, through whom all things came and through whom we live."[12] Paul had anticipated this reference earlier by gratuitously mentioning "lords" along with gods twice and by identifying the one God as "the Father," which for Paul is a shorted form of the formula "the God and Father of our Lord Jesus Christ."[13] The statement itself assures the reader that the Lord whom they worship and with whom they commune in the Lord's Supper (see 1 Cor 10:14-22) is truly Lord, unlike the so-called lords or demons.[14] Our Lord holds the title by right, for he is one "through" whom all things were created and "through" whom we live. The *means* or instrument of creation is just as essential as its origin, and the *means* of life is just as important as the goal of life. It is just as important to avoid making Christ jealous (1 Cor 10:22) by participating in an idol's table as it is to make God jealous by worshiping images (Ex 20:5). God the Father and the Lord Jesus Christ are inseparably united in the economy of creation and salvation. Just as identifying God as "the Father of our Lord Jesus Christ" makes clear God's identity and fills out the character of God, identifying Jesus Christ as the means by which God creates and sustains the world gives him the authority and power of the true Lord.

The role of Christ in creation is not Paul's overriding concern in this text, though of course Paul makes this assertion.[15] It is never far from Paul's mind that the Lord Jesus Christ, "through whom all things came and through

---

[12]According to Fee, *Pauline Christology*, pp. 89-90, 1 Cor 8:6 is Paul's reworking of the Septuagint version of Deut 6:4, in which he distinguishes between *theos* and *kyrios*, *theos* being the Father "from whom all things come" and *kyrios* being Jesus Christ "through whom all things come."

[13]Paul characteristically refers to God as "the God and Father of our Lord Jesus Christ" (1 Cor 1:3; Gal 1:3; Phil 1:2). In Paul, "God, the Father" is a shortened form of the full formula "the God and Father of our Lord Jesus Christ." This formula identifies God in much the same way that Pentateuch narratives identity God as "the God of Abraham, Isaac and Jacob." It places God in a narrative that fills out the divine character. The history of Jesus Christ identifies and characterizes God.

[14]Demons were a pervasive feature of Hellenistic religion and served to bridge the gap between the highest God and the human world. For thoughts on how Jewish and Hellenistic demonology illuminates 1 Cor 8–10, see McDonough, *Christ as Creator*, pp. 158-70.

[15]Cox, *By That Same Word*, pp. 159-60. Paul does not seem to be concerned here with issue of ontology. In the Hellenistic context, in Middle Platonism, in Hellenistic Judaism (especially Wisdom of Solomon and Philo of Alexandria) and perhaps among Paul's Corinthian opponents the issue was the nature of the intermediaries that bridged the gap between the transcendent God and the material world (for that story see ibid., pp. 1-140).

whom we live," is the crucified one (1 Cor 2:2); in *that* role Christ is the "power of God and the wisdom of God" (1 Cor 1:24) and thus is the model for an ethic of love, humility and unity in the Christian community. Nevertheless, asserting that the crucified one is the instrument of creation plays a vital role in the larger argument. Paul's opponents bragged about their wisdom (1 Cor 4:10) and knowledge (1 Cor 8:1), which seems to indicate that they viewed salvation through the lens of speculative cosmology more than through the historical-eschatological event of Jesus Christ.[16] Paul corrects his opponents by reversing the relationship, depriving speculative cosmology of its interpretive primacy by asserting that Jesus Christ the crucified and risen one is the agent of creation. And in that reversal Paul calls for a dramatic shift in the way we understand creation, salvation and ethics. Everything—God and creation, beginning and end, salvation and damnation—must be understood in view of the life, death and resurrection of Jesus Christ.

Colossians 1:15-20.

> The Son is the image of the invisible God, the firstborn over all creation. For in him all things were created: things in heaven and on earth, visible and invisible, whether thrones or powers or rulers or authorities; all things have been created through him and for him. He is before all things, and in him all things hold together. And he is the head of the body, the church; he is the beginning and the firstborn from among the dead, so that in everything he might have the supremacy. For God was pleased to have all his fullness dwell in him, and through him to reconcile to himself all things, whether things on earth or things in heaven, by making peace through his blood, shed on the cross.

Like other New Testament creation texts, these verses are set in a context that emphasizes the salvation brought by Christ, so that the creation theme

---

[16]Cox interprets 1 Cor 8:6 against the background of the Wisdom/Logos speculation in Hellenistic Judaism, which leads him to conclude that Paul's opponents likely viewed Wisdom as God's agent of creation and salvation. Paul replaces Wisdom with Jesus Christ and thus corrects his opponents by turning them away from cosmological speculation toward the crucified and risen one (*By That Same Word*, pp. 151-58). Cox's theory seems a bit speculative and underdetermined by the text under consideration. There is no evidence in the text that the Corinthians viewed Wisdom as a personified agent of creation. Nevertheless, it seems clear that Paul's opponents are enamored of the enlightenment brought by some form of speculation ("wisdom" and "knowledge") not rooted in the life, death and resurrection of Jesus Christ. Paul attempts to correct this error by bringing the focus back to the concrete historical-eschatological event of Jesus Christ.

is subordinate instead of independent.[17] Immediately preceding Colossians
1:15, Paul thanks the Father for rescuing Christians "from the dominion of
darkness" and placing them in the kingdom of "the Son he loves, in whom
we have redemption, the forgiveness of sins" (Col 1:14). Then follow Colossians 1:15-20, which some scholars argue is a hymn composed prior to being
incorporated into Colossians. The hymn naturally breaks into two stanzas,
Colossians 1:15-17 and Colossians 1:18-20. The first stanza extols the Son's
intimate relationship to God and his role in creating and governing the
world, while the second stanza shifts to focus on the saving work of the Son.
The Son, in whom the fullness of God dwells, achieves supremacy in this
sphere as well. Hence the Son's role as the instrument through which and
the sphere in which God creates and governs the world is surrounded by
references to the salvific work of the Son.[18]

Colossians 1:15 designates the Son as the "image" or *eikōn* of the invisible
God and the "firstborn" over all creation. These two qualifications indicate
a close ontological relationship between the Son and the Father and an ontological gulf between the Son and creation. The only other New Testament
text in which the Son is designated the "image" of God is 2 Corinthians 4:4,
where Christ the image of God is the place where God is revealed truly in
this world.[19] The revelatory function may also be behind the "image" title in
Colossians 1:15, for otherwise why speak of God as the *invisible* God?[20]
Designating the Son as the "firstborn" (*prōtotokos*) over creation indicates
not that he is the chronologically first creature but that he is superior in rank
to all creatures. According to McDonough the title "firstborn" may "speak
of an inextricable link between the Messiah and creation. . . . He was always
its intended ruler. It was made for his purposes, according to his
specifications."[21] The following verse explains why the Son deserves such an
exalted status. *All things* were created in him (*en auto*); then follow refer-

---

[17]McDonough, *Christ as Creator*, p. 189: "Colossians 1 provides the clearest evidence that Christ's
role in creation was developed in light of his role in redemption. . . . In this sense, the doctrine
of creation serves to undergird the doctrine of redemption."

[18]Since my concern is the NT theology of creation I will limit my further comments to the first
stanza (Col 1:15-16).

[19]The same concept is expressed in Heb 1:3, where the Son is said to be "the exact representation
of his [God's] being."

[20]Cox argues that it is likely "that *eikon* is an established technical term denoting the relationship
of the Son vis-à-vis God, a term the Colossians passage received from its Hellenistic Jewish
Vorleben" (*By That Same Word*, p. 173)

[21]McDonough, *Christ as Creator*, p. 185.

ences to every sphere of created being to emphasize the comprehensiveness of his authority and activity. The first stanza ends with "all things have been created through him and for him" (Col 1:16), repeating the first assertion of Colossians 1:16 and adding the prepositional phrase "for him" (*eis auton*). This addition intensifies the already intense focus on the Son's comprehensive rule over creation.[22] The Son is not only the instrument *through whom* all things were created but also the end *for whom* they were created. To sum up, cosmologically the Son's supremacy extends from the beginning to the end of creation and over every sphere of created being during all phases and throughout every dimension of its existence. And, as Colossians 1:17-20 makes clear, the Son who rules all creation is identified as the crucified and risen savior. Hence as in other New Testament creation texts, creation and salvation are brought into a dialogic relationship of mutual interpretation within Christology. For Paul, then, "salvation is itself the full flowering of what was made in the beginning."[23]

Hebrews 1:1-4.

> In the past God spoke to our ancestors through the prophets at many times and in various ways, but in these last days he has spoken to us by his Son, whom he appointed heir of all things, and through whom also he made the universe. The Son is the radiance of God's glory and the exact representation of his being, sustaining all things by his powerful word. After he had provided purification for sins, he sat down at the right hand of the Majesty in heaven. So he became as much superior to the angels as the name he has inherited is superior to theirs.

In these tightly compacted verses we see the ontological closeness of the Son to God and the revelatory, salvific and cosmological work of the Son blended in eloquent anticipation of the rest of the book. Clearly the revelation brought by the Son outranks and supersedes the piecemeal revelations that came through Moses and the prophets, a theme that the author pursues throughout the book. Jesus and the new order he brought are said to be "better" than or "superior" to other mediators and the old order (Heb 7:22;

---

[22]Nowhere else in the NT is the Son spoken of as the end of creation (*eis auton*). In Rom 11:36 (cf. 1 Cor 8:6) God is said to be the one "from whom and through whom and for whom" (*eis auton*) all things exist. See Fee, *Pauline Christology*, p. 302, for his comments on the implications for Christology of using *eis auton* with reference to Son's relationship to creation, given that this phrase is used only of the Father elsewhere.

[23]McDonough, *Christ as Creator*, p. 189.

8:6; 9:23; 11:40; 12:24). When the author asserts that God has appointed the Son heir of all things, it reminds us of the "for him" of Colossians 1:16 examined above. The Son possesses authority over "all things," which befits the Son and heir of God. The next phrase turns our attention to the beginning of creation. The Son is the one "through whom" (Heb 1:2) God made the ages (*aiōnas*). Given other texts about the work of Christ in creation we would expect "all things" (*ta panta*) rather than "ages" as the object of creation. Perhaps the writer wishes to bring not only Christ's past work under the category of creation but also his future work.[24] Clearly the Son is understood as the instrument of creation. And in Hebrews 1:3 the author refers to another of the Son's cosmological functions, that of sustaining "all things by his powerful word." Significantly, though God made the worlds "through" the Son (Heb 1:2), the Son himself sustains "all things" by his own powerful word (Heb 1:3). The author thus attributes to the Son an active, living and cooperative role in creation. The Son is not merely a passive tool.

The book makes two other references to creation. In proving that the Son is superior to the angels (Heb 1:10), the author quotes Psalm 102:25 as saying,

> In the beginning, Lord, you laid the foundations of the earth,
>   and the heavens are the work of your hands.

The Hebrews writer takes the "Lord" of this verse to apply to the Son and hence attributes creation directly to the Son without using the deferential "through whom" to indicate that the Son is God's instrument. The second reference is Hebrews 11:3, which echoes Genesis 1, and makes no reference to the instrumental agency of the Son. I do not think we should consider these two texts as undermining what the author asserts in Hebrews 1:2-3. In Hebrews 1:10 the author quotes the Septuagint Greek, in which the "through whom" is absent, and in Hebrews 11:3 the subject is not the superiority of Christ or creation as such but faith, and faith believes what is said in Genesis 1.

Having made reference to the unprecedented revealing, redeeming and cosmological work of the Son, the writer now speaks ontologically about the Son (Heb 1:3), showing why he is able and qualified to engage in such work. The Son is the "radiance" of God's glory and the "exact representation" of God's being. The glory of God here is synonymous with the reality of God

---

[24]McDonough, *Christ as Creator*, pp. 200-202.

understood under the metaphor of light. The Son embodies and radiates and communicates the light that is God. The Son is an exact representation (*charaktēr*) of God's being (*hypostaseōs*). The Son is light emanating from light and a visible representation of the invisible being of God. Although the Son's nature is clearly presupposed to be in the closest ontological relationship to God, the images here call to mind the revelatory function of the Son. His ontological resemblance to God coupled with his ability to become visible and audible to us enable him to serve as the definitive revealer of God.

Although Hebrews 1:1-4, understood in light of the whole book, aims primarily to bring out the ontological and revelatory superiority of the Son to all creatures and all other means of revelation, the text nevertheless places creation, salvation, revelation and the person of the Son in mutually supportive relationships. Certainly Hebrews gives no grounds to separate creation from the character of the Son through whom it came to be and by whom it is maintained or from the end toward which it is directed.

Summary. As these four texts witness, the New Testament identifies the Creator as the Father of the Lord Jesus Christ. The One revealed in the life, teaching, death and resurrection of Jesus is the Creator of heaven and earth. This identification not only highlights the universal significance of the work of Jesus Christ but also sheds new light on the character and purpose of creation. The divine act of creation anticipates and prepares the way for the divine act of salvation through Jesus Christ. The New Testament brings the biblical themes of creation and salvation into a new intimacy. In Jesus Christ we see the revelation of the Jesus-character of creation and the Creator-character of salvation.

3

"CREATOR OF HEAVEN AND EARTH"

THE FIRST SENTENCE OF THE BIBLE makes a monumental affirmation. "In the beginning God created the heavens and the earth" (Gen 1:1). The world with all its beauty and terror is not simply a brute fact we can take for granted or an illusion we can dismiss or a divine reality we must worship. It is not eternal or self-explanatory; rather, it has an origin and a beginning in *God*. The word "God" refers not to a god or gods, not to a mere divine force or principle, but to Israel's covenant God, the guide, savior and judge of Jacob, the God of Jesus Christ, the Father, Son and Holy Spirit. The rest of the Bible's teaching on creation and the doctrine of creation developed by the church explains, elaborates and defends this great assertion.

As we unfold the Christian doctrine of creation we will let this sentence guide us. God is the subject whose action creates the object, the heavens and the earth. The action of the subject takes place "in the beginning." Hence we have adverb, subject, verb and object, and we will reflect on each of these.[1] (1) Our first priority is identifying the subject. Who is God? (2) Next we will consider the verb, the action of creating. When we speak of God "creating" what kind of act is this, and what kind of relation between the Creator and the creature does this act indicate? (3) What is the nature

---

[1] I am not arguing that this sentence contains the whole doctrine of creation. It does not mention the means through which or the end for which God created the world, both essential components of the doctrine of creation. Later interpreters found mention of the means (the Word) in the very next verse: "And God said . . ." (Gen 1:2). That God created the world for a reason is presupposed because creation is described as the intelligent action of an intelligent being, which is always teleological. But the specific end comes into view only as the history of creation and salvation unfold. Only at the end (*eschaton*) will the end (*telos*) be realized and hence fully known.

of the object that results from God's act of creating?[2] What is a creature, and what kind of relation does it possess to the Creator? (4) With respect to the adverb, what does the relation of God's act of creating have to time, and what does the fact that creation has a beginning say about the nature and destiny of the creature?[3]

ONE GOD BEGINNING AND END

What reality does the word "God" identify in our master sentence? Most Christian theologians agree that we should understand the God named in Genesis 1:1 as the one God of Israel and the Father of Jesus Christ, whom Christians worship. However, such affirmations are not often carried through to their conclusions. Theologians often write about creation and the Creator as if he were an anonymous, transcendent cause of the universe and the topic of creation could be treated in relative isolation from the other topics in theology. In contrast to these approaches, I believe we should understand the word "God" in this sentence in light of the entire economy of salvation. The Creator of heaven and earth is also Lord, preserver, judge and savior of creation. The one God is not only the beginning but also the goal of creation. The beginning anticipates the end, and the end completes the beginning. The omnipotent Creator is also the loving, merciful and just ruler of his creatures. God always has been, is and always will be Father, Son and Holy Spirit.[4]

The Christian doctrine of creation places central focus on God's act of creation and the product of that act without losing sight of the existence and attributes of God and his other acts and their results. The doctrine of God places its central focus on the existence and attributes of God, who is Creator, preserver, judge and savior of creation. Other doctrines focus on other

---

[2]I will not deal in this book with the constitution of human nature, that is, with issues surrounding the relationship between soul and body or their natures. Nor will I enter into discussion of the ethics of the human relationship with other creatures and nature as a whole.

[3]John Webster divides the doctrine into three parts: "The Christian doctrine of creation treats three principal topics: the identity of the creator, the divine act of creating and the several natures and ends of created things. These topics are materially ordered: teaching about the identity of the creator governs what is said about his creative act and about what he creates" ("Love Is Also a Lover of Life: Creatio ex Nihilo and Creaturely Goodness," *Modern Theology* 29 [2013]: 157).

[4]Wolfhart Pannenberg, though he follows the traditional pattern of treating creation as a distinct topic of theology, agrees that "in a broader sense the fulfillment of the creature might be included in the concept of creation" (*Systematic Theology*, trans. Geoffrey Bromiley [Grand Rapids: Eerdmans, 1991], 2:8).

divine acts and their fruits. Hence although there are many topics of doctrine there is only one God who is the subject of them all. In the doctrine of creation, when speaking of the Creator, we are speaking of the one whose existence and attributes are the central focus of the doctrine of God. We must not allow the fact that we are focusing on one of God's acts, which is especially associated with divine omnipotence, blind us to the fullness of the divine being and action. Hence it is appropriate to remind ourselves here of the Christian doctrine of God.

In my book, *Great Is the Lord: Theology for the Praise of God*, I explained and defended what is often called the "traditional" doctrine of God.[5] This is the teaching about God's existence and attributes that has been held from the early centuries to today by most theologians in nearly all churches. The triune God is one, simple, spiritual, eternal, independent, immortal, immutable, incomprehensible, ineffable, impassible, omnipotent, omnipresent, omniscient and glorious, but also triune, loving, merciful, just, wise and patient. A theologian who thinks of God in these terms will construct a doctrine of creation consistent with the traditional doctrine. For, as I contended above, the doctrine of God and the doctrine of creation speak about the same God. And a doctrine of creation developed by a theologian who rejects or modifies the traditional doctrine of God will correspondingly fall in line with this modified teaching about God. Indeed many disagreements among theologians in the areas of creation and providence resolve into disagreements about the doctrine of God. And many of those disagreements turn on differences in relating God's transcendence to his immanence. How can a wholly other, infinite, completely independent God relate to a finite, changing, temporal world? Answers to this question can range from one extreme to another. To guard God's transcendence, some deny that God is related to the world in any sense. The divine act of creation as presented in the Bible is impossible without compromising the transcendence and perfection of God. Aristotle, Plotinus and the Gnostics fall into this extreme deist category.[6] Those at the other extreme of total immanence actually agree with extreme deism that the divine act of creation cannot be conceived without giving up divine transcendence. Their view of extreme divine im-

---

[5]Ron Highfield, *Great Is the Lord: Theology for the Praise of God* (Grand Rapids: Eerdmans, 2008).
[6]See David Sedley, *Creationism and Its Critics in Antiquity* (Berkeley: University of California Press, 2007), for the ancient debate about creation.

manence leads them to the edge of atheism or to what often amounts to atheism: pantheism. Of course most Christian theologians fall somewhere between these extremes, with one group sacrificing an element of transcendence to secure some immanence or others giving up some immanence to preserve a degree of transcendence. In contrast, traditional Christian theologians intuitively resist placing transcendence and immanence over against each other as opposites. These theologians argue that the very same nature that makes God so transcendent and perfect enables him to be absolutely immanent. The contrast between theologians who view divine transcendence and immanence as opposed to each other and those theologians who view them as complementary will come into play repeatedly in this study of the doctrine of creation.[7]

Simplicity. Central to the traditional doctrine of God is the doxological principle that we ought to speak of God as unsurpassable in greatness and that we ought never think or speak of God in a way that indicates imperfection. Indeed, doxology presses us beyond the thinkable or expressible toward silent wonder in the presence of the ineffable and the incomprehensible God. To illustrate how differences in the doctrine of God play out in the doctrine of creation we will examine briefly four traditional divine attributes: simplicity, eternity, immutability and omnipotence. The notion of divine simplicity is an important part of the doxological doctrine of God.[8] It contends that composition is a mark of imperfection. Anything that exists because other things are set into relation with each other is derivative; its

---

[7]Kathryn Tanner makes this division among theologies of creation an interpretive focus of her study of the doctrine of creation in contemporary theology. She sees a fundamental division between theologians who view divine and creaturely action as "contrastive" or "competitive" and those who view them as noncompetitive (*God and Creation in Christian Theology* [Minneapolis: Fortress, 2005]).

[8]For my treatment of this topic, see *Great Is the Lord*, pp. 261-74. For a book-length study of divine simplicity, see Frederik Immink, *Divine Simplicity* (Kampen: Kok Pharos, 1987). For a short defense of divine simplicity, see Steven R. Holmes, "Something Much Too Plain to Say: Towards a Defense of the Doctrine of Divine Simplicity," in *Listening to the Past: The Place of Tradition in Theology* (Grand Rapids: Baker Academic, 2002), pp. 50-67. See also Katherin Rogers, *Perfect Being Theology* (Edinburgh: Edinburgh University Press, 2000), p. 27; and Rogers, "The Traditional Doctrine of Divine Simplicity," *Religious Studies* 32 (1996): 165-86. Modern critics of divine simplicity include Alvin Plantinga, *Does God Have a Nature?* (Milwaukee: Marquette University Press, 1980); Christopher Hughes, *On a Complex Theory of a Simple God: An Investigation in Aquinas' Philosophical Theology* (Ithaca, NY: Cornell University Press, 1989); and Nicholas Wolterstorff, "Divine Simplicity," in *Philosophy of Religion*, ed. James Tomberlin, Philosophical Perspectives 5 (Atascadero, CA: Ridgeview, 1991), pp. 531-52.

cause lies outside itself. But nothing that depends on a cause outside itself can be thought of as God. Composite things are by nature subject to decomposition and death. But God is immortal, that is, God *cannot* die. Nothing that is composite can know itself perfectly because not all of its being is its own act. But God knows all things. I could continue this line of reasoning indefinitely, but it should be clear that the cost of denying divine simplicity is God's very deity. The tradition is quite clear on this point. Divine simplicity was defended by Augustine of Hippo (354–430)[9] and innumerable other Christian thinkers, ancient and modern, Roman Catholic and Protestant, who would agree with Gregory of Nyssa (ca. 330–ca. 395), who said, "We believe that the most boorish and simple-minded would not deny that the Divine Nature, blessed and transcendent as it is, is 'single.'"[10] John of Damascus (ca. 675–ca.749), says,

> The Deity is simple and uncompound. But that which is composed of many and different elements is compound. If, then, we should speak of the qualities of being uncreate and without beginning and incorporeal and immortal and everlasting and good and creative and so forth as essential differences in the case of God, that which is composed of so many qualities will not be simple but must be compound. But this is impious in the extreme.[11]

Thomas Aquinas (ca. 1225–1274) stated what, for him and his contemporaries, was crystal clear: "For there is neither composition of quantitative parts in God, since He is not a body; nor composition of matter and form . . . nor His essence from His existence; neither is there in Him composition of genus and difference, nor of subject and accident. Therefore, it is clear that God is nowise composite, but is altogether simple."[12]

Critics of divine simplicity argue that the doctrine represents an invasion by Greek metaphysics into Christian theology and is alien to the biblical worldview,[13] that the doctrine implies that God is not a person and hence

---

[9]Augustine, *City of God* 11.10 (*NPNF*[1] 2:210-11).

[10]Gregory of Nyssa, *Against Eunomius* 1.19 (*NPNF*[2] 5:57).

[11]John of Damascus, *Orthodox Faith* 1.9 (*NPNF*[2] 9:12).

[12]*SumTh* 1.3.7 (Anton C. Pegis, ed., *Basic Writings of Saint Thomas Aquinas* [New York: Random House, 1945], 1:34).

[13]Albrecht Ritschl, "Metaphysics and Theology," in *Albrecht Ritschl: Three Essays*, trans. Philip Hefner (Philadelphia: Fortress, 1972); William Herrmann, *Die Metaphysik in der Theologie* (Halle: Niemeyer, 1876); Adolf von Harnack, *History of Dogma*, trans. Neil Buchanan, 3rd ed. (New York: Russell & Russell, 1958), vol. 1; Emil Brunner, *The Christian Doctrine of God*, vol. 2, *Dogmatics*, trans. Olive Wyon (Philadelphia: Westminster, 1949), p. 243.

cannot know anything or do anything,[14] and that it denies what the Bible plainly teaches, that God is dynamic and relational.[15] Responding in detail to the critics of divine simplicity goes beyond the scope of this chapter and would repeat what I have done elsewhere.[16] I shall, rather, show its relevance to the doctrine of creation. Critics think that denying divine simplicity will enable us to conceive how God can change from the state of not being the Creator to being Creator. It allows us, they argue, to think of God as being able to enter into a real dynamic relation to creation, to love, to experience the temporal lives of his creatures, to respond to prayer and to suffer grief and pain—just the kind of relationships and interactions narrated in Scripture. Two observations: (1) These advantages are purchased at the price of complete divine transcendence and perfection, precisely the strategy I rejected above, making transcendence and immanence mutually exclusive. (2) What the critics seek is rational understanding of the mediation between God and the world. It is precisely this type of rationalism the doctrine of divine simplicity was designed to prevent. The doctrine of divine simplicity serves a Christian theology of creation by standing in the way of efforts to comprehend divine action in rational terms. As Brian Davies puts it, "From first to last the doctrine of divine simplicity is a piece of negative or apophatic theology and not a purported description of God."[17] Hence I will not seek a rational understanding of how God creates the world or acts in history and nature or answers prayer or works his will in all things.

Adhering to divine simplicity influences the doctrine of creation in another way. We must think of God's attributes or properties as finding their referent in the one simple being of God. As Dionysius the Areopagite (ca. 500) says, "all the Names [that is, attributes] proper to God are always applied in Scripture not partially but to the whole entire Godhead, and that they all refer indivisibly, absolutely, and unreservedly, and wholly to the

---

[14]Plantinga, *Does God Have a Nature?*

[15]Clark Pinnock, *Most Moved Mover: A Theology of God's Openness* (Grand Rapids: Baker Academic, 2001), pp. 68-78.

[16]Highfield, *Great Is the Lord*, pp. 268-74.

[17]Brian Davies, "Classical Theism and the Doctrine of Divine Simplicity," in *Language, Meaning, and God: Essays in Honor of Herbert McCabe*, ed. Brian Davies (London: Cassell, 1987), p. 59. David Burrell, in discussing Thomas Aquinas's view of simplicity, argues that Aquinas is not attempting to peer into the divine being but to guide our language so that we will know what not to say about God. Simplicity is a wholly negative doctrine (*Aquinas: God and Action* [Notre Dame, IN: University of Notre Dame Press, 1979], pp. 16-17).

wholeness of the whole and entire Godhead."[18] When Anselm of Canterbury
(ca. 1033–ca. 1109) speaks of simplicity he says, "For this Being it is the same
to be just that it is to be justice; and so with regard to attributes that can be
expressed in the same way: and none of these shows of what character, or
how great, but what this Being is."[19] The Reformed theologian Benedict
Pictet (1655–1724) defines divine simplicity as "nothing more than the intimate connection and entire unity of all the attributes of God and their
oneness or identity with the divine essence itself."[20] Because all God's attributes refer to God's simple essence, we must think of all God's actions as
expressions of the whole divine being. It makes no sense to say that God is
at one time just, at another loving and at another powerful or omnipresent.
Adhering to divine simplicity excludes viewpoints that make a particular
divine action arise from one aspect of God's character in isolation or in
tension with others.[21] Hence God's act of creation cannot be understood
merely as an act of power; it is also characterized by love, mercy, grace and
all the rest of the attributes. Creation cannot be attributed to the Father
alone, with reconciliation seen as a work of the Son alone and redemption
given over exclusively to the Spirit. All God's acts are acts of the whole
Trinity. God's act of creation cannot be separated from his saving and redeeming actions. Indeed, as I will explain later in this study, these "acts"
should not be thought as separate acts in too close analogy with human acts,
which must succeed one another in time. God's act is one even if the results
are manifold.

Eternity. The relationship between time and eternity plays an important
role in the doctrine of creation. Thinkers that differ in their understanding
God's eternity will carry those differences into their doctrines of creation.
Many modern theologians reject the classic notion of divine eternity and
argue that God is everlasting, that is, God lives a temporally sequenced but
infinite life, with no beginning or end.[22] Some advocate everlastingness in

---

[18]Dionysius the Areopagite, *Divine Names* 2 (C. E. Rolt, trans., *Dionysius the Areopagite on the Divine Names and The Mystical Theology* [Berwick, ME: Ibis, 2004], p. 65).

[19]Anselm of Canterbury, *Monologium* 16 (Saint Anselm, *Basic Writings*, trans. S. N. Deane, 2nd ed. [LaSalle, IL: Open Court, 1968], p. 64).

[20]Benedict Pictet, *Christian Theology*, trans. Frederick Reyroux (London: Seeley and Burnside, 1834), 2.8 (p. 99).

[21]The classic example of this error is the supposed tension the fall created between God's justice and his mercy, which made atonement through the death of Christ necessary.

[22]For my critique of everlastingness, see *Great Is the Lord*, pp. 302-6.

the interest of making God's relationship to creation conceivable. Since God lives in a series of moments or phases, God can relate to the temporal creation because God shares a common moment of time—the present—in which to meet with creation.[23] As one can readily see, this viewpoint uses time as a medium that encloses both God and creation in a kind of common space where a relationship between the two becomes possible.[24] The traditional doctrine of God, however, views time as the mode of existence of imperfect beings that must actualize their potential in successive stages and in dependence on external reality. They are never wholly with themselves in their full actuality. To conceive of God as temporal, even if everlasting, would force us to think of God as a being always in the process of *becoming* but never actually *being* God. God would never be identical to himself, simultaneous with himself, in control of himself, knowing himself completely. God's existence would never equal his nature in greatness. God might be conceived as a god but not God in the doxological sense.[25] In contrast to imperfect creatures, the tradition views God as fully actual and simultaneously and wholly present and transparent to himself. Boethius (ca. 480–ca. 524) gave classic form to the definition of God's eternity, and Christian thinkers, Roman Catholic and Protestant, have returned to it again and again for insight:

> The common opinion, according to all men living, is that God is eternal. Let us therefore consider what is eternity. . . . *Eternity is the simultaneous and complete possession of infinite life.* This will appear more clearly if we compare it with temporal things. All that lives under the conditions of time moves through the present from the past to the future; there is nothing set in time

---

[23]Nicholas Wolterstorff, "God Everlasting," in *Contemporary Philosophy of Religion*, ed. S. Cahn and D. Shatz (Oxford: Oxford University Press, 1982), pp. 77-98; and Wolterstorff, "Unqualified Divine Temporality," in *Four Views: God and Time*, ed. Gregory E. Ganssle (Downers Grove, IL: InterVarsity Press, 2001), pp. 186-213. William Lane Craig also defends divine temporality. However, in Craig's view God becomes temporal only "after" creation. See his *Time and Eternity: Exploring God's Relationship to Time* (Wheaton, IL: Crossway, 2001); and Craig, "Timelessness and Omnitemporality," in Ganssle, *Four Views: God and Time*, pp. 129-60. Alan G. Padgett views divine eternity as limitless duration, which is not the same concept as the Boethius's "simultaneousness." See Padgett's *God, Eternity and the Nature of Time* (New York: St. Martin's, 1992); and Padgett, "Eternity as Relative Timelessness," in Ganssle, *Four Views: God and Time*, pp. 92-110.

[24]We see here again what we noticed when discussing divine simplicity: the quest for a rational understanding of mediation between God and the world. And as we shall see, such views neglect the Christian answer to the problem of mediation, the Trinity.

[25]Clearly, viewing God as temporal goes hand in hand with denying divine simplicity. God's life would be composed of moments of time.

which can at one moment grasp the whole space of its lifetime. It cannot yet comprehend tomorrow; yesterday it has already lost. And in this life of today your life is no more than a changing, passing moment. And as Aristotle said of the universe, so it is of all that is subject to time; though it never began to be, nor will ever cease, and its life is co-extensive with the infinity of time, yet it is not such as can be held to be eternal. For though it apprehends and grasps a space of infinite lifetime, it does not embrace the whole simultaneously; it has not yet experienced the future. What we should rightly call eternal is that which grasps and possesses wholly and simultaneously the fullness of unending life, which asks naught of the future, and has lost naught of the fleeting past; and such an existence must be ever present in itself to control and aid itself, and also must keep present with itself the infinity of changing time. Therefore, people who hear that Plato thought that this universe had no beginning of time and will have no end, are not right in thinking that in this way the created world is coeternal with its Creator.[26]

Approaching the doctrine of creation through the lens of the traditional understanding of divine eternity as "*the simultaneous and complete possession of infinite life*" will significantly shape the doctrine. On one level the doctrine of creation becomes much more difficult and mysterious. For how can God go out of himself to create the world if nothing exists—no space or time—external to God to serve as medium within which to create? How can the Eternal be meaningfully said to act when the very concept of an act seems to presuppose time? And if God's act of creation, like himself, is eternal, must creation also be eternal? How can an eternal God experience the world as temporal? These and other difficult questions do not arise if one assumes that God and creatures share the temporal mode of being. But as I pointed out above, this advance for rational transparency is purchased with a retreat of God's divinity and praiseworthiness.

Immutability. Traditional theology defended the idea of divine immutability,[27] that is, since God is eternally perfect and fully actual, there is no need and no possibility for God to change.[28] Change implies imper-

---

[26]Boethius, *Consolation of Philosophy* 5, trans. W. V. Cooper (London: J.M. Dent, 1902), pp. 160-61.

[27]Immutability and impassibility—God's freedom from change in passion or emotion—are closely related. Immutability rejects change of any kind, while impassibility rules out change in emotional state. For my thoughts on impassibility, see *Great Is the Lord*, pp. 375-91.

[28]For my much fuller explanation and defense of the doctrine of divine immutability, see ibid., pp. 258-75. For book-length defenses, see Thomas G. Weinandy, *Does God Change? The Word's*

fection, potentiality and time, none of which applies to God.[29] Augustine explains, "For that which is changed does not retain its own being; and that which can be changed, although it be not actually changed, is able not to be that which it had been; and hence that which not only is not changed, but also cannot at all be changed, alone falls most truly, without difficulty or hesitation, under the category of Being."[30]

The tradition argued that God is eternally fully actual, which is doxologically optimal.[31] For it allows us to praise God as the One in whom every possible good is infinitely actualized eternally. As was the case with simplicity and eternity, critics of the traditional doctrine of immutability argue that it must be modified to secure God's ability to relate to a temporal and changing creation.[32] If God could not change, God could not create, for creating is an act of establishing a new relation. God changes from not being related to being related to creation, from not being Creator to being the Creator. Moreover, were God unable to change, God could not respond to changing circumstances within creation. If God could not change, praying the Lord's Prayer asking God to "give us our daily bread" or "lead us not into temptation but deliver us from evil" would make no sense.

From a traditional perspective, giving up divine immutability to make God's relationship to creation rationally transparent surrenders a central tenet of the Christian doctrine of creation, that is, that the Creator of our world is unsurpassably perfect and absolutely praiseworthy. In their less diplomatic moments, traditional theologians accuse critics of immutability of believing that only an imperfect and less than doxologically maximal God could have created the world we experience. In their even less charitable moments, defenders of immutability place the doctrine's modern critics in the camp of Marcionites, Gnostics and Manichees, who, for all their differ-

---

Becoming in the Incarnation (Still River, MA: St. Bede's, 1985); and Michael Dodds, *The Unchanging God of Love: Thomas Aquinas and Contemporary Theology on Divine Immutability*, 2nd ed. (Washington, DC: Catholic University of America Press, 2008).

[29]*SumTh* 1. 9. 1 (Pegis, *Basic Writings*, 1:70-71).

[30]Augustine, *On the Holy Trinity* 5.2.3 (*NPNF*¹ 3:88).

[31]For Augustine the superiority of immutability over change is obvious: "Now, no one is so egregiously silly as to ask, 'How do you know that a life of unchangeable wisdom is preferable to one of change?'" (Augustine, *On Christian Doctrine* 1.9 [*NPNF*¹ 2:525]).

[32]For an evangelical open theist perspective, see John Sanders, *The God Who Risks: A Theology of Providence*, 2nd ed. (Downers Grove, IL: IVP Academic, 2007); Pinnock, *Most Moved Mover*. For a critique from a process philosopher, see Charles Hartshorne, *The Divine Relativity* (New Haven, CT: Yale University Press, 1948), pp. 18-22.

ences from each other and modern critics of immutability, begin their systems on the very same premise. The Gnostics say, "The highest God is perfect and immutable, so the highest God did not create the world of our experience." The modern critics of immutability say, "The highest God created the world of our experience, so the highest God cannot be immutable and perfect." Both agree that immutability and absolute perfection are not compatible with creation. They differ only in which horn of the dilemma they choose.

In contrast to both Gnostics and the modern critics of divine immutability, traditional theology asserts both that God is absolutely perfect *and* that God created the world of our experience. It refuses to treat the profound mystery at the heart of the Christian doctrine of creation as an elementary logical mistake. And it joyfully ventures ever deeper into the mystery of God's relationship with creation, seeking greater clarity and deeper intimacy with that which cannot finally be resolved into propositions and their logical relations. In contrast to the caricature of the immutable God as cold, aloof, dead, unresponsive and static, the traditional doctrine understands God as unchangingly active and eternally fully alive. The term *static* applies only to something that could be active but is not. Those who defend immutability assert that God is pure act.[33] God's immutability does not make God aloof but guarantees that God is always and already there for us as our good in every situation.[34]

Omnipotence. Perhaps omnipotence is the divine perfection most closely associated with the doctrine of creation. The first sentence of the Nicene Creed (381) reads: "We believe in one God the Father all-powerful, Maker of heaven and earth, and of all things both seen and unseen."[35] The creed's connection between divine power and creation seems quite natural and was followed in all subsequent theology. The universe itself presents an awesome display of power.

---

[33]See Theodore J. Kondoleon, "The Immutability of God: Some Recent Challenges," *The New Scholasticism* 58 (1984): 293-315. Kondoleon argues the term *static* should not be applied to the immutable God (p. 298). For defenses of the traditional doctrine of immutability, see Dodds, *Unchanging God of Love*; Gerald Hanratty, "Divine Immutability and Impassibility Revisited," in *At the Heart of the Real* (Dublin: Irish Academic Press, 1992), pp. 135-62; and Weinandy, *Does God Change?* For a historical sketch of the doctrine of immutability in post-Reformation Reformed dogmatics, see Richard Muller, *PRRD* 3:308-20. According to Muller, modern critics of immutability and impassibility are not simply beating a dead horse, they are beating a "nonexistent horse" (p. 310).

[34]See my extensive treatment of divine immutability in *Great Is the Lord*, pp. 358-75.

[35]Jaroslav Pelikan and Valerie Hotchkiss, eds., *Creeds and Confessions of Faith in the Christian Tradition* (New Haven, CT: Yale University Press, 2003), 1:163.

How much more should we attribute power to the Creator! The traditional doctrine of divine omnipotence can be summed up in three theses: God is his power, God can do all things and God empowers all other powers. Dionysius the Areopagite touches on all three affirmations when he explains that

> God is power because in His own Self He contains all power beforehand and exceeds it, and because He is the Cause of all power and produced all things by a power which may not be thwarted nor circumscribed, and because He is the Cause wherefrom Power exists whether in the whole system of the world or in any particular part. Yea, He is Infinitely powerful not only in that all Power comes from Him, but also because He is above all power and is Very Power, and possesses that excess of Power which produces in infinite ways an infinite number of existent powers. . . . And, in short, there is nothing in the world which is without the Almighty Power of God to support and to surround it. For that which hath no power at all hath no existence, no individuality, and no place whatever in the world.[36]

In treating the first thesis mentioned above—God is his power—traditional theologians argue not only that God's power knows no limit but also, in keeping with the doctrine of simplicity, that God *is* his power. God does not merely have power or use power but is power, just as God *is* wisdom and love. And just as God's wisdom and love cannot be limited or measured in terms of quantity—because God himself cannot be limited or divided— God's power cannot be limited or measured. God does not deplete or strain or alter in anyway his power by creating. The second thesis affirms that God can do anything, anything that is really a "thing" and not a mere contradiction. Thomas Aquinas addresses this thesis:

> God is called omnipotent because He can do all things that are possible absolutely. . . . For a thing is said to be possible or impossible absolutely, according to the relation in which the very terms stand to one another, possible if the predicate is not incompatible with the subject, as that Socrates sits; and absolutely impossible when the predicate is altogether incompatible with the subject, as, for instance, that a man is a donkey.[37]

---

[36]Dionysius the Areopagite, *The Divine Names* 8 (Rolt, *Divine Names and the Mystical Theology*, pp. 154-57). Dumitru Staniloae uses Dionysius as his main example of the Greek fathers' teaching on the subject of divine omnipotence. See Staniloae, *The Experience of God*, vol. 1, *Revelation and Knowledge of the Triune God*, trans. and ed. Ioan Ionita and Robert Barringer (Brookline, MA: Holy Cross Orthodox Press, 1994), pp. 184-85.
[37]*SumTh* 1.25.3 (Pegis, *Basic Writings*, 1:263).

In grasping this thesis it is very important to keep clear the difference among a real existing thing, a common noun denoting a simple property or essence, and mere juxtaposition of words that form a nonsensical term. It is sometimes objected that the tradition did in fact limit God by denying that God can make a thing or perform an action that can be described in words only with contradiction. Can God make a "square circle"? The proper answer to this question is not yes or no. The proper answer is that the question is nonsense. A square-circle is not a thing hard to make or even a thing impossible to make. It is not a thing at all, because a square-circle is not a concept at all but a pseudo-concept, words put side by side to form a term that corresponds to nothing. Each word alone denotes a clear concept or property, and together they possess a meaningful grammatical form, a common noun preceded by an adjective. However, considered as a term it says nothing at all. Therefore nothing at all can be said in response.

The third aspect to the traditional doctrine of omnipotence, the thesis that nothing can happen apart from God's power, will play an important part in the doctrine of providence in the second half of this book. It is sufficient here to point out that God's act of creation is not a mere past event with the world continuing to exist and act by its own power. The act of creation and the power exerted in that act apply to the entire history of creation. The end of the beginning of creation in no way signifies a withdrawal of God's power or cessation of his act of creating the world.

Traditional theologians see creation as God's quintessential act of omnipotence. Calling this beautiful, vast, varied and complex universe into being from nothing demonstrates God's incomparable power. If God can create the universe simply by commanding it to exist, there can be no limit to his power. Nevertheless, there are many theologians from a variety of schools of thought that either deny God's omnipotence outright or argue that the existence of our world is incompatible with the *exercise* of omnipotence. Process theism falls into the first category. Charles Hartshorne argues that the very existence of anything other than God imposes limits on God, and since Hartshorne denies the traditional doctrine of creation from nothing, he contends that God has always been accompanied by the world. Hence God cannot be omnipotent.[38] The second group does not want to go

---

[38]Charles Hartshorne, *Omnipotence and Other Theological Mistakes* (Albany: State University of New York Press, 1984); David Ray Griffin, *God, Power and Evil* (Philadelphia: Westminster,

as far as Hartshorne but accepts a central tenet of process thought, that is, the tension between divine omnipotence and creaturely freedom.[39] They argue that though God is omnipotent in that God can do anything except something that involves contradiction, God must limit himself to create a world like ours. Clark Pinnock says, "Of course God is omnipotent. . . . [But God] exercises the kind of omnipotence which is compatible with his own decision to create a world with free agents."[40] Pinnock "understands God to be voluntarily self-limited, making room for creaturely freedom. Without making God finite, this definition appreciates God's delighting in a universe which he does not totally control."[41] Speaking for open theism, David Basinger explains, "We do not believe that God can unilaterally ensure that all and only that which he desires to come about in our world will in fact occur. . . . God voluntarily forfeits control."[42] So for these thinkers God's act of creation is at the same time an act of self-limitation or self-restriction that results in loss of complete control.

We shall deal with these and other challenges to omnipotence when we study God's act of creating in detail. Even this short anticipation shows that writing a doctrine of creation consistent with the traditional doctrine of God must show that creating and accompanying the world is an expression of divine omnipotence rather than an abandonment of that power.

THE TRINITY AND CREATION

We cannot remain satisfied to identify the Creator only by the attributes of simplicity, eternity, immutability and power; nor would it complete the task if

---

1976); and Griffin, *Evil Revisited: Responses and Reconsiderations* (Albany: State University of New York Press, 1991).

[39]See my discussion in *Great Is the Lord*, pp. 352-57. See also Ron Highfield, "Divine Self-Limitation in the Theology of Jürgen Moltmann: A Critical Appraisal," *Christian Scholar's Review* 32 (2002): 47-71; Highfield, "The Function of Divine Self-Limitation in Open Theism: Great Wall or Picket Fence?," *Journal of the Evangelical Theological Society* 45 (2002): 279-99; Highfield, "Does the World Limit God: Assessing the Case for Open Theism," *Stone-Campbell Journal* 5 (2002): 69-92; and Highfield, "The Problem with the 'Problem of Evil': A Response to Gregory Boyd's Open Theist Solution," *Restoration Quarterly* 45 (2003): 165-80.

[40]Clark Pinnock, "God Limits His Knowledge," in *Predestination and Free Will: Four Views of Divine Sovereignty and Human Freedom*, ed. David Basinger and Randall Basinger (Downers Grove, IL: InterVarsity Press, 1986), p. 153.

[41]Clark Pinnock, "Systematic Theology," in *The Openness of God: A Biblical Challenge to the Traditional Understanding of God*, ed. Clark Pinnock et al. (Downers Grove, IL: InterVarsity Press, 1994), p. 117. See also Pinnock, *Most Moved Mover*, pp. 92-96.

[42]David Basinger, "Practical Considerations," in Pinnock et al., *Openness of God*, p, 159.

we add such attributes as love, grace, wisdom and righteousness. Christianity identifies the God it worships and serves as the Father of our Lord Jesus Christ, as Father, Son and Spirit.[43] It knows no other God. Robert Jenson rightly says:

> Father, Son and Holy Spirit is simultaneously a very compressed telling of the total narrative by which Scripture identifies God and a personal name for the God so specified. . . . The church is the community and a Christian is someone who, when the identity of God is important, names him "Father, Son, and Holy Spirit." Those who do not or will not belong to some other community.[44]

Christianity does not think of God's triunity as an isolated factor that could be removed without damage to the doctrine of God. Nor is it true that Christianity and other monotheist religions share everything in doctrine of God except belief that God is triune. The doctrine of the Trinity is fundamental to Christianity and all-pervasive in Christian thinking. In Thomas Torrance's words, the doctrine of the Trinity constitutes the "fundamental grammar of Christian dogmatic theology."[45] It determines the way every other Christian teaching is framed.

The Christian identification of God as Father, Son and Holy Spirit receives its decisive impetus from the life, death and resurrection of Jesus Christ from the dead and from the sending of the Holy Spirit on the church.[46] Jesus Christ becomes the center and goal of history, revealer of God, savior of humanity and Lord and Judge of the world. From the apostles' and the apostolic church's experience of God's revelation in Christ arises the conviction that Christ is the center of the comprehensive divine movement toward the world, which flows "from the Father *through* the Son in the Spirit, and correspondingly in the trinitarian movement of faith and devotion in the Church, *in* the Spirit *through* the Son to the Father."[47] Christ is the center

---

[43]While the theological task of the doctrine of God includes examining whether the church's doctrine of the Trinity is grounded in Scripture, is coherent in itself and is consistent with other Christian doctrines, theologians of creation must presuppose the trinitarian doctrine in their work on creation. Otherwise they would be forced to begin all over again with every new topic. For my chapter on the Trinity, see *Great Is the Lord*, pp. 104-38.

[44]Robert Jenson, *Systematic Theology*, vol. 1, *The Triune God* (New York: Oxford University Press, 1997), p. 46.

[45]Thomas Torrance, *The Christian Doctrine of God: One Being Three Persons* (Edinburgh: T & T Clark, 1996), p. 82.

[46]There was of course a larger context for interpreting the significance of Jesus in the Old Testament and in contemporary Judaism. Telling this story falls outside the scope of this book.

[47]Torrance, *The Christian Doctrine of God*, p. 99.

of God's economy of creation and salvation both as it flows from the Father and is perfected by the Spirit, and as it returns to the Father. Because the church receives the gift of salvation from the Father through the Son in the Spirit and finally returns perfected in Spirit through the Son to the Father, the New Testament church confessed, worshiped, thanked and glorified the Son and the Holy Spirit along with the Father. The conviction that the true God became known and acted through Christ and the Holy Spirit gave rise to the identification of God as Father, Son and Spirit; God is in and for himself what God is for us. To use Thomas Torrance's words again, "The focal point is the Lord Jesus Christs. . . . What *Jesus Christ* is toward us in love and grace, in redemption and sanctification, in the mediation of divine life, he is inherently in himself in his own divine Being . . . [and] what *God* is toward us in Christ Jesus, he is inherently and eternally in himself in his own Being."[48] The economic Trinity is the immanent Trinity *for us.*[49]

The doctrine that God is triune does not simply explain and deepen our understanding of God's oneness by helping us to see its threefold unity. It's not a matter of metaphysical curiosity. Since the doctrine of the Trinity is Christocentric, it teaches us to view God through the lens of the life, work, character, death and resurrection of Christ. In Christ we know the power and wisdom and love of God and we experience the grace, mercy and patience of God. Everything we say about God must be spoken from this center and take into account God's triune life and activity.[50] And this "everything" includes God's act outside of himself, creation. To be genuinely Christian a doctrine of creation must be thoroughly Christocentric and trinitarian. On the metaphysical level this means that the multiform "problem of mediation"—the problem of the "one and the many" or the relation between transcendence and immanence or the problem of the "other"—must not be addressed by using eternal ideas or concepts or other nondivine things as mediating archetypes to bridge the gap between God and the world. Insight into the mystery of God's relation to the nondivine "other" begins with the Father's eternal relation to the Son and the Spirit,

---

[48]Ibid., p. 99.

[49]This is what I take as the meaning of Karl Rahner's famous epigram: "The 'economic' Trinity is the 'immanent' Trinity, and the 'immanent' Trinity is the 'economic' Trinity" (*The Trinity*, trans. Joseph Donceel [New York: Crossroad, 1997], p. 22).

[50]In my *Great Is the Lord*, I attempt to reflect in a trinitarian way on all the traditional divine attributes.

whose otherness of relation and unity of being with the Father are eternal and constitute the being of God. At the level of character or identity, creation can be understood as the act of the "God and Father of our Lord Jesus Christ." The Son of God incarnate as Jesus Christ is the one through whom creation was accomplished. Hence the act of creation is not merely an act of power and a revelation of glory but also an act of love, grace, generosity and condescension. Creation emerges in our thinking as the beginning of salvation and salvation the completion of creation. Just as creation came *through* Christ, salvation occurs *in* Christ. Christ is the end and meaning of creation.[51]

DIVINE FREEDOM

We must deal with one last issue before we leave the subject of the God who creates to examine his act of creation. As we have already noted, God is willing and able to create. God's power is infinite, and his gracious outreach in Jesus proves that his love is without limit. It is important to emphasize now that God is under no necessity or compulsion to create. God is totally independent and self-sufficient in every respect. There is no lack within God and no good outside of him that can explain why God created. In the words of the early Reformed Protestant theologian Wolfgang Musculus (1497–1563), God is "sufficient of himself, that not only has he all of himself, but he also suffices himself, that he has no need either of any of the things which he has, and which he has made and created. He is sufficient to himself through all things and unto all things. . . . He should not have been a point the richer, albeit he had created a thousand worlds."[52]

God is the infinite source of every finite good and possesses every good, every possible good, within his eternal life. "In his freedom, God is complete and possesses himself completely. God is alive, aware, and powerful—all in the absolute freedom of his self-determination. In himself and from himself and to himself, God is beautiful and bountiful, joyful and glorious. God is

---

[51]Jonathan R. Wilson devotes a chapter in his book on creation to applying the "trinitarian grammar" to the act of creating and perfecting the world. See *God's Good World: Reclaiming the Doctrine of Creation* (Grand Rapids: Baker Academic, 2013), pp. 71-96. The proper grammar can be stated as follows: "The Father initiates the work of creation, the Son implements that work and the Spirit completes it" (p. 85). Wilson maps the most common errors concerning creation as errors of trinitarian grammar.

[52]Wolfgang Musculus, *Loci Communes* 43, quoted in *PRRD* 3:369.

his own infinite goodness and God perfectly possesses this goodness as his own act and being."[53] Nevertheless from his overflowing goodness and unbounded freedom God created the world. Though God is free and uncompelled we should not think of God's act of creation as arbitrary, whimsical or experimental, because there is nothing in God's being arbitrary or accidental to serve as a foundation for a capricious act. David Fergusson's words hit the mark: "There is no natural necessity in the creation of the world. Nothing within the divine being requires that the world come into existence or implies that it must possess a particular form. As a free act, creation is gratuitous. God does not need to make the world; its appearance is itself a sheer act of divine grace."[54]

In light of God's love and power revealed in Jesus Christ we can affirm, if not comprehend, that God's act of creation, uncompelled yet not arbitrary, is fully consistent with and expressive of his eternal nature and character. Given its uncompelled and free character, the act of creation can be understood only as an act of divine love rooted in and reflective of the Father's eternal love for the Son in the Holy Spirit. With this thought, however, we are already venturing into the next topic, God's act of creation.

---

[53]Highfield, *Great Is the Lord*, p. 227.
[54]David Fergusson, *Creation*, Guides to Theology (Grand Rapids: Eerdmans, 2014), p. 20.

4

How God Creates
the World (Part One)

Probing a Mystery or Solving a Problem?

W HAT DOES IT MEAN THAT GOD *created* the heavens and the
earth? What kind of act is this? What sort of transition occurs? By
what type of causality does the created world come to be? How does it
come about that alongside God, whose existence is necessary, there exists
the world, whose existence is not necessary? As the subtitle of this chapter
adumbrates, we can approach these questions as mysteries to be entered
into at ever deeper levels or as problems to be solved and left behind.[1] In
his Gifford Lectures, Gabriel Marcel argued that modern philosophy triv-
ialized the depth of the human person by treating human nature as a
problem to be solved in the way modern science attempts to solve its
problems. We will miss the meaning of personhood, friendship, freedom,
love and other human experiences if we treat them as phenomena to be
comprehended on our own terms.[2] How much more would treating God's
act of creation as a problem rather than a mystery prevent us from entering into the depths of the God-creature relationship? In the spirit of
Marcel, then, I aim in this chapter to explore the mystery of God's gracious
act of creating the world rather than attempt to comprehend rationally
how God created it.

---

[1]Gabriel Marcel, *The Mystery of Being*, vol. 1, *Reflection and Mystery* (London: Harvill, 1950), pp.
204-19.
[2]Ibid., pp. 204-19.

CREATION AS DIVINE ACTION

Scripture and tradition speak of God's creating the universe as an act. We can see this in our master sentence. God "created" the heavens and the earth. This sentence possesses a grammatical structure similar to sentences that describe human acts, specifically acts of making. Solomon *built* the Jerusalem temple, the Israelites *made* a golden calf and Karl Barth *wrote Church Dogmatics* are examples of human acts of making. However, given who acts and what is accomplished in the act of creating the universe, it is clear that God's act differs qualitatively from human acts. Hence a major task of the doctrine of creation is clarifying the ways God's act of creating resembles and differs from human acts of making.[3]

The Scriptures provide the rules and limits for dealing with this issue but do not develop a systematic and comprehensive explanation of the act of creation. They do not address many of the questions that have arisen since they were written, questions posed by philosophy and natural science. As we saw in our study of the biblical theology of creation, the Scriptures assert that God is sovereign and free in his relation to creation, that the world that results from the divine act of creating is good (that is, it is in fact what it was intended to be) and that the divine act was accomplished *through* the Word, or Son. We turn now to those latter-day questions not addressed explicitly by Scripture.

What is an act? If Scripture speaks of creation as like but also unlike human acts of making, it makes sense for us to begin by reflecting on the nature of action.[4] First, the idea of an act rules out a static state of affairs where nothing happens but envisions, rather, transition from one state to another. Something happens and a change occurs. But this is also true of an event or a process. When we speak of an event we envision a segment of

---

[3]Hermann Cremer made divine action central to his discussion of the divine attributes (*Die christliche Lehre von den Eigenschaften Gottes* [Gütersloh: Bertelsmann, 1897]). Wolfhart Pannenberg, while acknowledging Cremer's contribution, criticizes him for not clearly differentiating the action of creatures from that of God (*Systematic Theology*, trans. Geoffrey W. Bromiley [Grand Rapids: Eerdmans, 1988], 1:368-70).

[4]William P. Alston attempts to find an overlapping univocal core in the concepts of human and divine action. See his essay "Divine and Human Action," in *Divine and Human Action: Essays in the Metaphysics of Theism*, ed. Thomas V. Morris (Ithaca, NY: Cornell University Press, 1988), pp. 257-80. Michael J. Dodds, *Unlocking Divine Action: Contemporary Science and Thomas Aquinas* (Washington, DC: Catholic University of America Press, 2012) criticizes much contemporary theology and philosophy of divine action for assuming that divine action is univocal with human action.

time, which can be of any length, within which one state of affairs transitions to another. If nothing happens we cannot speak of an event. To account for an event we must describe the initial conditions and the final state, and then account for the transition from one to the other. Examples are abundant: a heart attack, a chemical reaction or a solar eclipse. A process is a syndrome of interconnected events thought of as one. It is a cascading flow of causally related events that moves from one complex state to another. We use the term *process* rather than *event* when we want to emphasize the manifold nature of the middle states between the initial state and the resultant one. Event, in contrast to process, ignores mediation or middle states between the two states and treats the transition as if it were instantaneous. There is, however, a decisive difference between an event or process and an act.

To speak somewhat loosely for a moment, an act is an event or process attributable to an actor, an agent. Elements in events and processes cannot be considered agents except in a metaphorical sense. A physical law, a molecule or a force is not an agent in the normal sense of that term. Strictly speaking, an agent is a self or a person from whom action can originate.[5] To be an agent a subject must possess itself in self-knowledge so that it can reason, evaluate and set goals. Insofar as it is an agent it must be free from impersonal processes, for there is no such thing as a necessary or accidental act. Finally, an agent must possess power to effect change. An act, then, is a transition from one state of affairs to another originating in and directed by an agent toward an end. To account for a human act we must learn the initial state of affairs the agent wishes to change, the intentions or desired ends, the powers possessed by the agent, the means used by the agent to effect the desired transition and the actual state of affairs that results from the agent's efforts.[6]

---

[5]In the Aristotelian and Thomistic way of thinking there is teleological action in nature as well as in human beings. Things in nature are more than the sum of their parts. The reality that integrates a thing's constituent parts into the whole thing is its substantial form. Substantial form is the idea needed to account for a thing's existence as the particular thing it is with its qualities, powers and actions. Each thing in nature exercises its particular influence based on the kind of thing it is. Not every instance of causal influence is an act, though every event is a movement from a potential to an actual state. I think it is best to reserve the term *act* for human, angelic and divine action. My decision here coheres with the view that causality is known first and best by knowing ourselves as causes. Only by extension may we apply causality to relations between natural objects. See Dodds, *Unlocking Divine Action*, pp. 18-27.

[6]For Thomas Aquinas's thinking about an act, see *SumTh* 1-2.6.1: "Is there anything Voluntary in human acts?" Aquinas speaks of acts on a scale from imperfect to perfect. The more the prin-

To illustrate the nature of a human act we will examine the act of building a bookshelf. Let's begin our account by noting the feeling of displeasure and inconvenience that arises when we see our valued books lying on the floor or scattered around the room on surfaces meant for other purposes. We then envision an orderly arrangement of our books on a shelf and begin to desire that good end. We form the intention to build a bookshelf. Assuming that we have some experience in carpentry and after calculating the height, depth and width of shelf we need, we draw up the design plan for the shelf. As a next step we acquire the materials and tools needed to construct the desired article. Finally, following the blueprint and using the tools properly we build the shelf and begin using it for the intended purpose. We now look at our work and assess our success and its perfection or lack thereof.

Several aspects of this human act stand out as relevant to our discussion of the similarities and differences between the human act of making and the divine act of creating. (1) A human act begins in the agent's perception of a lack within the agent, which in the case above results in distress. Something is missing. Human acts not only have intentions but also motives, something that urges us to act. (2) The agent imagines something that can meet the agent's need and relieve the distress. (3) Though the agent's ability to perceive a lack and imagine something to fill that lack are significant powers, these powers alone cannot supply the lack. The agent must look to the external world to provide the material and the tools necessary to construct the thing it needs. (4) In its act, the human agent not only effects change in the external world but is also changed by its interaction with the thing it acts on (the patient). The agent expends time and energy, learns, gets tired and moves from lacking to possessing something. (5) The intentions of the human agent are never perfectly clear, and its ends are never fully realized in the product, because other factors outside its control have a say in the final outcome. Furthermore, human acts always produce unintended consequences, because, once we set in motion causal chains within the world, they escape our control. Our lack of understanding of ourselves or the initial conditions or the laws of nature at work or our inability to predict human

---

ciple of the act is internal and voluntary and based on knowledge of an end, the more it approaches perfection. For discussion of this point, see Eleonore Stump and Norman Kretzmann, "Being and Goodness," in Morris, *Divine and Human Action*, pp. 281-312.

reactions guarantees that our acts will soon be alienated from our intentions and diverted from our chosen ends.

What is a divine act? Clearly God's act of creating possesses some similarities to human acts of making. God is a kind of agent. God possesses himself in self-knowledge and is free from all impersonal processes and forces.[7] Indeed, God, being absolutely free, supremely wise and omnipotent, is the perfect agent. The divine act of creating the world can be thought of as effecting a transition from the created world not existing to existing, a transition that originates in and is directed by God. And of course, God's act of creating is not irrational or arbitrary but is accompanied by God's intentions and is directed toward an end. Despite these similarities, the differences between God's act of creating and a human act of making are profound and decisive. Even the similarities mentioned above are not exact correspondences. God is the perfect agent, and human agents are imperfect. Human beings are themselves products brought into being by something outside and prior to themselves. They do not possess perfect freedom and self-knowledge. Their wisdom is derived and limited, and their power is likewise limited and dependent. The divine act of creation does not begin with a state of affairs in the normal sense of that term. "Absolutely nothing" is not a state of affairs, and therefore the notion of change or transition does not strictly apply to the coming into being of creation.

The differences between the divine act of creating and the human act of making are so profound that we must conclude that they are not merely quantitatively different but qualitatively of a different kind. This becomes clear when we attempt to apply to divine creation the categories uncovered when analyzing human creation: (1) First, we consider the idea of motivation. Because God lacks nothing there can be no distress in the divine life. The theological tradition beginning with Athanasius and the Cappadocians in the East and Ambrose and Augustine in the West distinguished between the inner acts of God in the eternal trinitarian relations and the external act of the Trinity in creation. Since God is eternally active in the trinitarian relations we need not conceive of God as needing to create in order to become active. Even though the two types of acts are profoundly different because the internal acts are constitutive of God's triune life whereas the external act

---

[7]I addressed this subject in the previous chapter.

is not, the two "are not wholly different."[8] The idea of motive in the sense of a desire for a good that God does not possess merits no place in our thinking about God and creation. However, motivation must be distinguished from intention. Intention is the subjective side of the end for which God creates and need not arise from a lack. God's intention has to do with blessing creatures with the good that God is eternally.[9]

(2) The idea that God imagines a world that does not exist, which then provides the model for his act of creating, needs careful examination. God is the creator of all that is not God, and in God's life essence and existence are identical. Hence everything that is real in any sense is either God or God's creation. This means that concepts, possibilities, ideas, essences, possible worlds, or logical and physical laws are either created or are the finite mind's abstractions from God's being.[10] The human imagination constantly changes and is populated by images of things that exist independently of it. Human beings need to perform an act of imagination on the way to making a thing because they need a model or a design plan to guide their act of making. But why should God need to perform an act of imagination as an intermediary step before his act of creating? The images or models thus obtained, if they are not God's own being, must be either created or uncreated. In either case, to think this way would make God dependent on nondivine means to accomplish his act of creation. I shall return to this issue for a fuller treatment in the section on creation from nothing and in the discussion of the role of the Word in creation. Briefly, the trinitarian answer to the problem of mediation just formulated is that God (the Father) eternally images himself in the Word, or Son. But the Word is not a created image or a projected abstraction. The Word is God, the absolutely perfect and living image of God. It is *through* the Word that God creates. The world is a created image of the uncreated image of God. God creates through God and needs no other mediators.

(3) Apart from God's act of creating there would be nothing but God. A

---

[8]For a discussion of this issue, see Pannenberg, *Systematic Theology*, 2:1-9.

[9]We will deal with God's purpose and intention for creation in greater detail in a later section.

[10]Some argue that God thinks in concepts or propositions so that the divine mind and the human mind could think the same concept in common. I don't think it makes sense to speak as if God's way of "thinking" were knowable to us simply by knowing our own minds. Thomas Aquinas, in a modification of Plato and a departure from Aristotle's hylomorphism, speaks of God's being containing the "exemplary causes" of creatures. God, in this sense, is the formal cause of creatures. See Dodds, *Unlocking Divine Action*, p. 13.

human act of making presupposes the existence of form and matter and consists in bringing these two together through the intelligent exertion of power, which Aristotle called efficient causality. God's act of creating presupposes nothing outside God, for apart from God's will nothing else exists. In divine creating there are no nondivine mediators, no necessary connection between means and ends, no middle stages and no other causes.[11] We will discuss creation from nothing in its own section. But we can say here that every rational effort to bridge the gap between absolutely nothing and something is doomed to failure. Nor can reason explain how God creates apart from help from other causes, formal or material. This gap shatters the correspondence between human making and divine creation and forces us to make a qualitative leap.[12] Rationality can see relationships only between existing things. In analyzing an act of making, reason mentally puts things together or takes them apart to understand their relations. But there is no relation between nothing and something, and there are no middle stages between them. Reason can understand making only as imposing an ordered set of relations on matter. Divine creating is something else altogether.

(4) Whereas the human maker is changed by the act of making, God is not changed by his act of creating. This affirmation follows first from our view of God as simple and immutable. God is purely actual and hence possesses every possible good in infinite measure. Change occurs in imperfect beings because they possess unactualized potential and exist in a temporal process of becoming. In addition, changeable beings are changeable precisely because they are subject to being acted on by other powers. And changeable and becoming beings are by nature subject to death or dissolution because they are not the cause of their own existence and hence cannot oppose nonbeing with absolute sovereignty. None of this applies to God. When we consider God's act of creating, we can see another reason to reject the idea of God

---

[11]Pannenberg distinguishes the relation between means and ends within the order God creates and a supposed order of means and ends to which God's act of creation must conform. The existence of the latter order would "make God a needy and dependent being" (*Systematic Theology*, 2:16).

[12]Brian Robinette, "The Difference Nothing Makes: *Creatio ex Nihilo*, Resurrection, and Divine Gratuity," *Theological Studies* 72 (2011): 525-57. "Far from making the origin and ground of creation accessible to full comprehension, the statement [about *creatio ex nihilo*] requires the work of an apophatic discourse that opens up human understanding to the utter gratuity of creation. Nothing is necessary about creation at all. It derives wholly from the incomprehensible mystery of the creator God whose relationship to creation remains one of loving freedom and fidelity" (p. 529).

being changed by his act. Prior to God's act of creating, no other power or cause exists to act on God. Absolute nothingness can offer no resistance to God's act, and the creature must actually exist before it possesses causality of any kind. Hence God is not changed physically by creating. I will address in a later section the issue of whether it should be considered a change that God becomes the "Creator" only because and as God creates.

(5) As I observed above, a human act of making is imperfect at every stage. It arises from motives, its intentions are clouded and its ends are never perfectly realized and always involve unintended consequences. But for God none of the imperfections we experience are present. God knows himself and his intentions with perfect clairvoyance. His act of creating presupposes no causes or recalcitrant material forces to impede or divert his work or alienate the results from his intentions. God knows how to create exactly what God intends.

It should be clear by now that, although we cannot avoid speaking of God's creating heaven and earth as an act, we must be very careful to avoid imposing on God the imperfections of acts of making as we experience them. Theology is an exercise in affirmation and negation, an attempt to discipline our language so that we may speak in a way that honors God. Like all affirmations about God, when we say, "God acted to create the world," we must also add, "but not as human beings act to make things within the world." Many mistakes in the theology of creation arise from pressing the analogy between human making and divine creating too far. We know that the analogy has been pressed too far when any aspect of divine creating is considered to be "univocal" with creaturely making.[13]

---

[13]William Alston argues that in order to speak meaningfully about divine action we must assume that there is a univocal core of meaning that is shared between a human and a divine act. He denies that this contention forces him to accept the idea that God acts in exactly the same way as human beings act. His example is divine speech. One can say truly that God speaks without asserting that God possess lungs, a larynx and a tongue. The core meaning of speech shared by divine and human speech may have to do with communicating meaning from one mind to another mind whatever the method or "how" ("How to Think About Divine Action: Twenty-Five Years of Travail for Biblical Language," in *Divine Action: Studies Inspired by the Philosophical Theology of Austin Farrer*, ed. Brian Hebblethwaite and Edward Henderson [Edinburgh: T & T Clark, 1990], p. 68). Michael Dodds argues, correctly I believe, against any univocity between divine and human action. Such univocity would imply that God acts sometimes on the same level as created causes and this type of action would displace creaturely action from its natural place at the point where God acts. It would place God and creatures in competition. See Dodds, *Unlocking Divine Action*. Kathryn Tanner, applying a linguistic and functionalist approach to theological work, articulates the rules that govern statements about God's relationship to cre-

THE HOW OF DIVINE CREATING

As I hinted above, many discussions of creation and providence turn on some rational notion of *how* God acts to create and sustain the world.[14] And many pseudo-problems have been produced in the search for this elusive "how." This is especially the case in the "science and religion" debates and in the debates about divine providence. In the end I will show that there is no "how" comprehensible by human reason.[15] As a first step toward this conclusion we will ask why the quest for a "how" is so hard to give up.[16]

When seeking to understand an action by which one state of affairs is transformed into another by an agent, one of the first questions we ask is how?[17] For routine acts we take the how for granted. When we hear a familiar voice on the phone or the waiter sets a piece of lemon pie in front of us or someone drives up in front of our house, we generally don't ask, "How did you do that?" But if you are watching a magic act and the magician appears to make his assistant disappear or pulls a rabbit out of his hat, our amazement may be expressed by the words, "How did he do that?" Asking

---

ation found in first-order Christian talk and practice. Until certain modern developments subverted it, Christian talk observed the rule that one should never oppose divine action to human action so that they are mutually exclusive or indirectly proportional to one other. Divine action and properties should never be treated as univocal with creaturely action and properties (*God and Creation in Christian Theology* [Minneapolis: Fortress, 2005]).

[14]Not every writer who speaks about the "how" of divine action intends to make divine action transparent to human reason. David Burrell, in his essay "Divine Practical Knowing: How an Eternal God Acts in Time," in Hebblethwaite and Henderson, *Divine Action*, pp. 93-102, answers the question "how?" by saying that God, who is pure act, acts in time in and through temporal creatures by giving them being. Notice that Burrell's explanation of how God acts does not dispel the mystery of divine being and action; rather he states the mystery of the Creator-creature relation in such a way that given who God is and what "to create" means, we can see that we must express divine action in this way.

[15]God is his own "how" and needs nothing from outside. Since God is incomprehensible to us, his "how"—since he is his "how"—is also incomprehensible. See also Thomas F. Tracy, "Narrative Theology and the Acts of God," in Hebblethwaite and Henderson, *Divine Action*, pp. 173-210. Tracy distinguishes among different levels of answer to the question of how God acts. If we were asked how God saves human beings, we could name all sorts of instrumentalities: by guiding Israel, by inspiring the prophets, by sending Jesus, by raising Jesus from the dead and by sending the Holy Spirit. These answers address the "how" but not on the most basic level. But we cannot explain "how" God raised Jesus from the dead. We can call the former means "secondary" and the basic means "primary." Tracy denies that we can finally make the mode of divine action clear at its most basic level.

[16]Ian Ramsey, *Models of Divine Activity* (London: SCM, 1973). Ramsey proposes ways to give the notion of divine action and presence an experiential content.

[17]This question can also be asked of events and processes, but we won't pursue that possibility here.

for the how is asking for the means through which an agent is enabled to achieve a task. When I tell you I lifted my four-thousand-pound car six inches off the driveway to change a tire, you will assume that I used a jack as a means. A jack uses the laws of physics having to do with levers and fulcrums or fluids to concentrate the force I can generate with my body. This means enables me to do things indirectly I cannot do directly. The jack transforms my action of moving the lever up and down rapidly into a slower but more forceful upward action on the car to lift it. To generalize, a means is necessary when the agent's inherent power is not in a form suitable to the task to be done.

Our bodies generate physical force, but they must be guided by our minds. Through intelligence we gain knowledge, and knowledge guides the construction of tools to multiply or refine the physical power generated in our bodies. It is easy to see how a tool can transform our bodily action into a useful form. Physical energy is of the same nature whether it is generated in the human body, by an electric generator or by the sun. But it is also clear to everyone but the most entrenched materialist that knowledge, information, concepts, numbers and a host of other familiar things are not bodily or physical in nature. In addition, we are clear that our minds know, think, gain understanding and possess these nonmaterial things. Hence mind, too, cannot be physical in nature. Finally, everyone knows as a matter of experience that our minds direct our bodies according to the knowledge and wisdom they possess. But how does mind direct physical energy toward an end?

This way of looking at the human being treats the mind as the agent and the body as the means through which the mind effects change in the world. The mind uses both its knowledge and its body in its act of making. Knowledge guides and body moves other bodies so that they become informed with the knowledge possessed by mind. The human mind cannot by itself create anything or act in the world. On one side, it needs to be connected to an intelligible world of ideas, archetypes and concepts that can serve as models or forms to guide its activity, and on the other it must be interfaced to a physical body so that through its body it can interact with the physical world and bring the intelligible to bear on the material world. The individual mind is the origin of neither the intelligible nor the physical world. Its power resides in bringing the two together to form an actual object.

ANCIENT THEORIES OF MEDIATION

The reason this analysis of the "how" of human action is important is because it surfaces two very different analogies often probed for insight into how God creates and exercises providence. The first analogy is used naively in ancient myths of creation.[18] Many ancient creation myths picture the gods as possessing bodies that can interact with the physical material out of which they make the world. This picture makes the means of creation readily understandable to common sense, since it reflects the same use of means for the interaction of bodies to which we are accustomed.

The pre-Socratics and Plato. The philosophical problem of mediation begins with the tension between the process thought of Heraclitus (fl. 504–500 B.C.) and the philosophy of unchanging being taught by Parmenides (b. ca. 515). For Heraclitus the world has always existed and "has not been made by any god or man, but has always been, is, and will be."[19] Everything is flux, conflict and change. Nothing remains unchanged except change. Temporary stability in things is created by the balance among opposing forces. Unity and stability, then, are secondary and are reducible to the multiform and never-ceasing process. Fire seems to be at the root of this process of change. "All things are exchanged for fire, and fire for all things, as goods for gold and gold for goods."[20] Note that Heraclitus's philosophy places trust in the information received by the senses, which perceive constant plurality and change but only temporary unity and stability in the physical world. Parmenides and his followers in the Eleatic school held a position opposite to Heraclitus. For Parmenides, genuine being is simple and unchanging. In his philosophical poem he has the goddess instruct in these words: "What Is has no beginning and never will be destroyed: it is whole, still, and without end. It neither was nor will be, it simply is—now, altogether, one, continuous."[21] Do not follow the senses, warns the goddess: "Do not let custom, born of everyday experience, tempt

---

[18]Such stories are also used knowingly in religious teaching to guide the imagination, but they are always accompanied by a negation, explicit or implicit, to keep the hearer from taking the story too literally.

[19]Philip Wheelwright, ed., *The Presocratics* (New York: Odyssey, 1966), p. 70.

[20]Heraclitus, fragment 90, quoted in David Furley, *The Greek Cosmologists*, vol. 1, *The Formation of the Atomic Theory and its Earliest Critics* (Cambridge: Cambridge University Press, 1987), p. 35. Furley sees the function attributed to fire as an anticipation of the later notion of efficient cause (p. 36).

[21]Wheelwright, *The Presocratics*, p. 97.

your eyes to be aimless, your ear and your tongue to be echoes."[22] Instead, the divine guide urges her listeners to follow the way of reason, which says that only being can be. The assertion that nonbeing is, or that being both is and is not, is self contradictory.

The tension between these two powerful perspectives challenged philosophers to defend one or the other or seek a middle way, which is the problem of mediation. Are we trapped in the following dilemma or is there a way through? Either unity and stability are fundamental while plurality and change are mere appearance, or plurality and change are fundamental while unity and stability are mere appearance.[23] Plato (ca. 429–347 B.C.) sought to take into account the force of Heraclitus's thought while remaining true to the Parmenidean preference for being.[24] Plato and his followers perceive within themselves and the world as a whole a sharp distinction in kind and quality between intelligible things and the world of material stuff. There is no way to conceive them as deriving one from the other or to reduce them to a common factor.[25] Hence both must have always existed. Yet in the world and in our own existence we experience the intelligible and the material worlds in relation. The human soul plays the mediating role between the intelligible world and the material world in us. Plato sees in the human mind/soul and body a microcosm mirroring the order of the large cosmos.[26] In his dialogue *Timaeus* Plato explains the genesis and structure of the cosmos. Timaeus first lays down the principle that what is eternal and unchanging is superior to that which once did not exist and is always becoming. And as the eternal is superior to the temporal,

---

[22]Ibid., p. 96.

[23]Despite their differences, Heraclitus and Parmenides hold in common two presuppositions that are inimical to a Christian theology of creation: (1) there is a linear continuity between human existence and the ultimately real, so that the human mind has natural access to metaphysical reality, and (2) the world had no beginning and will have no end. These presuppositions continue foundational to subsequent attempts at mediation.

[24]Plato (*The Sophist*) found a way to affirm both being and nonbeing by using the concept of "the different" or "other" instead of "is not." See the discussion in David Furley, *Formation*, pp. 42-44.

[25]There were monistic philosophies such as Stoicism that understood the world to be reducible to the same fundamental reality, the difference between soul/spirit/mind and matter being only a matter of refinement.

[26]See Plato, *Timaeus* 44d-e, where this relationship can be seen in the way that the Maker of the world created the human soul in imitation of the world soul and placed it in "our spherical heads in imitation of the way the world soul occupies, and rotates through, the spherical heaven" (David Sedley, *Creationism and Its Critics in Antiquity* [Berkeley: University of California Press, 2007], p. 98).

so knowledge of the eternal is superior to belief and opinion about the changing order. The account that follows, Timaeus admits, since it concerns the changing order, is not knowledge but belief, a "likely story" or "reasonable account"[27] (29d). Given the changing nature of the world we must conclude that the visible order has come to be through some cause. This cause is of course "the maker and father of the universe" (28c). But the next sentence shows that Timaeus will be focusing on the nature of the world rather than the nature of God (28c):[28]

> Now to find the maker and father of this universe is hard enough, and even if I succeed, to declare him to everyone is impossible. And so we must go back and raise this question about the universe: Which of the two models did the maker use when he fashioned it? Was it the one that does not change and stays the same, or the one that has come to be?

Even though the dialogue's central theme is not the nature of the divine, there are good arguments for the position that for Plato the "maker and father" of the world transcends the world and stands decisively on the intellectual side of the divide between the intelligible and the material realms.[29] From this position, the Maker creates the world in the likeness of the perfect living creature, thus making it an ever-living god. The Maker himself creates the world soul, the souls of the heavenly gods and their heavenly bodies while delegating the creation of human beings and the rest of the cosmos to these newly created gods. The Maker in his perfection must keep his distance from the earthly realm; for were he to create creatures there they too would be gods and immortal. The dialogue continues to describe the

---

[27]See Miles F. Burnyeat, "Eikôs Mythos," *Rhizai: A Journal for Ancient Philosophy and Science* 2 (2005): 143-65, for his argument in support of this translation of *eikos logos/mythos*.

[28]Plato, *Timaeus*, trans. Donald J. Zeyl, in *The Complete Works of Plato*, ed. John M. Cooper (Indianapolis: Hackett, 1997), p. 1235.

[29]Using coherence with Plato's overall cosmology as her criterion, Sarah Broadie makes a strong case that the Maker in the *Timaeus* must transcend the cosmos (*Nature and Divinity in Plato's Timaeus* [Cambridge: Cambridge University Press, 2011], pp. 16-24). For discussion of the modern debate over whether the Maker is metaphorical or literal, see Donald J. Zeyl's discussion in his introduction to Plato, *Timaeus*, trans. Donald J. Zeyl (Indianapolis: Hackett, 2000), pp. xx-xxv. See also Thomas Kjeller Johansen, *Plato's Natural Philosophy: A Study of the Timaeus-Critias* (Cambridge: Cambridge University Press, 2004), pp. 79-83. On the other hand, if we assume that the demiurge is a symbol for the immanent "operation of Reason in the universe," the realm of forms functions to mediate the material world of plurality and change with the unchanging and simple highest principle of being, the Good or the One. See Frederick Copleston, *A History of Philosophy*, vol. 1, pt. 1, *Greece and Rome* (Garden City, NY: Image, 1962), p. 203.

structure and contents of the cosmos in terms of the physics, biology and mathematics of its day. But further description of the cosmology in the dialogue is not necessary for my purpose.

We can see the analogy of human making at work in this picture. In Sarah Broadie's terminology we see "cause, product and materials" or "user, used and product."[30] The Maker is the cause, matter is the material and the cosmos is the product. The Maker brings together the two principles (*archai*), or fundamental constituents, the intellectual and the material, to form the world. As I pointed out earlier, the *Timaeus* seeks to develop cosmology rather than metaphysics; so Plato does not labor to distinguish the Maker from the intellectual model in whose likeness he makes the cosmos. However, by having the created gods make the lower creatures Plato shows that he is conscious of the problem of mediation between the transcendent Maker and the material world. He sets up the problem for those who follow him, but he does not make the problem of mediation central to his creation story.

Middle Platonism and Neo-Platonism. After a period of decline and skepticism in Hellenist philosophy, the school of Middle Platonism (80 B.C.–A.D. 220) was founded by Antiochus of Ascalon (ca. 130–68 B.C.).[31] Antiochus wished to overcome the skepticism of his immediate predecessors by combining the thought of Plato, Aristotle and the Stoics into a harmonious system. Middle Platonists perpetuated Plato's distinction between the principles of the intelligible and the material; however, they tended to emphasize more than Plato had the transcendence of highest principle and to give the mediating realm greater specificity and definition. Hence Middle Platonists invoked *three* principles to explain the cosmos rather than Plato's two. Eudorus of Alexandria (fl. ca. 50–25 B.C.), for example, postulates a higher One above a lower One, the lower One serving as an intermediary between the transcendent One and the world of matter. The higher One is turned away from the world and contemplates only its own perfection. Because the lower One is positioned between the higher One and the material world, it can transmit the creativity of the highest principle without compromising its transcendence. Middle Platonists des-

---

[30]Broadie, *Nature and Divinity in Plato's Timaeus*, pp. 9-10.
[31]For a study of Middle Platonism, see John Dillon, *The Middle Platonists: 80 B.C. to 220 A.D.*, rev. ed. (Ithaca, NY: Cornell University Press, 1996). Middle Platonism lasted until its evolution into Neo-Platonism in the work of Plotinus (A.D. 205–269/270).

ignated the mediating principle variously: second god, second One, Heavenly Mind, world soul, logos or Maker. While the name differs from thinker to thinker the function of the intermediary principle remains the same. Dillon concludes, "Whatever the differences in detail, however, it is common ground for all Platonists that between God and Man there must be a host of intermediaries, that God may not be contaminated or disturbed by too close involvement with Matter."[32] We can take a statement by Alcinious, author of a second-century handbook of Platonic doctrines, as typical of Middle Platonism's feel for the issue of mediation:

> [The first principle] is Father through being the cause of all things and bestowing order on the heavenly intellect and the soul of the world [the second principle] in accordance with himself and his own thoughts. By his own will he has filled all things with himself, rousing up the soul of the world and turning it toward himself, as being the cause of its intellect. It is this latter that, set in order by the Father, itself imposes order on all of nature in this world (Epit. 10.3, 164.40–165.4).[33]

Plotinus (204–270), the most famous representative of Neo-Platonism, so emphasized the transcendence of the first principle that he considered it beyond thought, being and expression. Middle Platonism had assumed that the first principle, the One, was also Thought or Mind without noticing the contradiction between absolute unity and the subject-object distinction necessary for thought. Plotinus considered it necessary to remove Thought or Mind from the One and make thought the second principle.[34] The second principle, Thought or the "Intellectual-Principle," emanates out of the One and mediates the undifferentiated richness of the One in the form of the ideas of all possible things.[35] As the Intellectual-Principle contemplates the One in eternal concentration it generates the "All-Soul." Likewise, as the

---

[32]Dillon, *Middle Platonism*, p. 47. Thomas H. Tobin voices a similar judgment: "The emphasis on the transcendence of the Supreme One creates the need for an intermediate realm in which one finds the proximate principles or causes of things" (*The Creation of Man: Philo and the History of Interpretation*, p. 15). Quoted in Ronald Cox, *By that Same Word: Creation and Salvation in Hellenistic Judaism and Early Christianity* (Berlin: Walter de Gruyter, 2007), p. 37, n. 30.

[33]Quoted in Cox, *By that Same Word*, p. 36.

[34]John Dillon, introduction to *The Enneads*, by Plotinus, trans. Stephen MacKenna (London: Penguin, 1991), pp. xcii-xcviii.

[35]See John Bussanich, "Plotinus's Metaphysics of the One," in *The Cambridge Companion to Plotinus*, ed. Lloyd P. Gerson (Cambridge: Cambridge University Press, 1996), pp. 38-65, for explanation of the difficult idea of Intellect's necessary but uncompelled flow from the ungrudging perfection of the One.

All-Soul contemplates the Intellectual-Principle of which it is the image, it produces the world patterned on the ideas it finds there. In Plotinus we find a hierarchical system flowing downward from the absolute perfection of the One to matter, which is almost, but not quite, nothing.[36] Though all being flows from the One and strives to return to the One, the highest principle is protected by the Intellectual and Soul principles from direct contact with anything other than its own perfection.[37]

John Polkinghorne—A Contemporary Theorist of Mediation

The terms, presuppositions and scientific understanding of modern thinkers differ dramatically from those of the ancient world; yet the problem of mediation between God and the world remains. I have chosen to focus on John Polkinghorne, a study of whose work is well suited to sharpen our understanding of what is at stake in the quest for the "how" of divine action in creation and providence.[38] Polkinghorne has played a leading role in this discussion for the past thirty years, and his theological position is closer to traditional Christian theology than many others in the discussion. Polkinghorne argues that believing theologians and scientists should not give up the quest to find a way to conceive how God acts in the world. The rationality of theology's talk about God is at stake. Polkinghorne contrasts his view with that of Austin Farrer, whom he quotes as saying, "God's agency must actually be such as to work omnipotently on, in and through creaturely agencies, without either forcing them or competing with them."[39] In this sentence Farrer articulates his well-known theory of

---

[36]Dominic J. O'Meara objects to the term *hierarchical* because applied to Plotinus it is anachronistic, since it was first used in the sixth century by Pseudo-Dionysius. O'Meara suggests speaking of a series of types of priority/posteriority as a way to grasp the leveling that the term *hierarchical* attempts to describe. See "The Hierarchical Ordering of Reality in Plotinus," in Gerson, *Cambridge Companion to Plotinus*, pp. 66-81.

[37]See Tanner, *God and Creation in Christian Theology*, pp. 36-48, for a concise treatment of Greek metaphysical theories of mediation and their differences from the Christian doctrine of creation.

[38]John Polkinghorne, "Kenotic Creation and Divine Action," in *The Work of Love: Creation as Kenosis* (Grand Rapids, Eerdmans, 2001), p. 98. Polkinghorne uses the term "scientist-theologians," which refers largely to participants in the dialogue of theology and science whose primary expertise and research lies in natural science but who also have a working knowledge of philosophy and theology. Below I treat science-and-theology issues in greater detail.

[39]Austin Farrer, *A Science of God*, p. 76, quoted in John Polkinghorne, *The Faith of a Physicist: Reflections of a Bottom Up Thinker* (Minneapolis: Fortress, 1996), p. 81.

double-agency, in which every event and act has at its root both divine and creaturely causality operating on levels appropriate to each agent. Polkinghorne replies, "I find it an unintelligible kind of doublespeak."[40] Without entering on an extensive defense of Farrer's view of double-agency, it seems that so casual a dismissal of so celebrated a thinker's most praised contribution to theology should not be mistaken for an intelligent critique.[41] Moreover, in rejecting Farrer's formulation of double-agency Polkinghorne dismisses a venerable theological tradition that includes Thomas Aquinas, Protestant Orthodoxy and Karl Barth.

In many essays and books Polkinghorne participates in the search for phases in the world process, at the quantum level or at the initial conditions of chaotic systems, where God could determine a system with input of information rather than energy. In analogy to human freedom in relation to bodily movements, he argues that God uses (or it is conceivable that God uses) "the inherent incompleteness" and "resultant flexibilities" in the physical processes to act in particular ways in the world.[42] We could conceive of God's "interaction with creation . . . purely in the form of active information."[43] Polkinghorne admits that the term "active information" (mind or form?) is rather vague but argues that it may escape the requirement true of passive information, such as that stored on a computer hard drive or in a DNA molecule, that it requires energy for the process of transference from one medium to another. Polkinghorne avoids making strong claims about the method of divine interaction with creation. He seems content with

---

[40]Polkinghorne, *Faith of a Physicist*, pp. 81-82. In "Kenotic Creation and Divine Action," p. 97, he calls the theory of double causation "double talk."

[41]For a study that shows the similarities between Karl Barth's understand of the creature-Creator relation with that of Austin Farrer, see Darren M. Kennedy, *Providence and Personalism: Karl Barth in Conversation with Austin Farrer, John Macmurray and Vincent Brümmer* (New York: Peter Lang, 2011), pp. 131-90. Kennedy concludes that Farrer's and Barth's views on double-agency are "remarkably similar in form" (p. 190).

[42]John Polkinghorne, *Science and Providence: God's Interaction with the World* (Boston: Shambhala, 1989), p. 34. Polkinghorne varies in the level of confidence with which he asserts these claims. Sometimes he asserts God's way of acting in the world without qualification: God acts "by means of causal joints hidden within the unpredictability of process" (p. 34). And sometimes he hedges his claims with qualifiers: "it *seems conceivable to suppose* that there is scope for action" or "God's purposive will *may be* exercised within his creation" (p. 34 [emphasis added]).

[43]John Polkinghorne, "The Metaphysics of Divine Action," in *Chaos and Complexity: Scientific Perspectives on Divine Action*, ed. Robert John Russell, Nancey Murphy, and Arthur R. Peacocke (Vatican City and Berkeley, CA: Vatican Observatory Publications and the Center for Theology and the Natural Sciences, 1997), p. 155.

showing the possibility of such action and, hence, with defeating strong claims about the impossibility of divine action. Put in its most conservative form, he wishes to demonstrate the rationality of religiously motivated belief in divine providence or at least that such faith is not irrational. But should we accept even this minimum requirement as a condition for the rationality of theology?

Apparently, Polkinghorne considers belief in divine action irrational or at least nonrational, apart from knowledge of flexibilities and incompleteness in the *physical* world, places in physical systems where God could exert causal influence. But does Polkinghorne's theory of the "how" of divine action really set theology on a better rational foundation than Farrer's theory of double-agency? At best his theory shows the physical world's receptivity to information at certain levels. But he does not explain *how* information can be communicated from the divine mind to the physical world, which is the very question that drives the quest. His vague notion of "active information" gets us no closer. Despite the sophistication of the physical theories he invokes and the resulting impression of profound insight, our understanding of the how of divine action never gets beyond the analogy of human action.[44] And we don't need to know anything about quantum physics or chaos theory to experience our own free action.

But even if we grant that Polkinghorne has secured theology's rationality, such rationality is purchased at considerable cost. The quest to find an open place where God can act presupposes that God's ability to act is conditioned by creation's physical laws. It seems to grant Pierre Laplace's (1749–1827) mechanistic determinism its premise: God cannot act through and within a system that is determined with respect to physical causes.[45] Since Polkinghorne accepts the Copenhagen interpretation of quantum indeterminacy—that our inability to measure both the position and the momentum of events at the quantum level is due to their ontological indeterminacy—he does not think quantum events are determined by previous deterministic

---

[44]To make human action comprehensible, Polkinghorne advocates "dual aspect monism," that is, mind and body are two manifestations of the same underlying reality. This view of the human being seems a bit too facile and does not really explain anything; but applied to the God-world relation it results in pantheism or at least panentheism.

[45]It is reported that after reading Laplace's five-volume *Celestial Mechanics*, Napoleon Bonaparte asked Laplace where God was in his system. Laplace replied, "I have no need of that hypothesis" (see C. B. Boyer, *A History of Mathematics*, 2nd ed. [New York: Wiley, 1968], p. 538).

causes; rather, they happen by chance. God, then, replaces chance to determine quantum states. Nevertheless, Polkinghorne seems to agree with
Laplace that God cannot act in the world without taking the place of some
natural process, in his case that process works in a nondeterministic way.
Hence to affirm that God acts in the world to achieve his purposes while
also avoiding anything like Farrer's double-agency, Polkinghorne postulates
that God acts in a causal way that replaces natural causes, that is, in a way
univocal with natural causes, albeit in ways undetectable by empirical
methods. This assumption makes sense only when God and God's action
are placed on the same ontological plane as the physical world and when
empirical science is mixed with metaphysics or theology in a confused way.
In Michael Dodds's judgment, Polkinghorne treats divine causality as univocal with the causality of creatures. Speaking with reference to Polkinghorne's idea that God must limit his action to make "metaphysical room for
creaturely action,"[46] Dodds says, "Here, God is clearly conceived as a univocal cause who must limit his causality to make 'room' for the causality of
creatures."[47] Steven Dale Crain passes similar judgment on Polkinghorne's
approach: "I believe that the claim that God transcends the world as the
creator renders highly suspect attempts like Polkinghorne's to argue that
God must exploit a built-in physical feature of the world in order to act in
the world."[48]

Polkinghorne makes divine action empirically falsifiable without making
it verifiable. If future scientists begin to believe that quantum events are
determined by previous states and deterministic chaotic systems become
measurable in practice, divine action will again become mysterious if not
doubtful. But no amount of indeterminacy and openness can ever show that
any particular event is a divine act. Sheer chance is always an option and
would probably pass the Ockham's razor test (simple explanations are to be
preferred over complex ones) with higher marks than divine action. The
only way I know that my acts are mine is that *I* am doing them. No empirical
examination can add to that certainty, though it might reduce it through
obfuscation. And the only way that we could ever know that an act is *God's*

---

[46]John Polkinghorne, *Belief in God in an Age of Science* (New Haven, CT: Yale University Press,
      1998), p. 13.
[47]Dodds, *Unlocking Divine Action*, p. 232; the Polkinghorne quote is from *Belief in God in an Age
      of Science*, p. 13.
[48]Steven Dale Crain, "Divine Action in a World of Chaos," *Faith and Philosophy* 14 (1997): 58.

action is if God allows us to participate in his self-knowledge of his action.

A second theologically troubling problem is Polkinghorne's competitive view of divine power and action.[49] In advocating what he calls a "kenotic" view of divine creation he assumes that God must change and withdraw in order to create and that God must suffer and restrain his action in order to allow for human freedom.[50] According to Polkinghorne, "It has been an important emphasis in much recent theological thought about creation to acknowledge that by bringing the world into existence God has self-limited divine power by allowing the other truly to be itself. . . . God has stood back, making metaphysical room for creaturely action."[51] God must withhold action and restrain power, live in time, and dwell in ignorance of the future. For Polkinghorne, divine self-limitation is necessary if there is "to be an interweaving of providential and creaturely causalities."[52] Divine action and creaturely action cannot occupy the same causal space.[53] Of course, Polkinghorne is correct if causal "interweaving" is necessary for divine providence to make any sense. As I argued above, however, it is precisely the attempt to understand the method of God's action on the world that is the fundamental mistake that makes necessary Polkinghorne's drastic revisions in the doctrines of God, creation and providence. Such projects fall victim to the ancient and ever-persistent intuition that God's active presence in the world can be increased only insofar as his otherness from the world is decreased, that in order to be able to relate to the world God must have some properties in common with the world. In arguing for their kenotic view of creation, Polkinghorne and other scientist-theologians derive unwarranted

---

[49]See my section on divine omnipotence in *Great Is the Lord*, pp. 332-57, for an extensive discussion and refutation of the competitive view of divine power.

[50]Polkinghorne, "Kenotic Creation and Divine Action," pp. 90-106.

[51]Polkinghorne, *Belief in God*, p. 13.

[52]Polkinghorne, "Kenotic Creation and Divine Action," p. 105.

[53]Taede A. Smedes voices this concern. "The category mistake is thus a confusion between natural causality and divine action: divine action is supposed to work on the same level as natural causality. In other words, divine action is considered to be limited by the same factors as is human action, thereby blurring the distinction between the order of creation and its inherent limits, and the transcendent order of its Creator. This confusion results in a conceptual 'devaluation' of God, i.e. a reduction of the Creator to his creation" (*Chaos, Complexity, and God: Divine Action and Scientism* [Leuven: Peeters, 2004], p. 198). Tanner, *God and Creation in Christian Theology*, also address this mistake of making divine and creaturely action "contrastive" or "competitive." See also James R. Pambrun, "Creatio ex Nihilo and Dual Causality," in *Creation and the God of Abraham*, ed. David Burrell et al. (Cambridge: Cambridge University Press, 2010), pp. 192-220.

conclusions from Christian belief in the incarnation: that the incarnation is an aspect of a wider divine self-limitation or emptying necessary if creatures are to exist and exercise freedom.[54] This view seems to imply that the Word was made flesh to enable God to relate to the world when the Scriptures witness to the opposite reason: to enable the world to relate rightly to God.

[54]Polkinghorne, "Kenotic Creation and Divine Action," p. 104.

5

How God Creates
the World (Part Two)

Assessing the Quest

In our quest to discover how God creates the world we've examined the idea of act and means. In this chapter we continue this theme by looking at two more concepts that have been used to gain insight into God's creation of the world: causation and relation.

Creation and Causation

Is God's act of creation an exercise of causality? If so, how does our understanding of causality within the world relate to God's creation of the world? Causality is built into our understanding of our experience of ourselves and the world. We want something to change, and we set about to effect that change. In considering the way we use our bodies and other tools as means, we need the concept of causality to explain our acts fully. To wish things were different is not equivalent to effecting the change. As our analysis of an act revealed, the soul's power and freedom to move itself, though necessary, is not sufficient to complete an external act. Fortunately, the soul's movement possesses the power to move the body, which then causes movement in the world outside the body. The relationships between the various means and the end we seek involve changes produced by the transmission of energy and information. The communication between agents, means and products they produce we call causality.

It seems to me that the notion of causality is derived centrally from experience of our bodies' interactions with the external world and from our ability to persuade other people into action.[1] From this focal point we trace the trail of causality inward to the power of the soul to move the body. However, the power of the soul to move the body is called causality in a modified sense from that used to describe the communication between our bodies and physical means we use to effect change in the world. This is so because the soul is not a physical force that causes movement by contact. Here we face the same problem we faced in our discussion of mediation: what kind of communication can the intelligible world have with the material world, and how does the soul link the two? In the other direction, we observe events and processes in the world unconnected to human acts. Things happen and transformations occur, and the power of one thing to effect change in another we also call causality.

Aristotle's theory of causality played an important role in theological thinking about God's action on and in the world, especially in the Middle Ages and in Protestant Orthodoxy. For Aristotle, too, the central model of causality is a human act, which then is extended to cover all change, whether effected by human agents or interactions among things. Any transformation in the world can be explained by discovering its four causes: material, formal, efficient and final. Clearly Aristotle's causes correspond to our everyday experience of our minds, bodies and the world.[2] If you come upon an object that bears the marks of intelligence, such as a house or a living thing, and you wish to explain the factors that contribute to its existence, you will need to answer four questions. (1) What is the possibility of a thing's being any particular thing whatever? (2) What causes it to be the particular thing it is with its properties and activities? (3) Who or what made it? (4) For what purpose is it made, or for what ends does it aim? Remove any one of these four causes and the artifact or natural object ceases to exist. Aristotle's contribution to understanding is to bring our commonsense experience of making things under one category, causality.

---

[1]R. G. Collingwood, *An Essay on Metaphysics* (Chicago: Regnery, 1972), pp. 285-387, notes three types of causality: (1) one human being brings about the conditions in which another human being freely acts; (2) a human being acts on nature to achieve a certain end; and (3) things act on other things impersonally to bring about a new state of affairs. Collingwood considers the human use of persuasion to evoke the free action of another human being as primary; the other senses are derived. In the minds of such early modern scientists as Galileo and Newton the third sense has completely escaped anthropocentrism. But Collingwood's main object in these chapters is to show that an anthropomorphic core remains in the third sense.

[2]Aristotle, *Metaphysics* 1.3.

Aristotle's comprehensive understanding of causality was rejected by the architects of modern natural science. After Galileo, Descartes and Isaac Newton, mathematics was deemed the only way to comprehend the real relationships of things in nature to each other. The book of nature, according to Galileo, "is written in the language of mathematics, and its characters are triangles, circles, and other geometric figures without which it is humanly impossible to understand a single word of it; without these one wanders about in a dark labyrinth."[3] And Newton observed, "The moderns, rejecting substantial forms and occult qualities, have endeavored to subject the phenomena of nature to the laws of mathematics."[4] Causality was narrowed to the efficient causes of things, that is, to causes that move physical objects by a transfer of physical energy. Formal and final causality cannot be articulated in mathematical categories; hence they were put aside as unhelpful to scientific understanding or denied real causal effect.[5] The logic and actual history of this movement leads directly to mechanistic materialism, deism or atheism, and loss of a way to conceive of divine action in the world; or at least it makes it more difficult to conceive of God as acting in creation without acting as "a force or energy that moves the atoms of the universe."[6]

According to Michael Dodds, new developments in natural science, specifically quantum theory, emergence or top-down causality, and the fine-tuning of the universe or the anthropic principle, have refuted the mechanistic view of causation. These developments call for broadening the concept of causality to include again Aristotelian and Thomistic formal and final causality. For example, the notion of emergence takes account of the unity and holistic identity of complex natural objects. The properties and activities of living things or even molecules cannot be explained by reducing them to their parts and inner relationships. Something new comes into view when these objects emerge, and the whole exercises a new causal influence

---

[3]Galileo Galilei, excerpts from the Assayer, in *Discoveries and Opinions of Galileo*, trans. Stillman Drake (Garden City, NY: Doubleday, 1957), pp. 237-38.

[4]Isaac Newton, *Mathematical Principles of Natural Philosophy*, in *Sir Isaac Newton's Mathematical Principles of Philosophy and His System of the World*, trans. Andrew Motte, rev. Florian Cajori (Berkeley: University of California Press, 1946), p. xvii; quoted in Michael Dodds, *Unlocking Divine Action: Contemporary Science and Thomas Aquinas* (Washington, DC: Catholic University of America Press, 2012), p. 49. Rejection of the relevance of formal causality ("substantial form") to natural science was characteristic of the new science.

[5]See Dodds, *Unlocking Divine Action*, p. 54, on Bacon's setting aside, but not denying, final and formal causality.

[6]Ibid., p. 50.

over the parts of which it is composed. Dodds points out that some reality must be postulated to account for this new causal power, and this explanation bears a striking resemblance to the Aristotelian substantial form, which causes prime matter to be a particular thing. For if we answer the question, "What makes the whole to be a distinct entity and not just a hodge-podge of parts?" by saying, "the structure of the parts," we simply return to reductionism. To avoid reductionism we must also account for the structure of the parts through the causality of the whole. Dodds concludes correctly, "If priority is given to the whole over the part, it must also be given to that principle that accounts for the distinctive identity of the whole."[7] Since new developments in science imply that something like the Aristotelian substantial form is a necessary explanatory principle of physical entities in the world, divine action may now be conceived more readily as a type of formal causality.[8]

On the issue of final causality, biologists seem to be driven by their study of living things to explain the functioning and activities of organisms in terms of something resembling final causality.[9] Some cosmologists see the production of such thinking and self-conscious beings as human beings as the final explanation of the basic laws of physics.[10] The reintroduction of final causality or teleology into modern science facilitates the possibility of again considering God as exercising a type of final causality in relation to creatures. Specifically, God is the supreme and original good toward which all creatures strive as they endeavor to realize their natures. Dodds explains as follows:

---

[7]Ibid., p. 184.

[8]Dodds uses Aquinas's Platonic notion of extrinsic exemplar causes existing in the divine mind. Every created substantial form bears a likeness to God. See ibid., p. 186, where he quotes Aquinas: "The form which is a part of the thing is a likeness of the first agent, flowing from him. Thus, all forms are traced back to the first agent as to the exemplar principle" (quoting *Sentences* 2.1.1.1 and 5).

[9]"We can conclude that, from the point of view of the present scientific world view, the existence of teleological dimensions in our world—not only in the biological, but also in the physico-chemical—is a plain fact. Until now the state of the sciences did not provide sufficient grounds for it: only the scientific progress of the last decades of the twentieth century has made it possible to reach this vantage point" (Mariano Artigas, *The Mind of the Universe: Understanding Science and Religion* [Philadelphia: Templeton Foundation Press, 2000], p. 130, quoted in Dodds, *Unlocking Divine Action*, pp. 102-3). Also, "Importantly, then, the onset of third-order emergence defines the onset of telos on this planet and for all we know, in the universe. Creatures have a purpose, and their traits are for that purpose" (Ursla Goodenough and Tarrence W. Deacon, "From Biology to Consciousness to Morality," *Zygon* 38 [2003]: 804; quoted in Dodds, *Unlocking Divine Action*, p. 103).

[10]See Michael J. Denton, *Nature's Destiny: How the Laws of Biology Reveal Purpose in the Universe* (New York: Free Press, 1998).

The influence of God's final causality is present in the least act of every creature. Each one acts in accordance with its nature. In doing this, it is consciously or unconsciously seeking some good, namely the fulfillment of its nature. Its nature, however, bears some likeness to God since God is the creator of all things, and each thing in some way resembles its creator. The fulfillment of its nature, therefore, also bears some likeness to God, since that fulfillment is proportionate to its nature and its nature is like God. In seeking its fulfillment as a good proportionate to its nature, therefore, it is also seeking God or divine goodness since the fulfillment of its nature bears a likeness to God. In its every act, therefore, the creature is in some way seeking the goodness of God as its final cause.[11]

As I noted earlier, Dodds rejects the use of particular interpretations of these new theories in a univocal sense to explain the possibility of divine action. However, he argues that the expanded notions of causality, if used analogically, can advance our understanding of divine action. These developments make possible a retrieval of the Aristotelian and Thomist understanding of causality as material, formal, efficient and final. Using this expanded understanding of causality along with care to preserve the transcendence of God in relation to creation, according to Dodds, yields an understanding of divine action that preserves the integrity of creaturely action and the intimate and universal presence and action of the God of love in creation and providence.

CREATION AND RELATIONALITY

Since the concept of a relation is used by many modern authors to shed light on God's relation to the world in creation and providence—especially by those who wish to demonstrate that God must limit himself to be able to engage the world—we must clarify its meaning.[12] The concept of relation is more comprehensive than causality. Causality is one type

---

[11]Dodds, *Unlocking Divine Action*, p. 181.

[12]The concept of relation and its application to God in a univocal sense as articulated by Charles Hartshorne in *The Divine Relativity: A Social Conception of God* (New Haven, CT: Yale University Press, 1948) has exercised influence on many contemporary theologians. Open theist William Hasker, for example, credits Hartshorne with convincing him that God is really related to creatures as opposed to being related logically ("An Adequate God," in *Searching for an Adequate God*, ed. John B. Cobb Jr. and Clark H. Pinnock [Grand Rapids: Eerdmans, 2000], pp. 216-17). Hasker says, "That God is really related to the creatures is a genuine and important point of agreement between process theism and the open view of God" (p. 217).

of relation but not the only kind. There are logical relations: entailment, contradiction, contrariety and others. Interpersonal relations seem to differ from both causal and logical relations, although they are also causal. Perhaps there are others. But what is a relation? Again, we face a difficulty we faced when discussing causality, for the concept of relation is so primitive that we find it impossible to explain in other terms. But we can say something about it. The concept of relation presupposes the existence of at least two things between which there can be a relation. One thing considered only in its unity can possess no relations. A relation also presupposes a space that encompasses both things and enables them to relate. Is the relation between two things a third thing?[13] Already we see similarities between the concept of mediation and the concept of a relation. Both involve a bridge between two different things. The question about relations that is relevant to our discussion of creation is: Does a's relation to b always count as a property of a, so that removing the relation would change a? To put it in theological terms: Does God's relation to creation always count as a property of God, so that apart from God's relation to creation God would be different? Do God's relations of creator, sustainer, knower and guide count as accidental or essential properties of God?

For Aristotle, a relation is not a real thing two other things share by mutual participation but an accident that appears only when two things are considered together. Relations are not substances. Aristotle seems to have in mind primarily logical relations among propositions, and such relations are indicated most often by the prepositions "of" and "to." Consider the relation of being taller than. John is taller than Samuel. Being "taller than" is not a substance that can exist on its own. But neither does it seem to be an accidental property that inheres in John. It appears only in relation to Samuel or something else shorter than John. But the master/slave relation assumes an accidental property in both the master and the slave. The master and slave are interdependent since the one cannot exist without the other.[14]

---

[13]See John Heil, "Relations," in *The Routledge Companion to Metaphysics*, ed. Robin Le Poidevin et al. (New York: Routledge, 2009), pp. 310-21, for his typology of seven views of the ontological status of relations from "flat out anti-realism" to "Hyper-realism." Heil labels the view that sees relations as "ontologically fundamental" as "Hyper-realism" (p. 312).

[14]Aristotle, *Categories 7*, in *The Basic Works of Aristotle* ed. Richard McKeon (New York: Random House, 1941), p. 20.

Medieval philosophers debated whether relations exist only in the mind or have a real foundation in the related substances. The idea that relations exist only in the mind seems to have been a minority position.[15] There must be something in the related things that makes the relation hold, some sort of real correspondence. The question naturally arose concerning the nature of God's relation to creation. What sort of relation is this? For medieval thinkers it was unacceptable to think of God as acquiring or losing accidental properties or to attribute any change to God. But how can the Creator-creature relation be conceived apart from such change? We will return to this question.

Early twentieth-century British philosophers distinguished between internal relations and external relations.[16] F. H. Bradley and Alfred North Whitehead asserted that all things are related internally, while G. E. Moore and Bertrand Russell insisted that all relations are external.[17] In *Appearance and Reality* (1893), Bradley asserts that "a relation must at both ends affect, and pass into, the being of its terms" (p. 364). Also: "Every relation essentially penetrates the being of its terms, and is, in this sense, intrinsical" (p. 392). And again: "To stand in a relation and not to be relative, to support it and yet not to be infected and underdetermined by it, seems out of the question" (p. 142).[18] And Whitehead says of every event: "Each unit has in its nature a reference to every member of the community, so that each unit is a microcosm representing in itself the entire all-inclusive universe."[19] The assertion of the all-encompassing relativity of the universe finds a home in the tradition of philosophy that begins with Plato and continues with Plotinus and Hegel and the absolute idealist F. H. Bradley. That all things are

---

[15]Jeffrey Brower, "Medieval Theories of Relation," in *Stanford Encyclopedia of Philosophy* (Spring 2014 ed.), ed. Edward N. Zalta, http://plato.stanford.edu/entries/relations-medieval/.

[16]For a concise summary of historical and contemporary debates about the nature of relations, see Heil, "Relations," pp. 310-21. See Jonathan Schaeffer, "The Internal Relatedness of All Things," *Mind* 119 (2010): 341-76, for a defense of monism and the internal relatedness of all things.

[17]The ancient atomists and the modern thinker Bertrand Russell argue that all relations are external. F. H. Bradley and Alfred North Whitehead argue that all relations are internal. See G. E. Moore, "External and Internal Relations," *Proceedings of the Aristotelian Society* 20 (1919–1920): 40-62; See also John Macquarrie, *Twentieth Century Religious Thought: The Frontiers of Philosophy and Theology 1900–1980*, rev. ed. (New York: Charles Scribner's Sons, 1981), pp. 28-29, 228-32.

[18]Each of these quotations is quoted in Moore, "External and Internal Relations," 40.

[19]Alfred North Whitehead, *Religion in the Making* (Cambridge: Cambridge University Press, 1926), p. 79, quoted in John Macquarrie, *In Search of Deity: An Essay in Dialectical Theism* (New York: Crossroad, 1987), p. 145.

internally related implies that "everything is what it is in virtue of its relations with everything else."[20]

Other thinkers argue that at least some relations are purely external. An
external relation is a relation the removal of which would leave the related
things unchanged. Moore gives the example of the part/whole relation.[21]
A whole constituted of particular parts would be different if it were made
of different parts; but the reverse is not true. Take for example the bricks
that constitute a building. The bricks of which a building is constructed
are externally related. A brick is a brick before and after it becomes a part
of a building, and when the building is demolished the brick's essence
remains unchanged. The relation of knowing is another. Seeing that the
moon is in total eclipse affects the knower, but this knowledge does not
affect the moon.[22]

Process philosopher Charles Hartshorne contends that God and the
world are really related and are mutually constitutive of each other.[23] God
would not be what God is without the God-world relation. God is in one
aspect absolute and nonrelative but in another God is supremely related. If
God were wholly nonrelative, God could not know us or love us: "All our
experience supports the view that the cognitive relation, still more obviously,
if possible, a relation such as love, is genuinely constitutive of the knower or
the lover."[24] Schubert Ogden asserts that in order for God to relate to creation "God must enjoy real internal relations to all our actions and so be
affected by them in his own actual being."[25] According to David Griffin, the
idea of the internal relatedness of all things "is the basis for understanding
causation as incarnation, for regarding the presence of God in all things and

---

[20]Macquarrie, *In Search of Deity*, p. 145. Douglas Pratt argues that Macquarrie's "dialectical theism" and Hartshorne's "neoclassical theism" share significant common features, specifically the
relationality of God (Douglas Pratt, *Relational Deity: Hartshorne and Macquarrie on God* [Lan-
ham, MD: University Press of America, 2002]).

[21]Moore, "External and Internal Relations," p. 51.

[22]For a historical and analytic study of relations concentrated on the distinction between internal
and external relations, see Richard Rorty, "Relations, Internal and External," in *The Encyclopedia
of Philosophy* (New York: Macmillan, 1967), 7:125-33.

[23]Hartshorne, *Divine Relativity*. For Hartshorne, God is both absolute and relative. Even though
God's identity is not dependent on the actual existence of any contingent creature, God is not
independent of the possibility of contingent creatures as such. God is by eternal nature the
creator and is "bound to create, provided he is not thus bound with respect to any given creature
or set of them" (p. 74).

[24]Hartshorne, *Divine Relativity*, p. 17.

[25]Schubert Ogden, *The Reality of God* (New York: Harper & Row, 1963), p. 47.

the presence of all things in God as fully natural."[26] And according to William Hasker open theism agrees with process theology that "God is really related to his creatures, where 'really related' means that it makes a difference to God how things are with creatures."[27] Open theist John Sanders designates his approach to theology as "relational theism" and argues that the biblical picture of God's interaction with creation makes no sense unless there are "genuine give-and-take relations between God and humans such that there is receptivity and a degree of contingency in God."[28]

Old Testament scholar Terence Fretheim applies the modern idea of internal relations to the God-world relation in the Old Testament. In *God and World in the Old Testament: A Relational Theology of Creation*, after voicing criticisms of a covenantal-monarchial view of God's relation to the world, Fretheim asks, "What if we took the word *relationship* seriously?"[29] In the Old Testament narrative, the world "affects God" so much that God's relationship with his people "is constitutive of the divine identity" (p. 20). In the relationship with creation, God is "interdependent" with the world (p. 27). Because God has established this relationship, God is "vulnerable" and must take "risks" (p. 38). When God makes promises, this "entails an eternal self-limitation regarding the exercise of divine freedom and power" (p. 83). In relating to creation, God becomes "dependent upon both human and nonhuman in the furtherance of God's purposes in the world. All creatures have a God-given vocation within God's creation-wide purposes; in other words, God has freely chosen to rely upon that which is not God to engage those purposes" (p. 270). Fretheim dismisses without argument the idea that God's relationship to the world could be "logical" (an oblique reference to Thomas Aquinas's view, which I discuss below) and simply assumes that all

---

[26]David Griffin, "Process Theology and the Christian Good News: A Response to Classical Free Will Theism," in Cobb and Pinnock, *Searching for an Adequate God*, p. 5. Griffin's assertion in this quote presupposes the Whiteheadian axiom "God is not to be treated as an exception to all metaphysical principles, invoked to save their collapse. He is their chief exemplification" (Alfred North Whitehead, *Process and Reality: An Essay in Cosmology* [New York: Macmillan, 1929], p. 521). Griffin simply assumes that God is subject to the laws that govern the nature and actions of the things we experience and that the language we use of creatures can be used univocally of God.

[27]Hasker, "An Adequate God," p. 216.

[28]John Sanders, *The God Who Risks: A Theology of Providence* (Downers Grove, IL: InterVarsity Press, 1998), p. 12.

[29]Terence Fretheim, *God and World in the Old Testament: A Relational Theology of Creation* (Nashville: Abingdon, 2005), p. 16 (emphasis original). Subsequent in-text page references are to this work.

relations must be real and mutually constitutive of the related things. Apparently this is what he means by taking the word *relationship* "seriously." Fretheim seems to assume that we must choose between the concepts of internal (real) relations and external relations, both taken in a univocal sense, to express the God-creature relation.[30]

THE QUEST FOR HOW—AN ASSESSMENT

Does the quest for an answer to the question of how God created the world advance our understanding of God's act of creation? Do the ideas of act, means, mediation, cause and relation help us in this quest? We discovered in our discussion of action that all acts with which we are familiar involve an actor using means to bring about an end. The human actor needs means to achieve the end. Given the sovereignty with which God creates and acts within the world, we cannot view God's act of creation in the same way we view human acts. The God of Christian faith does not need means outside of his eternal being. Hence the rational understanding of an act as necessitating means must not be imposed on the divine act of creating.[31] The idea of a mediator is intimately related to that of means. In those systems of thought that seek to explain how the highest ontological principle relates to the physical world (Platonism and Gnosticism) we find that the mediator must exist on a lower ontological level than the highest reality. It must participate in (but not be identical to) the highest reality to relate to it, but it must participate in (but not be identical to) the lower reality to relate in it. The problem with these metaphysical theories of mediation is that once one posits the distinction between the two things to be mediated as an absolute qualitative difference—the one and the many, the perfect and the imperfect, eternity and time, good and evil, the spatial and the nonspatial, and the intelligible and the unintelligible—rational mediation is logically excluded. Any "slight" ontological lowering of the mediator represents a qualitative

---

[30]And Charles Hartshorne asserts that "theology (so far as it is the theory of the essence of the deity) is the most literal of all sciences of existence. . . . The pure theory of divinity is literal, or it is a scandal, neither poetry nor science, neither well reasoned nor honestly dispensing with reasoning" (*Divine Relativity*, pp. 36-37).

[31]Unlike a human actor God's active power does not need activating by a prior actor. God is "pure act" and hence needs no means to act. In David Burrell's words, it makes no sense to "ask how pure act acts" ("Divine Practical Knowing: How an Eternal God Acts in Time," in *Divine Action: Studies Inspired by the Philosophical Theology of Austin Farrer*, ed. Brian Hebblethwaite and Edward Henderson [Edinburgh: T & T Clark, 1990], pp. 93-102).

change that might as well be infinite and represents a mixing of categories, the quantitative and the qualitative. Anything less than perfect is not perfect at all, and there is nothing qualitatively between one and many and so on. We can always ask, Who mediates the mediator?

In the Christian doctrine of creation, the Word, or Son, who became incarnate as Jesus Christ serves as the means of creation and salvation. We will deal at length with the role of the Word/Son in creation later. In this context we must assert that Christian theology cannot accept the notion that the Word/Son exists on a lower ontological level than the Father. The distinction and otherness of the Son from the Father is a relation *within* the being of God, not a distinction *between* the being of God and the being of the Son. This understanding of mediation arises from reflection on God's self-revelation in the economy of salvation, not from rational reflection on the soul as mediator between the physical and intelligible worlds. The Christian understanding of mediation is beyond the power of rational self-reflection to achieve, and a Christian theology of creation must not allow itself to be judged by this understanding.[32]

In what ways may the concept of causality be helpful in understanding God's act of creating? Certainly, looked at from the perspective of the created world, creation owes its existence and form wholly to God and remains dependent on God. So the world is an effect of God's act. But does that make God the cause of the world? If so it must be in a very special sense of the word *cause*. Unlike other causes God is not changed by causing the world to exist. Nothing in God is transferred or communicated to the world. God cannot be the material cause of the world, for the world is not made out of God. Nor can God become the formal cause of the world because that would make the creature divine, an incarnation of God.[33] God can be thought of as the efficient cause of the world only in an attenuated sense because

---

[32]Perhaps we ought to take J. G. Fichte's judgment on the impossibility of conceiving creation seriously: the idea of creation is "the absolute and fundamental error of all false metaphysics and religious doctrine. . . . A creation cannot be conceived of in a respectable intellectual fashion at all . . . and no man has ever thought it out in such a way" (*The Way Towards the Blessed Life: or the Doctrine of Religion*, trans. William Smith [London: Chapman, 1849], p. 191, quoted in Otto Weber, *Foundations of Dogmatics*, vol. 1, trans. Darrell L. Guder [Grand Rapids: Eerdmans, 1981], p. 468 n. 18).

[33]Thomas Aquinas denies that God is the formal cause of the world (*SumTh* 1.3.8). According to David Burrell this identification would "directly foster pantheism" ("The Act of Creation: Theological Consequences," in *Creation and the God of Abraham*, ed. David Burrell et al. [Cambridge: Cambridge University Press, 2010], p. 45).

nothing exists for God to work on and the change happens only in the creature. According to Scripture, God is the cause of creation in that God wills it to be and for this reason it comes to be. But *how* God's willing brings about the existence of creatures we cannot say. Hence we must be wary of arguments that follow this pattern: God is the cause of the world. Causes are changed in the process of producing their effects. Therefore God is changed in his act of creating the world.

The concept of relation, like the concepts of act, means, mediation and cause, proves inadequate to the task of explaining how God creates, and we must resist efforts to force divine creating into the mold of the rationally derived concept of relation. Consider the three characteristics of a relation discussed above. (1) In conceiving God's "relation" to the world we must reject the notion that God and creation met in a space that contains both of them, because it would make God dependent on this larger space. The divine possibility for meeting creation is grounded solely in God. God is his own space and the space of creation is something altogether different.[34] (2) God's relation with creation cannot depend on the existence of creation, as the concept of relation seems to demand. God and the world are not interdependent. For the possibility of the existence of the creature lies in God alone; hence God's relation with the creature and the creature's relation with God are established unilaterally by God in his act of creation. (3) The relation itself cannot be considered mutual participation of God and creatures in a third thing, such as a property or a space or a substance. This aspect of a relation reminds us of the philosophical idea of mediation and is subject to the same critique. Such mutuality would imply a natural connection between God and creatures, which would compromise the independence of God and the allied doctrine of creation from nothing, with which we will deal below. A glance at the ideas of external and internal relations in relation to God will confirm how challenging it is to find a sense in which the notion of relation applies to God's creating. The God-creature relation cannot be thought of as purely external. This would mean that God and creatures relate like two bricks in a building put together within a common space and by another force. And the Scriptures clearly teach that God's act of creating arose from an inner decision and as an expression of his goodwill. On the

---

[34]On God's space, see Ron Highfield, *Great Is the Lord: Theology for the Praise of God* (Grand Rapids: Eerdmans, 2008), pp. 284-89.

other hand, it won't do to interpret God's relation to creation as an internal relation, on the divine side at least. This would make God dependent on creation, for God would not be God apart from creation. It would imply that God's act of creating was less than completely free and was partially an automatic process demanded by God's inner nature.

Thomas Aquinas felt these problems and addressed them in many contexts. In discussing the subject of creation in book 2 of *Summa Contra Gentiles*, Aquinas begins by saying that Scripture and tradition assert many things of God relatively: that God knows creatures and that God creates and moves them. Clearly God possesses some kind of relation to creatures, but what kind? The relation cannot be an accidental property, "since there is no accident in Him."[35] Nor can the relation be in God substantially since that would make God in some way dependent on creatures. The above relations Thomas calls "real" relations, that is, the relation inheres in the thing in a way that the thing would be different without it. It would need to change to acquire or to lose the relation. This cannot be true of God. But neither can it be said that God's relation to creatures exists outside of him; for then God would need other relations to relate to the external relations, and so on. Hence, argues Aquinas, God's relation to his creatures must be "rational" or "logical" in nature whereas the creature's relation to God is real. That is to say, if God had not created the world, God would still be God but *the world* would not exist at all. The relation between God and creatures can be called a "mixed relation." According to Thomas Weinandy, Aquinas understands a mixed relation to come about when two things of different ontological orders relate; for example Creator and creature, known and the knower, and the divinity and humanity in the incarnation. In each example just listed the first term would not be changed were it not related to the second term.[36] God is said to become the Creator not because of a change in him but because of the coming to be of creatures.[37] Clearly Aquinas assumes the sovereign transcendence of God as axiomatic or already proved, as he reasons about how to understand God's relationship to creation.

[35]*SCG* 2.12.2 (Saint Thomas Aquinas, *Summa Contra Gentiles, Book Two: Creation*, trans. James F. Anderson [Notre Dame, IN: University of Notre Dame Press, 1975], p. 43).

[36]Thomas Weinandy, "Aquinas: God Is Man: The Marvel of the Incarnation," in *Aquinas on Doctrine: A Critical Introduction*, ed. Thomas Weinandy, Daniel Keating and John Yocum (London: T & T Clark, 2004), pp. 76-77.

[37]Burrell, "Act of Creation," pp. 27-44.

Perhaps it is good to remind ourselves here that for Christian theology the doctrine of creation, like the doctrines of incarnation and Trinity, is a doctrine of faith. It is based on revelation and cannot be concluded from reason's reflection on the world. Thomas Aquinas and other theologians adapted such concepts as relation and cause to express as far as possible the church's faith in divine creation. But they did not allow that faith to be trumped by the limits of reason. Hence their use of concepts derived from metaphysical thought seems a bit fractured and inconsistent. But to insist as process theology and open theism do that such concepts be used of God and God's relationship to creation in the same sense as they are used of things within the world is to abandon the faith premise of the Christian doctrine of creation.

WHY GOD MADE THE WORLD

In our examination of human acts of making we noticed five aspects of agent action that demand our attention: motive, model, means, change and result. In each area we noted the dissimilarity between imperfect human action and perfect divine action. In that section we were disciplining our language for use in articulating Christian faith in God's creating and sharpening our abilities to perceive misuses of these concepts. In this section we will approach the subject positively by seeking insight into the central scriptural and traditional affirmations about God's act of creating, keeping in mind the same five aspects of agent action explored above. We will seek insight into Scripture's affirmations concerning (1) God's motives and intentions for creating; (2) the model and means for creating; (3) the change effected in creating; and (4) the result of God's act.

Why did God create the world? In dealing with this question we will examine both the motive and the goal for creating. Nowhere does the Bible tell us explicitly what motivated God to create the world or to what end. And if we insist on separating God's act of creation from those of salvation and consummation, we will find little guidance on the subject. If, for example, God's original motivation and purpose for creation was thwarted by the fall as some contend, we must exclude everything said in the Scriptures about God's motivation and purpose for his saving actions as irrelevant to the original creation. However, I have insisted from the beginning that creation must not be viewed as a divine act separate from salvation and consummation. The proof of this is that Jesus Christ, the one through whom God saves, is the very same one through whom God created. I shall assume this unity for the present and look for God's motive and purpose in creation in what is said about salvation and consummation. According to the New Testament, God's saving act centers on sending his Son to become human to suffer, die and be raised for our salvation. And the universal consensus is that God's love or mercy or kindness motivated this act. John is unambiguous on this point: "For God so loved the world that he gave his one and only Son . . ." (Jn 3:16; cf. Jn 16:27; 1 Jn 3:1; 4:9-11, 16). Paul sees in Christ's act the depths of divine love: "But God demonstrates his own love for us in this: While we were still sinners, Christ died for us" (Rom 5:8; cf. Rom 8:39; Gal 2:20; Eph 2:4; 5:2; Tit 3:4). In Ephesians 1:3-10 we see a clear connection between creation and the love of God expressed in Christ:

> Praise be to the God and Father of our Lord Jesus Christ, who has blessed us in the heavenly realms with every spiritual blessing in Christ. For he chose us in him before the creation of the world to be holy and blameless in his sight. In love he predestined us for adoption to sonship through Jesus Christ, in accordance with his pleasure and will—to the praise of his glorious grace, which he has freely given us in the One he loves. In him we have redemption through his blood, the forgiveness of sins, in accordance with the riches of God's grace that he lavished on us. With all wisdom and understanding, he made known to us the mystery of his will according to his good pleasure, which he purposed in Christ, to be put into effect when the times reach their fulfillment—to bring unity to all things in heaven and on earth under Christ.

In this text the eternal, the present and the future are fused into one grand movement from God to God. The present spiritual blessings experienced in Christ are traced to the eternal love of God that was already directed toward us in Jesus Christ "before the creation of the world." It will be completed only when all things "in heaven and on earth" are united under Christ. As Wolfhart Pannenberg observes, "The final ordering of creatures to the manifestation of Jesus Christ presupposes that creatures already have the origin of their existence and nature in the Son. Otherwise the final summing up of all things in the Son (Eph. 1:10) would be external to the things themselves."[38] The eternal Jesus Christ is not a nameless mediator but the one through

---

[38]Wolfhart Pannenberg, *Systematic Theology*, trans. Geoffrey W. Bromiley (Grand Rapids: Eerdmans, 1994), 2:25.

whose blood we have redemption. Colossians 1:15-20 also identifies the one through whom all things were created as the same one whose blood was "shed on the cross." Hebrews 1:1-3 makes the same identification, for the one through whom God made the universe is also the one who "provided purification for sins." Karl Barth's words appropriately sum up the message of these texts:

> In respect of His Son who was to become man and the Bearer of human sin, God loved man and man's whole world from all eternity, even before it was created, and in and in spite of its absolute lowliness and non-godliness, indeed its anti-godliness. He created it because He loved it in His Son who because of transgressions stood before Him eternally as the Rejected and Crucified.[39]

Thomas Torrance also brings out the motivational unity between God's act of creation and the salvation accomplished in Christ: "Creation arises out of the Father's eternal love of the Son, and is activated through the free ungrudging movement of that Fatherly love in sheer grace which continues to flow freely and unceasingly toward what God has brought into being in complete differentiation from himself."[40]

In answer to the question of what motivated God to create the world the only answer can be *his love for the world in his Son*. As we argued earlier, God's creating could not have been motivated by distress or need or lack of any kind. Clearly we are using the concept of a "motive" in an analogical way. Nothing outside God "moves" or changes God. Nor was it an event without a motive, a necessary overflow of divine goodness. Still less was creation an arbitrary or capricious or chance event. Though completely free and uncompelled, God willed to share his love and life with creatures. Jesus' prayer in John 17 gives us some insight into this love: "Father, I want those you have given me to be with me where I am, and to see my glory, the glory you have given me because you loved me before the creation of the world" (Jn 17:24). The Father's eternal love for the Son is in the Son turned toward the world and through the Son returns to the Father.

The other side of the question of why God created the world is the issue of the purpose or intention of that act. What was God's goal? It is often said

---

[39]*CD* III/1, pp. 50-51. For a similar view, see Otto Weber, *Foundations of Dogmatics*, trans. Darrell L. Guder (Grand Rapids: Eerdmans, 1981), 1:479-80, 485-86.
[40]Thomas F. Torrance, *The Christian Doctrine of God: One Being Three Persons* (Edinburgh: T &T Clark, 1996), p. 209.

that God's goal was his own glory. There is of course an element of truth in this claim. In my doctrine of God, I define the glory of God as *"the manifestation and perception of the greatness, splendor, and excellence of God's being and actions."*[41] God is glorious in his eternal being and hence glorious in all his actions, including his act of creation. So God displays his glory in creating. The theological rationale for considering God's glory the end of his creative action is desire to avoid giving God an end other than himself. For this would make God dependent on whatever that end is. This intuition is sound. But there is something not quite right about making self-glorification the end of God's actions. It's not only that we tend to judge God's actions by human standards so that self-glorification sounds self-centered and arrogant. More importantly, how can we think of the *purpose* of creation as God's glory while thinking of the *motivation* for that act as love? Love is other directed, and self-glorification seems directed inward. Augustine pointed out that the idea of doing something "for the sake of" something can be used in two senses that are often confused: (1) as the beneficiary of an action or (2) as the ultimate ground for an act. It is not inconsistent to believe that God created the world (1) to benefit and bless creatures and (2) for his glory, as long as one understands that God's glory is to love.[42] God's glory radiates in his act of being himself, that is, as Father, Son and Holy Spirit. God is acting as himself in his act of benefiting and blessing his creatures. Hence God glorifies himself in creating creatures that might know and enjoy him forever. But to disconnect God's glory from his eternal love for the Son and view it as an exercise in sheer willful power would greatly distort it.

If we return to Ephesians 1:3-10 we see that God's eternal goal for salvation, and hence for creation, is making his chosen ones "holy and blameless in his sight" and adopting them "to sonship through Jesus Christ." And along with them, to "bring unity to all things in heaven and on earth under Christ." In Paul's writings everything that precedes Christ aims toward him

---

[41]Highfield, *Great Is the Lord*, p. 391 (emphasis original).

[42]Jonathan Edwards deals with this question in his *Dissertation Concerning the Chief End for Which God Made the World*. He shows how God brings glory to himself in creating creatures. In valuing creatures he values himself, and in valuing himself he values creatures (*The Works of Jonathan Edwards* [1834; repr., Peabody, MA: Hendrickson, 1998], 1:94-121). Jonathan R. Wilson makes essentially the same point when he speaks of the "telos" of creation as "life," the fullness of life in the new creation as it was manifested in Jesus Christ (*God's Good World: Reclaiming the Doctrine of Creation* [Grand Rapids: Baker Academic, 2013], pp. 116-17).

and finds its fulfillment in him. The law of Moses was made necessary by sin and served as a "guardian until Christ came." (Gal 3:24; cf. Rom 5:20-21; 7:13). Creation "was subjected to frustration . . . in hope that the creation itself will be liberated from its bondage to decay and brought into the freedom and glory of the children of God" (Rom 8:20-21). And "all things have been created through him and *for him*" (Col 1:16). This last prepositional phrase is highly significant for our discussion. Christ is here the beginning and the end of creation. The purpose of the beginning reveals itself in the end. God created the world so that everything would be reconciled and brought to unity and completion in Christ.

6

JESUS CHRIST AS CREATOR

The New Testament views Jesus Christ as model and means for God's act of creation. In the section on the New Testament's theology of creation, we examined the four texts in which Christ is designated as Creator or the means through whom God created the world (Jn 1:1-5, 10-14; 1 Cor 8:4-6; Col 1:15-17; and Heb 1:1-4, 10). We concluded that salvation and creation are intimately related and inseparable in Christ and that Jesus Christ, who exists eternally in intimate relationship with God, is also God's instrument of creation. But we were left with many unanswered questions and unexplored implications that have fascinated Christian theologians from the patristic era until today. We take up those questions in this chapter.

HELLENISTIC JEWISH BACKGROUND

Some among the large Jewish community living in Alexandria in the first century B.C. found in Middle Platonism (discussed above) a system of thought within which to understand and communicate its faith in the one God to the Hellenistic mind. Philo of Alexandria (20 B.C.–A.D. 50) is the most outstanding and well-documented representative of this trend. Holding firmly to the monotheism of his ancestral religion, he attempted to show the harmony of the Hebrew Scriptures with the truth he found in philosophy. In this endeavor Philo developed a theology of mediation in considerable detail, if not consistency. Having inherited a Jewish tradition of speculation about wisdom's (Sophia) role as God's mediator in relating to the created world, Philo incorporates insights from Middle Platonism and Stoicism.[1] Philo uses a variety of names for the intermediary (Sophia,

---

[1]Ronald Cox, *By That Same Word: Creation and Salvation in Hellenistic Judaism and Early Christian-*

goodness, cherubim), but the Logos dominates the list and assumes the functions of the rest.[2] What is the Logos, and how does it function as a mediator? The idea of the Logos as the rational structure that gives meaningful order to the world derives from Heraclitus (fl. 504–500 B.C.) and became a central tenet of the Stoic worldview. Philo finds a connection between the structural Logos of philosophy and the spoken word by which God creates the world in Genesis. The Logos can be compared to the mind of God containing all the ideas and forms used as models for creation. The Logos is "beginning," "firstborn" and "eldest," having priority over all things but God himself. "Ontologically speaking, Philo holds the *Logos* to be the closest thing to God that is not God himself."[3] The functions of the Logos in mediation are described under three headings: image, instrument and divider.[4] It may be the case that the Logos carries out his functions as the instrument and divider precisely by being the image.[5] The visible world is a sensible image of the intelligible image of God.[6] Hence in Philo we find the same three principles we find in Middle Platonism: the transcendent One, the intelligible world that serves as orderer and shaper of the material world, and unformed matter itself.

Recalling our chapter on the biblical view of creation, we can see that the New Testament speaks about Jesus Christ in ways that remind us of Philo's doctrine of the Logos and Middle Platonism's intermediary principle. The four texts we examined previously (Jn 1:1-5, 10-14; 1 Cor 8:4-6; Col 1:15-17; and Heb 1:1-4, 10) speak of Jesus Christ as the one "through whom" or "for whom" creation was accomplished. But the overwhelming emphasis is on the saving work of Jesus, the crucified and risen one. Clearly the New Testament identifies Jesus with the instrument of creation about which phi-

---

ity (Berlin: de Gruyter, 2007), pp. 109-10. The apocryphal book Wisdom of Solomon (6:22–10:21) provides a good example of how personified wisdom of Proverbs 8 becomes near hypostatic as God's "breath, his emanation, his image" (ibid., p. 87).

[2]Ibid., p. 99.

[3]Ibid., p. 101.

[4]John Dillon, *The Middle Platonists: 80 B.C. to 220 A.D.*, rev. ed. (Ithaca, NY: Cornell University Press, 1996), pp. 158-61. By "divider" Philo means "orderer." By dividing undifferentiated matter by number and proportion, the Logos orders it.

[5]Cox, *By That Same Word*, p. 117.

[6]In commenting on Gen 1:27, Philo sees a distinction between God and the image of God after which God made the human beings. The human beings are images of the image. Apparently Philo wishes to generalize God's way of creating human beings to the whole creation process (Cox, *By That Same Word*, p. 119).

losophers and Philo speculated. However, rather than using Jesus Christ as a convenient cipher to be filled with philosophical content, the abstract idea of an intermediary is commandeered and transformed by being filled with the personal identity of Jesus Christ. Instead of Jesus Christ being depersonalized by being absorbed into a rational theory of the Logos, the speculative Logos is personalized by being transformed into the Son who came in obedience to the Father for our salvation and will come again for our redemption. The Son is indeed the "image" and "firstborn," but there is hardly a hint of the intellectualism of Philo and the philosophers. The emphasis falls squarely on the holiness and love of God mirrored in Jesus and set forth as an example for believers (Eph 5:1-2; Phil 2:1-11). And given the New Testament authors' conviction that Jesus' life and message had received validation by God's act of raising him from the dead, their identification of the incarnate Son as the agent of creation was legitimate. If Jesus is God's instrument of salvation and final redemption, he had to be God's instrument of creation.

TWO MODELS FOR A THEOLOGY OF CREATION

Before developing a systematic understanding of the issue of mediation, we will examine briefly two very different patristic-era theologians, Irenaeus of Lyons (ca. 130–ca. 202) and Origen (ca. 185–ca. 254).[7] According to Leo Scheffczyk, the Apostolic Fathers, the generation of Christian writers immediately following the New Testament era, were content to assert creation as a matter of passing on the received teaching and in contexts of praise.[8] The next generation, the Apologists, as their name suggests, attempted to show the truth of Christianity in terms that made sense to the contemporary philosophical mind. In Scheffczyk's opinion, the efforts of such second-century thinkers as Aristides, Theophilus of Antioch, Athenagoras, Justin Martyr and Tatian "to adapt the idea of the Logos to Middle Platonic ideas" were not altogether successful, for they "led to a subordination of the Logos and limited him to a purely cosmological role."[9]

Unlike the Apologists, Irenaeus did not concern himself with translating the gospel into philosophical terms acceptable to cultured pagans. He found

---

[7]The Gnostic model could be considered a third.

[8]Leo Scheffczyk, *Creation and Providence*, trans. Richard Strachan (New York: Herder and Herder, 1970), pp. 47-54.

[9]Ibid., p. 60. On Justin Martyr, see John Behr, *Formation of Christian Theology*, vol. 1, *The Way to Nicaea* (Crestwood, NY: St. Vladimir's Seminary Press, 2001), pp. 93-110.

himself troubled more by heresy from within than unbelief from without, that is, from the extremely Hellenized Gnostics claiming to have discovered the true gospel. The Gnostics, by no means a monolithic group, made use of Middle Platonist conceptual tools to rewrite Christianity into an anti-creation religious myth. They asserted the transcendence of the highest principle to such an extreme that the creation event was interpreted as the act of a lower, misbegotten, ignorant and malevolent god. Creation itself is malformed and evil; hence, salvation consists in the liberation of the spiritual essence of humanity from the material world within which it is trapped. To these speculative theories Irenaeus opposed a salvation-historical reading of the Old and New Testaments, demonstrating the identity of the God of salvation and the God of creation and uniting the history that connects the beginning to the end. The challenge, which Ireneaus met in his own way, faced the entire age. According to Paul Blowers, "In this period a strong continuity between apostolic and sub-apostolic Christianity was the crucial challenge of piecing together a coherent narrative of creation and redemption from the witnesses of the Hebrew Scriptures and the sacred texts coming into use in and among diverse Christian communities."[10]

Ireneaus integrates creation and salvation in this brief summary:

> So then the Father is Lord and the Son is Lord, and the Father is God and the Son is God; for that which is begotten of God is God. And so in the substance and power of His being there is shown forth one God; but there is also according to the economy of our redemption both Son and Father. Because to created things the Father of all is invisible and unapproachable, therefore those who are to draw near to God must have their access to the Father through the Son.[11]

Against those who see the material world as alien and external to the true God, Irenaeus argues that God would be less than omnipotent if there existed something outside of him or if he did not control all things:

> It is proper, then, that I should begin with the first and most important head, that is, God the Creator, who made the heaven and the earth, and all things that are therein (whom these men blasphemously style the fruit of a defect),

---

[10]Paul Blowers, *The Drama of the Divine Economy: Creator and Creation in Early Christian Theology and Piety* (New York: Oxford University Press, 2012), p. 96.

[11]Irenaeus, *Demonstration of the Apostolic Preaching* 47, trans. Armitage Robinson (London: SPCK, 1920), p. 112.

and to demonstrate that there is nothing either above Him or after Him; nor that, influenced by any one, but of His own free will, He created all things, since He is the only God, the only Lord, the only Creator, the only Father, alone containing all things, and Himself commanding all things into existence.[12]

In response to the question of how God could create the world apart from a series of mediators, Irenaeus responds:

> It was not angels, therefore, who made us, nor who formed us, neither had angels power to make an image of God, nor any one else, except the Word of the Lord, nor any Power remotely distant from the Father of all things. For God did not stand in need of these [beings], in order to the accomplishing of what He had Himself determined with Himself beforehand should be done, as if He did not possess His own hands. For with Him were always present the Word and Wisdom, the Son and the Spirit, by whom and in whom freely and spontaneously, He made all things, to whom also He speaks, saying, "Let Us make man after Our image and likeness."[13]

Irenaeus answered his Hellenized opponents not with attempts to show that Christian views of God, creation and redemption are consistent with popular philosophical theories but by showing the inner coherence of those views with the sweep of the biblical history and with the traditional teaching and practices of the church. In his well-known theory of recapitulation, he argued that in his life, death and resurrection Jesus Christ summed up "the seamless and purposive action of the Creator in and for the world."[14] Beginning, middle and end are integrated in Christ, so that salvation cannot be peeled off from creation.[15] On the issue of creation specifically, Irenaeus did not attempt to explain philosophically how God created the world;

---

[12]Irenaeus, *Against Heresies* 2.1.1 (*ANF* 1:359). Irenaeus asserts creation from nothing not as if it were a new doctrine but as if it were the accepted view: "While men, indeed, cannot make anything out of nothing, but only out of matter already existing, yet God is in this point pre-eminently superior to men, that He Himself called into being the substance of His creation, when previously it had no existence. But the assertion that matter was produced from the Enthymesis of an Æon going astray, and that the Æon [referred to] was far separated from her Enthymesis, and that, again, her passion and feeling, apart from herself, became matter—is incredible, infatuated, impossible, and untenable" (*Against Heresies* 2.10.4 [*ANF* 1:370]).

[13]Irenaeus, *Against Heresies* 4.20.1 (*ANF* 1:487-88).

[14]Blowers, *Drama of the Divine Economy*, p. 87.

[15]Eric Osborn, *Irenaeus of Lyon* (Cambridge: Cambridge University Press, 2001), pp. 97-98, finds nine ideas wrapped up in Irenaeus's concept of recapitulation: "Unification, repetition, redemption, perfection, inauguration and consummation, totality, the triumph of Christus Victor, ontology, epistemology, and ethics."

rather, he argued from the acknowledged fact of the salvation mediated to the church through Christ and the Spirit that God possesses "two hands" that enable him to act in the world as himself. Irenaeus's theological method involves what would later be called "faith seeking understanding" and the "analogy of faith."[16] Here we see a connection between theological method and metaphysical assumptions. Theologians who assume there is an ontologically graded continuity between God and the world practice theology as a religiously colored metaphysical system. Theologians who assume a sharp and qualitative ontological divide—but not opposition—between God and all other things approach theology as faith seeking understanding by discovering the analogy of faith in the Bible, understood as the record of God's gracious self-revelation.

Origen of Alexandra pursues theology in a way very different from Irenaeus. Origen inherits the Alexandrian style of allegorical interpretation of the Bible and philosophical argumentation pioneered by Philo and carried on by Origen's immediate predecessor Clement of Alexandria (ca. 150–ca. 215). Although Origen affirms the tradition that God created heaven and earth and connects creation and redemption, he interprets the first chapters of Genesis in a distinctly Platonic way.[17] The reference to "heaven and earth" indicates two different creations. First, God created a spiritual world containing "rational or intellectual creatures" (*First Principles* 9.1). Even though this world was created, it always existed, for it would be absurd to think that God became the Creator at some point in time (*First Principles* 1.4.3). Everything in Genesis 1, even Genesis 1:26, which speaks of the creation of humankind as male and female, refers to the first creation and is interpreted allegorically to harmonize with this view.[18] The material world as we know

---

[16]The expression comes from Anselm of Canterbury. See his *Proslogium* 1 (*Saint Anselm: Basic Writings*, trans. S. N. Deane, 2nd ed. [LaSalle, IL: Open Court, 1969], pp. 4-7). For the post-Reformation Reformed Orthodox use of the analogy of faith in Scripture interpretation, see *PRRD* 2:493-97. For Barth's use of the analogy of faith see *CD* I/1, pp. 243-47. For my treatment of the analogy of faith, see *Great Is the Lord*, pp. 144-48.

[17]Blowers points out that though Origen, unlike Irenaeus, uses Platonic categories he rejects Gnosticism just as thoroughly as Irenaeus does (See *Drama of the Divine Economy*, p. 90).

[18]As an example of allegorization, consider Origen's interpretation of male and female of Gen 1:26 as referring to Christ and the church. Peter C. Bouteneff, *Beginnings: Ancient Christian Readings of the Biblical Narratives* (Grand Rapids: Baker Academic, 2008), pp. 103-12, addresses Origen's view of history. He concludes that whether Adam and Eve are historical individuals or not—Origen does not go out of his way to deny this—"is of no consequence as such to his theological vision" (p. 112).

it was made necessary by the fall of these spiritual beings, the diversity and uneven distribution of good and evil in the world being a reflection of

> the diversity and variety in the movements and declensions of those who fell from that primeval unity and harmony in which they were at first created by God, and who, being driven from that state of goodness, and drawn in various directions by the harassing influence of different motives and desires, have changed, according to their different tendencies, the single and undivided goodness of their nature into minds of various sorts.[19]

God created the material world with its diversity and hierarchy to mirror the merits of fallen souls, for the purpose of their ultimate salvation. In his providence God guides, educates and persuades his creatures with the goal of returning them to their previous state of unity in the spiritual world.[20]

Lest it seem that I am being too hard on Origen, we must admit that he attempted to preserve and pass on the tradition that he had received. He begins *First Principles* with these affirmations:

> *First*, That there is one God, who created and arranged all things, and who, when nothing existed, called all things into being. . . . *Secondly*, That Jesus Christ Himself, who came (into the world), was born of the Father before all creatures; that, after He had been the servant of the Father in the creation of all things—"For by Him were all things made"—He in the last times, divesting Himself (of His glory), became a man. . . . Then, *Thirdly*, the apostles related that the Holy Spirit was associated in honour and dignity with the Father and the Son.[21]

In addition, Origen refuted the Gnostics' contention that explanation for the evil in the world is that the world was created by an ignorant and evil god (*First Principles* 2.9, 5-7). And he argues for creation from nothing against the Platonists (*First Principles* 2.1.4). We must also acknowledge that we may not possess a clear understanding of Origen's views in their larger context,

---

[19]Origen, *First Principles* 2.1; 9.6 (ANF 4:268, 292).

[20]Origen's views may be more complicated than Scheffczyk and other interpreters think. See Bouteneff, *Beginnings*, pp. 108-15, and the specialized literature he cites. Origen's commentary on Gen 1-4 is lost. C. P. Bammel marshals evidence from Origen's scattered references to Adam that supports the idea that Origen's teaching about two creations and a preincarnate fall is not incompatible with his also teaching that Adam was a real individual who experienced a historical fall. See C. P. Bammel, "Adam in Origen," in *The Making of Orthodoxy: Essays in Honour of Henry Chadwick*, ed. Rowan Williams (Cambridge: Cambridge University Press, 1989), pp. 62-93.

[21]Origen, *First Principles*, preface 4 (*ANF* 4:240).

since many of Origen's thousands of writings are lost and *First Principles* is preserved only in a Latin translation.[22] Nevertheless, on the topic of creation Origen provides a mostly negative example.[23] In his apologetic zeal to show the rationality and hidden depths of the teaching of Scripture and tradition on creation, he seems to read into them alien elements of speculative thought current in the Platonism of his day. Although Origen *correlated* philosophic ideas with the biblical history and the economy of salvation by means of his allegorical interpretation of biblical texts, he can hardly be said to have *derived them from* the economy.[24] In some areas of his theology Origen crossed the line between *explaining* the faith in metaphysical terms and reading an alien system into the Scriptures and traditional formulas of the church. I realize this is not an easy balance to maintain; yet in this study I shall follow the more cautious way mapped by Irenaeus rather than the bold one tread by Origen.

JESUS CHRIST IN MODERN THEOLOGY OF CREATION

In this section we will examine the work of four theologians who take seriously the ontological and epistemological unity of the economic and immanent Trinity and attempt to derive their doctrines of creation from the economy of salvation.

Karl Barth. Karl Barth begins his doctrine of creation by asserting that the first article of the Apostles' Creed, "I believe in God the Father Almighty, Maker of heaven and earth," is a statement of faith based on God's self-revelation in Jesus Christ, who is both the "noetic" and the "ontic" basis of

---

[22]Contemporary scholars differ widely in their interpretations of Origen's views. See Eric Osborn, "Origen: The Twentieth Century Quarrel and Its Recovery," in *Origeniana Ouinta*, ed. R. Daly (Leuven: Leuven University Press, 1992), pp. 26-39.

[23]Colin Gunton also contrasts Origen unfavorably with Ireneaus. Specifically, Gunton (1) objects to Origen's introduction of an "intermediate world between God and the creation," (2) objects to his depreciating "the importance of the material world," and (3) observes that "by speculating about the possibility of a plurality of possible worlds Origen sailed rather close to the wind of cyclical theories of the universe and, perhaps, more important, called into question the uniqueness of this one" (*The Triune Creator: A Historical and Systematic Study* [Grand Rapids: Eerdmans, 1998], pp 60-61).

[24]Blowers defends Origen on this point, saying that Origen's highly philosophical theology is nevertheless a vision "in which the biblically-grounded *oikonomia* of creation and redemption still has primacy" (*Drama of the Divine Economy*, p. 91). Blowers cites Lothar Lies, *Origenes' "Peri Archôn": Eine Undogmatische Dogmatik* (Darmstadt: Wissenschaftliche Buchgesellschaft, 1992), pp. 68-121, as having "demonstrated" this primacy in Origen.

creation.[25] Neither the existence of "God the Father Almighty" nor the genuine existence of "heaven and earth" can be known on any other basis. To a certain type of biblicism that proceeds as if the doctrine of creation can be established merely by citing the many biblical texts that attest to creation, Barth opposes a Christocentric biblicism that sees Jesus Christ as the unifying center of the Bible. We can know that God the Father Almighty created heaven and earth because God has in Jesus united himself to humanity and affirmed the existence of the creature. Scripture teaches us "to know the Father through the Son, the Creator through the redeemer."[26] Jesus Christ is God and God's creature united in one. He firmly establishes our knowledge of the reality and identity of the Creator: the Father of our Lord Jesus Christ. He manifests the character of the relationship between Creator and the creature as divine grace and human gratitude. And his existence guarantees the real and distinct being of the creature alongside God as one loved by him and destined for eternal life with him. Faith in Jesus Christ includes within it the decision to take "seriously the truth that God is the Creator of all the reality distinct from Himself . . . that this reality is at the disposal of God as the theatre, instrument and object of His activity . . . that God has controlled and does and will control it."[27]

Jesus Christ, however, is not only the basis of our knowledge of creation (the "noetic"), but also the basis of the existence of creation (the "ontic"). What does the New Testament mean when it declares that the Son, or the Word, is the means through whom God created all things? Barth gives a two-part answer to this question:

> In the same freedom and love in which God is not alone in Himself but is the eternal begetter of the Son. . . . He also turns as Creator *ad extra* in order that absolutely and outwardly He may not be alone but the One who loves in freedom. . . . The eternal fellowship between Father and Son, or between God and his Word, thus finds a correspondence in the very different but not dissimilar fellowship between God and His creature.[28]

However true the above statement is, Barth finds it inadequate because

---

[25]*CD* III/1, p. 28. That is to say, Jesus Christ is both the basis of our knowledge of the divine act of creation and, along with the Father and the Spirit, the cause and model of creation.

[26]Ibid., p. 24.

[27]Ibid., p. 32.

[28]Ibid., p. 50.

it does not yet identify the Son and Word with the incarnate Jesus Christ who died and rose again. When the New Testament refers to the Son and the Word as the one "through whom" or "in whom" God created the world, it is clear that it refers not to a *logos asarkos*, a preincarnate Word, but to Jesus Christ. In God's eternal counsel and will God determined that his Son should become human, suffer, die and rise again for the sake of humanity and all creation. God "created it because He loved it in His Son who because of its transgressions stood before Him eternally as the Rejected and Crucified . . . [and] as the Elected and Resurrected."[29] Given the identity of the Son, Barth can say that "it was not only appropriate and worthy but necessary that God should be the Creator," a necessity of course rooted in his loving decision. Barth concludes, "The fact that God has regard to His Son—the Son of Man, the Word made flesh—is the true and genuine basis of creation."[30]

However, for Barth, creation is the work of the whole Trinity and not just a matter between the Father and the Son. Barth points out that in the Nicene Creed the Spirit is called the "giver of life" and in Scripture the Spirit creates and renews life. The Spirit gives new birth and brings about a new creation. Through the Spirit, who is the bond of communion between the Father and the Son, God supports and gives life and space for the creature. And it is through the Spirit that the Father brings creatures into fellowship with the Father and the Son. Since this is how the Spirit works in the economy of salvation, we can say that "it is in God the Holy Spirit that the creature as such pre-exists. That is to say, it is God the Holy Spirit who makes the existence of the creature as such possible, permitting it to exist, maintaining it in its existence, and forming the point of reference for its existence."[31]

Jürgen Moltmann. In his *Trinity and the Kingdom*, Jürgen Moltmann asks what the Creator-creation relationship means for God and specifically for our understanding of the inner-trinitarian life of God. According to Moltmann, the Old Testament concepts of creation derive from Israel's experience of salvation in the exodus and her hope for the messianic future. The New Testament view of creation is shaped by the salvation already experienced in Jesus Christ and the expected transformation of all things in

[29]Ibid., pp. 50-51.
[30]Ibid., p. 51.
[31]Ibid., p. 56.

and through him. The world will be transformed into a place fitted to be God's dwelling, his home. If, through the death and resurrection of Jesus and the power of the Spirit, the world is destined to become united with and included in God, how must we conceive the act of creation?

Moltmann wishes to capture the truth in two extreme positions, which he labels "Christian theism" on the one hand and "Christian panentheism" on the other. Christian theism is so focused on protecting the freedom of God in creation that it makes the divine act of creation seem arbitrary and completely external. Christian panentheism roots creation so firmly in the divine nature and life that it absorbs the Creation-creature relationship into the inner-trinitarian Father-Son relationship. The world becomes the eternal Son and an aspect of the eternal life of God. Moltmann reclaims the truth in both positions by rejecting the dichotomy between necessity and freedom in God and between creation as an external work and creation as an outflow of the divine nature. God is freely his own goodness, and his free goodness is freely productive. "This is why," explains Moltmann, "the idea of the world is already inherent in the Father's love for the Son."[32] The Father loves the Son who is like him in essence but other in hypostasis or relation. Within the Father's love for "the Other in the like" (the Son) is included his love for the "like in the Other" (the world).[33] Since God has in fact communicated himself to the world in the suffering love of his Son, we know that there was no possibility that God could have refrained from creating the world, the object of his love. "In this sense God 'needs' the world and man. If God is love, then he neither will nor can be without the one who is his beloved."[34] For Moltmann, the New Testament teaching that God created "through" the Son can be understood as meaning that the Father's love for the Son includes his creative love for the world. The Son is the Logos or archetype of the world in the sense that in and through the Son the Father will ultimately receive the same responding and self-giving love from the world that the Son bestows on the Father eternally.[35] That is to say, the Son's uncreated relationship to the Father is the archetype for the world's created relationship to the Father.

---

[32]Jürgen Moltmann, *The Trinity and the Kingdom: The Doctrine of God*, trans. Margaret Kohl (Minneapolis: Fortress, 1993), p. 108.

[33]Ibid., p. 59.

[34]Ibid., p. 58.

[35]Ibid., p. 113.

However, for Moltmann, God's creative act should not be understood as an effortless, painless and immutable "Let it be." Such a view cannot be derived from the economy of salvation. On the basis of the suffering and death of the incarnate Son and the groaning of the Spirit (Rom 8:26), Moltmann argues that the Father's eternal love for the world in the Son corresponds to inner divine suffering. "Outward acts correspond to inward suffering, and outward suffering corresponds to inward acts."[36] These "outward acts" include creation, incarnation (including the cross) and consummation. To make room for the world God must restrict himself, let go control and endure creation's ingratitude. In the cross the Father abandons the Son and the Son experiences godforsakenness. In the resurrection and the consummation of all things, the negation within God and between God and the world will be overcome. Hence, even in the eternal love of the Father for the Son, given its dual aspect as love for the Son and love for the world in love for the Son, there is a negative moment:

> If we think about this external state of affairs [economic suffering and alienation], transferring it by a process of reflection to the inner relationship of the Trinity, then it means that the Father, through an alteration in his love for the Son (that is to say through a contraction of the Spirit), and the Son through an alteration in his response to the Father's love (that is, through an inversion of the Spirit) have opened up the space, the time and the freedom for that "outwards" into which the Father utters himself creatively through the Son.[37]

For Moltmann, then, the divine act of creating cannot be understood merely as the outward effect in time of a painless eternal resolve of the divine will. Creation's coming to be is caused by a change in the perfect self-giving and returning love among the Father and Son whose perfect union is sealed by the Spirit. The Father withdraws from the Son and so creates discontent within the divine life. The discontent and suffering within the divine life gives creation a reason to exist, a kind of necessity and therefore being; for creation now becomes the means of reuniting the Trinity by reconciling the world to God.

Wolfhart Pannenberg. Wolfhart Pannenberg responds to the charge that the Christian doctrine of creation views the act of creation as "pure caprice."[38]

---

[36]Ibid., p. 98.
[37]Ibid., p. 111.
[38]Wolfhart Pannenberg, *Systematic Theology*, trans. Geoffrey W. Bromiley (Grand Rapids: Eerdmans, 1988), 2:20. Pannenberg is responding to a charge made by Hans Blumenberg, *Die Legitimität der Neuzeit* (Frankford am Main: Suhrkamp, 1966), pp. 102-200.

This charge presupposes that the only alternative to pure caprice is pure necessity, that is, the world is an extension of the divine essence. Pannenberg adumbrates his answer to this dilemma by proposing that the act of creation must be "viewed as the expression of an intention to create this reality that is different from his own, which has its basis in the eternity of the Creator."[39] To meet this demand Pannenberg develops a trinitarian doctrine of creation in dialogue and criticism with patristic and medieval theology. He criticizes the approaches of Origen and Augustine because they tend to assimilate the role given to the Son in the New Testament to ideas in the divine mind. Such a strategy is inspired by the Platonic and Neo-Platonic understanding of mediation instead of the economy of salvation.[40] Pannenberg argues instead from the incarnate Son's self-distinction from the Father to the existence of creatures. He begins with the presupposition of all trinitarian thinking, that is, that in the Son the divine and creaturely perspectives combine so that the relationship of the incarnate Son to the Father reveals the eternal Son's relationship to the Father. In his earthly existence Jesus distinguished himself from the Father in a dual sense, as the eternal Son and as a creature. Pannenberg distinguishes between the noetic and ontic implications of this self-distinction. On its basis we know (the noetic) that the incarnate Son is the eternal Son in unity and distinction from the Father. But the self-distinction is also the ontic basis, the very possibility, for the existence of creatures. In this self-distinction of the Son from the Father in which the Son "moves out of the unity of the deity by letting the Father alone be God . . . the creature emerges over against the Father."[41] That is to say, when the Son freely wills to become a creature in harmony with the Father's sending, he becomes a human being. But his free decision to become the something other than God is at the same time the decision to create something other than God. In Pannenberg's words, "But if from all eternity, and thus also in the creation of the world, the Father is not without the Son, the eternal Son is not merely the ontic basis of the existence of Jesus in his self-distinction from the Father as the one God; he is also the basis of the distinction and independent existence of all creaturely reality."[42]

---

[39]Pannenberg, *Systematic Theology*, 2:20.
[40]Ibid., 2:25-26.
[41]Ibid., 2:22.
[42]Ibid., 2:23

Robert Jenson. Robert Jenson develops his doctrine of God by articulating the dramatic narrative of the Bible in which God is identified as Father, Son and Holy Spirit. These are the *"dramatis dei personae,* 'characters of the drama of God.'"[43] God can be known only in the dramatic relations among these three and in their acts toward the world. As Jenson says, "Any work of God is rightly interpreted only if it is construed by the mutual roles of the triune persons. This must first of all be true of creation."[44] Accordingly, Jenson views the trinitarian act of creation in dramatic terms. God creates by making room and time for creatures among the persons of the Trinity. When in their eternal conversation Father, Son and Spirit address others with their "Let it be," others come into being. This way of thinking makes sense only if we think of being not as monadic being-there but as being-in-relation, that is, personhood. To establish relationship is to establish being-in-relation. "Creatures occur as in this discourse others are commandingly mentioned or addressed beyond the three who conduct it. So and only so there are entities that truly are and are truly other than God."[45] The Father is the absolute source of creatures because the Father is the absolute source of all things. The Spirit frees the Father to create by opening a space within God that is not the Father and hence making possible other beings who are not the Father. In other words, if the Father were monadic, his natural generative nature could only extend itself, producing a divine outflow of the divine being. God could not get outside himself. But because God is the Spirit who unites the Father and the Son, the Father can produce creatures, beings that are other than God. Because the Son possesses the determinate character as the crucified and risen one, he "mediates between the Father's originating and the Spirit's liberating" and gives determinate actuality to creatures. In Jenson's words, "As the Father's love of the Son as other than himself is the possibility of all otherness from God, and so of creation, so the Son's acceptance of being other than God is the actual mediation of that possibility."[46]

---

[43]Robert Jenson, *Systematic Theology,* vol. 1, *The Triune God* (New York: Oxford University Press, 1999), p. 75.

[44]Robert Jenson, *Systematic Theology,* vol. 2, *The Works of God* (New York: Oxford University Press, 1999), p. 25.

[45]Ibid., p. 26.

[46]Ibid., p. 27.

CHRIST AND CREATION

In my previously published discussions of omnipresence and eternity I developed the concepts of divine space and time and of Christ as the place and time where God meets creatures.[47] God is not a spaceless monad. Adopting a relative view of space that Luco van den Brom defines as "the generalized next-to-each-other relation extended to subsume all possible objects," I argue that the trinitarian relations among the Father, Son and Spirit constitute divine space.[48] God does not need an external space within which to relate and move; for God is his own space constituted by the relations and mutual indwelling of the triune persons. Nor is God timeless. Regarding time as constituted by the before/after/simultaneous causal relations,[49] we take the begetting and being begotten relations between the Father and Son and the proceeding of the Spirit as quasi-temporal relations even though they are eternal. The triune relations are not reversible; hence the before/after relation retains its temporal and causal character even though there is no temporal separation of time between Father, Son and Spirit. God's eternity is not timelessness (and hence lifeless) but simultaneity.

These thoughts on divine space and time are not merely speculative; they are based on how God has acted in the economy of salvation. The God who sent his Son to become flesh and blood among us to die and rise again through the Spirit for us sinners cannot be spaceless, timeless and loveless in his eternal life. Since what we know of God derives from how God has been revealed in Jesus Christ, we need not think of the triune relations as characterless causal relations. Those relations are acts of self-giving love. The Father's fatherhood is his giving love for the Son in the Spirit and the Son's sonship is his returning love for the Father in the Spirit. The Spirit is the bond of love between Father and Son. The Trinity is an eternal community where action, presence, life, causality, simultaneity (and hence time), fruitfulness, love and otherness make sense. Hence in the eternal life of the triune God lies the possibility of creation. Since God is his own space and time and action and otherness, God can make room and time for creatures. Allowing creatures to share in the fellowship of the triune persons does not

---

[47]Highfield, *Great Is the Lord*, pp. 284-87, 307-10.

[48]Luco van den Brom, *Divine Presence in the World: A Critical Analysis of the Notion of Divine Omnipresence* (Kampen: Kok Pharos, 1993), p. 114.

[49]Time is no more an empty container within which things happen than space is an empty container within which things happen. Time is constituted by relations of causality and dependence.

contradict or pollute the divine nature. But what about the actual existence
of creatures and the act of creating? Can we say more?

Whatever we say must be based on the Scriptures interpreted in light of
their center, who is Jesus Christ. It cannot be based merely on the analogy
of human action whereby an agent uses means to effect change in the world.
In the economy of salvation the eternal Son of God became a creature. He
acknowledged and obeyed his Father even unto death, and was raised to life
by the Spirit. In thinking about the incarnation, the Son's assuming a human
nature, we tend to presuppose the existence of creation, indeed a fallen creation, as a given. We think of the incarnation as the result of a divine decision in time in response to the sin of humanity. Following the same trajectory it would seem that God's decision to create and his decision to send
the Son to become a creature are two separate decisions made for different
reasons. Furthermore, given this line of reasoning, there would be no
grounds for the incarnation had humanity not fallen into sin. God's eternal
plan, "the mystery of his will . . . to bring unity to all things in heaven and
on earth under Christ" (Eph 1:9-10), would to be relegated to plan-B status.[50]
But if we view salvation as the perfection and completion of creation, we will
come to a different conclusion. We will then be able to reason backward to
creation from the incarnation, death and resurrection, the end of which is to
"bring unity to all things in heaven and on earth under Christ." The decision
to unite all things to God in Christ entails the decision to become incarnate,
die and rise, and the decision to become incarnate entails the decision to
create. Hence Danish theologian Regin Prenter concludes: "Creation is the
beginning of redemption and redemption is the consummation of creation."[51]

In what way is the Son, or the Word, the means of creation? In consonance with the basic point of agreement among Irenaeus, Barth, Moltmann,
Pannenberg and Jenson, and in disagreement with the Logos theologians
and Origen, we must not begin with reflection on the *logos asarkos* in

---

[50]The negative consequences of placing such a gulf between salvation and creation are staggering.
We are plunged into darkness concerning the motive and purpose of creation. We are tempted
with Manichean and Gnostic explanations of the differences between the God of creation and
the God of salvation, between the law and the gospel, and between the material world and the
spiritual world. It becomes difficult to see the world of our experience as God's creation because
the fall is given such significance. Indeed, the fall becomes a kind of substitute act of creation
by a malevolent force.

[51] Regin Prenter, *Creation and Redemption*, trans. Theodor I. Jensen (Philadelphia: Fortress Press,
1967), p. 200.

eternity, which is really a sophisticated reflection on human action. Since salvation is the completion of creation, we should begin, rather, in the economy of salvation by asking about the salvation "through" and "in" Jesus Christ.[52] Creatures are saved by the incarnation, cross and resurrection of Christ. The Father sent and the Spirit conducted the Son into the world to assume human nature and become a human individual. Jesus Christ accepted this mission, emptying himself in obedience to the Father (Phil 2:7). Jesus Christ lived human life as a creature should live. He lived under the conditions of existence as the divine Son, affirming his eternal relationship to the Father in gratitude and self-giving love. His absolute love for the Father and his uncompromising love for us sinners led to his death in which he identified with mortal and dying sinners. As the eternal Son and as human he never broke faith with his God and Father. In his resurrection by the Spirit he now fills the whole creation thereby joining it to God in a new way; that is, creation joins in the returning love of the Son for the Father. In this one man creation fulfills its purpose to bring glory to God by receiving and returning itself to its Creator to affirm God as all in all. Through the Holy Spirit, Jesus Christ invites all persons to join him in turning to the Father in self-giving love. The Spirit's inviting address possesses transforming and empowering power. It gives what it requests. As Paul expressed it, "For Christ's love compels us, because we are convinced that one died for all, and therefore all died. And he died for all, that those who live should no longer live for themselves but for him who died for them and was raised again" (2 Cor 5:14-15).

What can we learn about God's act of creating from the divine act of salvation? The Gospel of John may give us a clue. In Jesus' final prayer for his disciples he said, "Father, I want those you have given me to be with me where I am, and to see my glory, the glory you have given me because you loved me before the creation of the world" (Jn 17:24). Why, then, did God create the world? Because Jesus loved us and wanted us to be with him forever, and the Father loves his Son. Karl Barth is surely correct when he says that God created the world because from all eternity he determined that the Son of God should become human and through suffering bring "many sons and daughters to glory" (Heb 2:10). How did God create the world?

---

[52]The same prepositions are used of Christ's relationship to God's act of creation.

God created the world the same way he does everything else. The Father loves his Son, and the Son loves us in his love for the Father. Just as in his act of salvation the Son's love "compels" us to love him in return, in creation it "compels" us to be (2 Cor 5:14-15). Robert Jenson's dramatic understanding of creation helps us to articulate the event of being created by love. When in their eternal conversation, which is the mutual self-giving and returning that constitutes God's life, Father, Son and Holy Spirit lovingly address creatures as if they exist, they come into existence. As Paul says in Romans 4:17, God is he "who gives life to the dead and calls into being things that were not." We exist because in his great love God called and continues to call our names . . . in expectation that we will echo his call. More than this we cannot say. But isn't this enough?

7

CREATION FROM NOTHING

Creation as an Act of Sovereign Generosity

T HE FORMULA *creatio ex nihilo*, or creation from nothing, functions today as a shorthand way to encapsulate the Christian doctrine of creation and differentiate it from alternatives such as dualism and pantheism. But like many sacred formulas it is more invoked than understood. To grasp its meaning at a deeper lever we need to (1) examine its form, (2) trace its history, (3) relate it to the biblical understanding of the God-world relation and (4) pursue its implications to their limit.

DEVELOPMENT OF THE DOCTRINE OF CREATION FROM NOTHING

The expression "creation from nothing" parallels other expressions of making. The builder built the house out of bricks or the car is made out of metal and plastic. Clearly in these expressions the two words "out of" indicate the material from which something is constructed. But the *ex nihilo* cannot possibly mean out of "nothing" in this sense. "Nothing" is not a material from which something can be built! The *ex nihilo* doesn't fit any of Aristotle's four causes. The "nothing" in question cannot be a formal, material, efficient or final cause. We can see why ancient philosophers, beginning with Parmenides, asserted *ex nihilo nihil fit*, "nothing comes from nothing." Hence we can conclude that the formula is either meaningless or it means something other than what its form would lead us to think. Only by looking into the history of the concept can we get a clear idea of what it means.

Gerhard May argues that the first clear expression of what came to be the definitive meaning of *creatio ex nihilo* can be found in Theophilus of Antioch (later second century).[1] In his apology, *Ad Autolycus*, he argues against the Hellenistic theory of world-formation and for the idea that "God has created everything out of nothing into being."[2] According to May the works of Theophilus exercised significant influence over Irenaeus, Hippolytus and Tertullian, and "played a decisive role in the breakthrough of the doctrine of *creatio ex nihilo* in Catholic theology which occurred as the second century drew to a close."[3] Irenaeus asserted the omnipotence and freedom of God against the world-formation theories current in his day.[4] Against Gnostics and Platonists Irenaeus asserts that God is the absolute and free origin of every dimension of creatures: "While men, indeed, cannot make anything out of nothing, but only out of matter already existing, yet God is in this point pre-eminently superior to men, that He Himself called into being the substance of His creation, when previously it had no existence."[5] Or again:

> It was not angels, therefore, who made us, nor who formed us, neither had angels power to make an image of God, nor any one else, except the Word of the Lord, nor any Power remotely distant from the Father of all things. For God did not stand in need of these [beings], in order to the accomplishing of what He had Himself determined with Himself beforehand should be done, as if He did not possess His own hands. For with Him were always present the Word and Wisdom, the Son and the Spirit, by whom and in whom, freely and spontaneously, He made all things, to whom also He speaks, saying, "Let Us make man after Our image and likeness"; He taking from Himself the substance of the creatures [formed], and the pattern of things made, and the type of all the adornments in the world.[6]

For Irenaeus, only a doctrine of creation that views God as the sole source,

[1]Gerhard May, *Creatio ex Nihilo: The Doctrine of "Creation Out of Nothing" in Early Christian Thought*, trans. A. S. Worrall (Edinburgh: T & T Clark, 2004). J. C. O'Neill argues for an earlier date for the fully articulate doctrine ("How Early Is the Doctrine of Creatio ex Nihilo?," *Journal of Theological Studies* 53 [2002]: 449-65).
[2]Theophilus, *Ad Autolycus* 1.4, quoted in May, *Creatio ex Nihilo*, p. 156. The term "world-formation" is May's way of designating the Platonic understanding of the coming together of form and matter to constitute the world of our experience.
[3]May, *Creatio ex Nihilo*, p. 159.
[4]Paul Gavrilyuk, "Creation in Early Christian Polemical Literature: Irenaeus Against the Gnostics and Athanasius Against the Arians," *Modern Theology* 29 (2013): 22-32.
[5]Irenaeus, *Against Heresies* 2.10.4 (*ANF* 1:370).
[6]Irenaeus, *Against Heresies* 4.20.1 (*ANF* 1:487-88).

power, archetype and end for creation can honor God as the Creator. According to May, not until Christian theologians realized that the Greek world-formation theory fundamentally undermines the freedom and omnipotence of God did they begin to formulate their doctrine of creation in dialectical antithesis to this theory. The doctrines of creation formulated by Theophilus and Irenaeus stand self-consciously in clear "antithesis to the world-formation idea of the Greek philosophers and calls in question the axiom 'Ex nihilo nihil fit.'"[7] Only with the work of these two theologians can we speak of a genuine doctrine of *creatio ex nihilo.*

With the definitive doctrine in mind we can now examine the earlier statements that seem to express *creatio ex nihilo.* The term is not used in the Old Testament. In the section on the Old Testament theology of creation, I derived five theses from the Old Testament references to creation. Two are relevant to the issue of creation from nothing. The first states that God is the absolute origin of all that is not God. Old Testament references to creation do not mention the origin of God or point to an eternal reality existing alongside of him. The second thesis states that in unhindered freedom and power God alone established the Creator-creature relation. Clearly the later doctrine of creation from nothing is implied by the Old Testament view of creation. We must recognize, however, that the later doctrine remains inchoate in the Old Testament itself.[8]

Hellenistic Judaism does not speak in one voice on the subject. Wisdom of Solomon 11:17 declares that God created the world "out of formless matter" (*ex amorphou hylēs*). Wisdom has not yet seen the contradiction between the world-formation theory and the biblical affirmation of the omnipotence and freedom of God. The earliest reference that sounds like creation from nothing is 2 Maccabees 7:28: "I beg you, child, to look at the heavens and the earth and see all that is in them; then you will know that God did not make them out of existing *things* [*ouk ex ontōn*]; and in the same way the human race came into existence." This text, however, contains no reference to the world-formation theory. In May's judgment, "The text implies no more than the conception that the world came into existence through the sovereign

---

[7]May, *Creatio ex Nihilo,* p. 163.

[8]David Fergusson makes a similar point, comparing the way the doctrine of creation from nothing is taught implicitly in Scripture to the way the doctrine of the Trinity is implied in Scripture but only fully developed in the fourth century (*Creation,* Guides to Theology [Grand Rapids: Eerdmans, 2014], p. 15).

creative act of God, and that it previously was not there."[9] Two New Testament texts are often cited as teaching *creatio ex nihilo*. Paul speaks of "the God who gives life to the dead and calls into being things that were not" (Rom 4:17), and the book of Hebrews asserts that "the universe was formed at God's command, so that what is seen was not made out of what was visible" (Heb 11:3). These texts witness to the omnipotent power and freedom of God, but are not set in antithesis to the Greek world-formation theory. Hence they cannot be said to teach *creatio ex nihilo* in the same sense as that taught by Theophilus and Irenaeus.

THE BIBLICAL AND THEOLOGICAL MEANING OF *CREATIO EX NIHILO*

Is *creatio ex nihilo*, then, a biblical teaching? Yes. The Bible teaches that in relation to the world God is completely free. There are no conditions that constrain or limit God's act of creating the world. When the biblical teaching is confronted by the world-formation theory of the Greeks, which understands the demiurge as forming the world using preexisting matter and eternal ideas, it must then become explicit in its rejection of the eternal existence of matter or ideas that are independent of God's will or power. The positive side of that rejection can be stated as *creatio ex nihilo*, though the "positive" formula retains its predominately negative meaning. *Creatio ex nihilo* denies the world-formation theory and affirms that God creates through his own power without help from anything else. The formula does not explain "how" God created the world. *Creatio ex nihilo* resembles such statements as "The builder built the house of wood" or "The potter molded the vase from clay" only in grammatical form. It means instead something like "God creates without help from other sources" or "God is the sole cause and source of all that is not God."

What are the implications of the teaching that God created the world *from nothing* and the biblical conviction that God creates in absolute freedom and power, which *creatio ex nihilo* attempts to express? Recall that the doctrine was developed in direct antithesis to the Greek world-formation theory. In Plato's theory there were two principles (or *archai*), the intelligible and the material, and in the Middle Platonists three principles,

---

[9]May, *Creatio ex Nihilo*, p. 7.

the intelligible forms, the second god (or world soul or demiurge) and matter. The intermediate principle uses the intelligible forms to which he has access through his mind as archetypes or design plans to shape chaotic and formless matter into an orderly world. In this theory the demiurge is the divine agent of creation, the counterpart of God in the Christian doctrine of creation. The intelligible forms exist independently of the demiurge, as it were externally or in their own space. Matter exists independently of the other two principles. All three principles are eternal and exist on their own. The demiurge does not create in absolute freedom and power. He depends on the other two principles. Hence the world-formation theory cannot serve as a model for how God created the world. The Christian doctrine of creation begins with the assumption that in creating the world God depends on nothing else. Matter and intelligible form must depend on God.

DIVINE IDEAS AND POSSIBLE WORLDS

The doctrine of *creatio ex nihilo* is not an explanation of how God creates. Quite the opposite; it is a rejection of this question. To affirm creation from nothing is to affirm that everything that is exists only because God wills it to exist. And this "everything" includes the material and intelligible realms. However, the analogy of human agency still exerts great attractive force even on theologians who accept the doctrine of creation from nothing. Merely to assert that God created the world from nothing while declining to speculate on how this is possible is greatly dissatisfying to the philosophical mind. *Ex nihilo nihil fit!*

Augustine of Hippo. Augustine, to avoid making God dependent on something outside himself and to guarantee that God created intelligently, asserted that the ideas or forms after which God made the world are located within the divine mind. According to Augustine,

> The ideas are certain original and principal forms of things, i.e., reasons, fixed and unchangeable, which are not themselves formed, and being thus eternal and existing always in the same state, are contained in the Divine Intelligence. And though they themselves neither come into being nor pass away, nevertheless, everything which can come into being and pass away and everything which does come into being and pass away is said to be formed in accord with these ideas.[10]

---

[10]Saint Augustine, *Eighty-Three Different Questions*, trans. D. L. Mosher, Fathers of the Church 70 (Washington, DC: Catholic University of America Press, 1977), p. 80.

Augustine explains why he locates archetypical ideas in the divine mind: "For it would be sacrilegious to suppose that he was looking at something placed outside himself when he created in accord with it what he did create."[11] Augustine is eager to make clear that, unlike the Platonic demiurge, God does not depend on an external world of forms to guide his creative activities. But we must ask whether simply moving the world of forms from outside to inside the divine mind really goes far enough to protect God's independence. Do these ideas possess a possibility of their own not founded on God? According to Frederick Copleston, Augustine does not see the archetypes of created things as separate things within the divine being. Instead God contemplates himself from eternity and "sees in himself all the possible limited essences, the finite reflections of His infinite perfection, so that the essences or *rationes* of things are present in the divine mind from all eternity as the divine ideas."[12] Copleston goes on to argue that, given Augustine's firm commitment to the doctrine of divine simplicity, we should not take these "ideas" as "'accidents' in God, ideas which are ontologically distinct from His essence."[13] If Copleston is correct about Augustine's theory, I think it is compatible with the doctrine of creation from nothing.[14] But Colin Gunton's judgment should give us pause. In his estimation, Augustine's move takes us "only half way to the doctrine of creation from nothing."[15] In an even stronger statement, Gunton objects, "We really must take extreme exception to a theology that speaks of entities other than God" that are eternal, fixed and unchangeable.[16]

---

[11]Ibid., p. 81.

[12]Frederick Copleston, *A History of Philosophy*, vol. 2, pt. 1, *Mediaeval Philosophy: Augustine to Bonaventure* (Garden City, NY: Image, 1963), p. 88.

[13]Ibid. These qualifications go a long way to protect God's independence and *creatio ex nihilo*; however one concern remains. If we think of God's essence or nature as given to God in the way the human essence is given to us, then the possible ways in which God's essence is imitable are also given in that way. If God's nature escapes his freedom, the possibilities for creatures rooted in his nature are also beyond his freedom.

[14]One may ask whether Copleston reads Augustine in the light of Thomas Aquinas.

[15]Colin Gunton, *The Triune Creator: A Historical and Systematic Study* (Grand Rapids: Eerdmans, 1998), p. 78.

[16]Ibid., p. 78 n. 29. For the debate about Thomas Aquinas's use of divine ideas in his doctrine of creation, see John Hughes, "Creatio ex Nihilo and the Divine Ideas In Aquinas: How Fair Is Bulgakov's Critique?," *Modern Theology* 29 (2013): 124-37. Aquinas's reasons for including the exemplar ideas within the divine mind are much the same as Augustine's reasons: to protect the freedom of God in creation without making creation purely arbitrary. Hughes, agreeing in a limited way with Sergius Bulgakov's criticism of Aquinas (*The Bride of the Lamb*, trans. Boris Jakim [Edinburgh: T & T Clark, 2002]), shows how the notion of divine ideas can actually imply

William Lane Craig. Many contemporary philosophically oriented theologians and theologically oriented philosophers continue to find Augustine's adaptation of Plato alluring. In the course of defending the doctrine of creation out of nothing, Paul Copan and William Lane Craig address the problem posed by the Platonist contention that there exists a realm of ideas or forms independent of God, to which God must look for models for creation. After analyzing the strengths and weaknesses of two anti-Platonist solutions, absolute creationism and fictionalism, Copan and Craig express a preference for conceptualism.[17] These authors trace conceptualism back to Augustine and understand its main feature to be its assertion that the exemplars for creation exist only as ideas dependent on the operation of the divine mind and not as "independently existing abstract objects."[18] Divine concepts must not be mistaken for divine thoughts, for a thought is a particular operation. A particular cannot serve as a model for creation. God's concepts are the "pure concepts, the sort of things that are the intentional objects of thought."[19] Copan and Craig do not speculate further in their book *Creation Out of Nothing* about the ontological status of divine concepts, their range, how God selects which of them to actualize or how they serve God as models for creation. Their conclusion is in fact rather modest: "In our view some sort of nominalist or conceptualist account seems to be the most promising solution. So long as some such account seems plausible, the doctrines of divine aseity and *creatio ex nihilo* need not be compromised to accommodate the metaphysical pluralism entailed by Platonism."[20]

In Craig's discussions of divine foreknowledge and providence, however, we find a robust use of divine concepts in creation.[21] Craig finds the theory

---

voluntarism. If God possesses an infinite number of ideas of things that he does not will to create, does this imply that God chooses in libertarian fashion which ideas to actualize? Is then creation arbitrary after all?

[17]Paul Copan and William Lane Craig, *Creation Out of Nothing: A Biblical, Philosophical, and Scientific Exploration* (Grand Rapids: Baker Academic, 2004), pp. 167-95. Absolute creationism contends that the abstract objects that serve as archetypes for creation are themselves created. Fictionalism considers sentences that seem to refer to abstract objects as useful fictions constructed by the mind for practical reasons.

[18]Copan and Craig, *Creation Out of Nothing*, p. 189.

[19]Ibid., p. 194.

[20]Ibid., p. 195.

[21]William Lane Craig, *The Problem of Divine Foreknowledge and Human Freedom from Aristotle to Suárez* (Leiden: Brill, 1980); Craig *The Only Wise God: The Compatibility of Divine Foreknowledge and Human Freedom* (Grand Rapids: Baker Books, 1987); Craig "Middle Knowledge: A Calvinist-Arminian Rapprochement," in *The Grace of God and the Will of Man*, ed. Clark Pinnock

of divine middle knowledge developed by Luis de Molina (1535–1600) persuasive and fruitful for the doctrines of creation and providence.[22] Since I will explore Craig's Molinism in detail in chapter 12 on the foreknowledge model of providence, I will focus here on matters directly relevant to creation from nothing. According to Craig, God knows what each free creature would freely do in every possible set of circumstances. Instead of God's will alone determining God's knowledge of what will happen, God's knowledge of what *would* happen also comes into play. By using middle knowledge "God can plan a world down to the last detail and yet do so without annihilating creaturely freedom, since God has already factored into the equation what people would do freely under various circumstances."[23] God's middle knowledge is not God's self-knowledge of his power and will but God's knowledge of "counterfactual" truths of human freedom.[24] And these counterfactual truths are made true by what free creatures *would* do, not by what God can do or will do. These truths exist in God's mind eternally. God cannot make them untrue or create a world that does not conform to them. If God chooses to create a world that contains free beings, God must choose from those worlds available and known in his middle knowledge. God cannot simply create the world he wants.

Now let's ask whether Craig has really broken as free of the spell of Plato as he thinks he has. The Platonic demiurge depends on the existence of ideas or forms to guide him in shaping the world of unformed matter. Unformed matter not only exists apart from the will of the demiurge but even resists his efforts to impose form. Matter's inherent resistance to being formed accounts for evil in the world. In Craig's theory God finds himself confronted with eternal truths not under his control, the so-called counterfactuals of creaturely freedom. His knowledge is eternally determined by something other than himself. Though these truths do not compel God to create a

(Minneapolis: Bethany House Publishers, 1989), pp. 141-64; and Craig "God Directs All Things: On Behalf of a Molinist View of Providence," in *Four Views on Divine Providence*, ed. Dennis W. Jowers (Grand Rapids: Zondervan, 2011), pp. 79-100.

[22]*Liberi arbitrii cum gratiae donis, divina praescientia, providentia, praedestinatione et reprobatione Concordia.* For a translation of the section on divine foreknowledge, see Luis de Molina, *On Divine Foreknowledge (Part IV of the "Concordia")*, trans. Alfred J. Freddoso (Ithaca: Cornell University Press, 1988). This translation contains an 81-page introduction by Freddoso. Molina's teaching is also summarized in chapter 5 of Craig, *The Problem of Divine Foreknowledge and Human Freedom.*

[23]"God Directs All Things," in *Four Views on Divine Providence*, pp. 79-100, at p. 82.

[24]Ibid., pp. 80-81, 95.

world, they do constrain God's choices should God choose to create. Now it should be clear why Craig sees in middle knowledge a "means" to help God choose which world to create.[25] Counterfactuals stand between God and the world ontologically; for they are neither God nor God's creatures.[26] Their power as a means consists in their power to bridge the gulf between God and creation. They have an ontological "foot" in both. On the divine side, they exist in God as his concepts, but on the creation side the entities to which these concepts refer and make knowable to God are *creatures* of possible creatures. They possess no virtual existence in God before (logically) they come to exist in the hypothetical free agents who create them. They are not created by God; rather they cause and determine God's knowledge. God knows them because they are true, not the other way around. When God creates a world, God does not create the reality (free decisions) that makes the eternal counterfactuals true. In creating the world, God merely surrounds these decisions with their facilitating but non-determining circumstances. Hence, God does *not* create the world from nothing, for the form of the created world is determined by free creaturely decisions that constitute the shape of that world.[27] God, like the demiurge of Platonism, is guided in his creating the world by something other than himself.[28]

---

[25]Seeking for rationally understandable means always involves pressing the analogy of human action. We use ideas whose truth exists independently of us to guide our physical action. If the ideas were not independent of us they would not qualify as means. Our action would be immediate. As the following line of reasoning will show, Craig's reasoning on middle knowledge as a means follows the human analogy quite closely. This is the only way to rationalize creation.

[26]See Mark Ian Thomas Robson, *Ontology and Providence in Creation: Taking Ex Nihilo Seriously* (London and New York: Continuum, 2008), for an extensive analysis and criticism of theories that resort to eternal divine concepts to give God knowledge of the creatures he could create. Robson argues that taking creation from nothing seriously and preserving God's complete freedom excludes the notion of preexisting determinate possible worlds or individual creatures either outside the divine mine (as in Platonism) or inside the divine mind as in Augustine. However, Robson's view of possibility and divine capacity as an indeterminate continuum drives him to deny that God knows what a creature will be until God creates it. Furthermore, since God does not know creatures until they exist, causing them to exist changes God's knowledge of what is. But doesn't this admission undermine Robson's project of defending divine independence and freedom?

[27]Molinism usually points to the amazing divine knowledge involved in knowing counterfactuals of creaturely free decisions. But it fails to take into account the significance of its admission that God's knowledge is determined by the existence of these counterfactuals.

[28]For my more extensive critique of Craig's view of creation and providence, see Ron Highfield, "Response to William Lane Craig," in *Four Views on Divine Providence* (Grand Rapids: Zondervan, 2011), pp. 114-122.

MATTER—THE THIRD PRINCIPLE

Matter is Platonism's third principle, or *archē*. When people speak of creation from nothing they usually take for granted the existence of the agent and the intelligible archetypes and think of the creation of matter. *Creatio ex nihilo* is taken as a denial of the eternal existence of matter, the stuff out of which the physical world is constructed. No one who defends *creatio ex nihilo* would defend the Platonic idea that matter is one of the uncreated components of the world. We must then examine this theme.

The Bible does not distinguish between matter and intelligible form, for that is a philosophical distinction requiring more abstraction than we find in the Scriptures. The Bible distinguishes between body and soul or flesh and spirit. Both are created by God, and both are good in their own sphere. The soul animates and empowers the body, which apart from soul is dead. Flesh is weak and mortal. In the Bible, God is not thought of as an immaterial object but as Spirit, alive and active. Soul is used of living beings, human and animal, as the life of the body. Just as creation from nothing could not be formulated until the abstract distinction between being and nothing was made clear, further philosophical development had to occur before the distinction between the intelligible forms and matter could be made. Thinking of form and matter as two irreducible principles, as Platonism did, requires a third mediating principle that can act as an agent to unite the other two. In the Bible the fundamental ontological distinction is between God and creatures. And God is understood to be alive and active by nature. No third principle is required.

In the commonsense way of thinking, matter corresponds to the senses and form or idea corresponds to the mind. But for Plato the concept of matter is derived by removing in thought the form of which the objects of sense experience are the images. Matter is what is left, and it accounts for the distinction between images and the originals that they imitate.[29] Aristotle arrived at his concept of matter by analysis of the process of change to discover the presuppositions necessary to account for its possibility. Change occurs when a substance such as a tree loses and gains a form; for example,

---

[29]Leonard J. Eslick, "The Material Substrate in Plato," in *The Concept of Matter in Greek and Medieval Philosophy*, ed. Ernan McMullin (Notre Dame, IN: University of Notre Dame Press, 1963), pp. 39-54. Aristotle used the term *hylē* (which means "timber") to designate the unformed principle. Plato called it the Receptacle.

its leaves change from green to red. Such change does not involve transformation of one thing (substance) into another kind of thing but merely of the accidental characteristics of the same thing. But change can also be substantial or a change in the kind of thing. In all change there must be something (a subject) that remains the same throughout the transition. If substantial change is possible, something Aristotle assumes, there must be a universal and unchanging "subject" that is not itself any kind of thing but is capable of receiving the substantial form of any substance. This subject is matter, which Aristotle conceives as positive but indeterminate; that is, matter is pure potentiality for any and every kind of thing.[30]

In general, form is the intelligible and matter is the unintelligible. Form is active and matter is passive. Matter is the unlimited and form is the determinate. Form is structure and matter is the unstructured. The different concepts begin with the different avenues through which we contact the external world, but they are refined dialectically by their mutual exclusion. Thus understood we don't seem to be able to explain one in terms of the other. If we think of form as the intelligible and the intelligible as structure and structure as a system of relations, clearly to make a concrete object we need *things* to stand in relation to one another and thus "inflate" the system into an object. Those things standing in external relation to one another are, for the purpose of experiencing the concrete object, not considered to have internal relations or structures. Before recent discoveries of modern physics atoms were considered to have external relations only, but now we know that atoms are very complex systems of relations. But the philosophical notion of matter is the idea that reality cannot simply be structures within structures to infinity. What "inflates" the structures? What makes the difference between a blueprint and a house or the idea of an atom and an atom?

As this little reflection on the nature of matter teaches us, there is no agreement about what matter is. It stands as a warning that we must not link the Christian doctrine of creation from nothing to a particular view of matter. Consequently it must reject the project of speculating about how God creates matter. The doctrine of creation asserts that God created the world we experience—the world of concrete objects, of ideas and sense ex-

---

[30]John J. FitzGerald, "'Matter' in Nature and the Knowledge of Nature: Aristotle and the Aristotelian Tradition," in McMullin, *Concept of Matter*, pp. 59-78; and Joseph Owens, "Matter and Predication in Aristotle," in McMullin, *Concept of Matter*, pp. 70-93.

perience, of minds and intelligibility and unintelligibility—through himself without help from any other power. It makes no difference to the doctrine of creation how concrete objects may be resolved into their constituents. The doctrine of *creatio ex nihilo* makes clear that the God-creature relation is unlike any creature-creature relation. Reason can hope to discover physical relationships—spatiotemporal, quantitative and causal—among creatures; and it may speculate about metaphysical constituents and relations within the world. It may even speculate about a metaphysical cause of the world based on the world's apparent lack of self-evident existence. But reason, metaphysical or scientific, cannot cross the gap expressed in the term *creatio ex nihilo*. For God's act of creating is a free act of the divine will, reasonable and beneficent for sure, but nevertheless not necessary.[31]

---

[31]In David Burrell's words, "If philosophical argument will suffice to 'ground' the universe in a creator, that creator can hardly be free. The argument will have to assume some form of emanation modelled in logical deduction, whereby the initial premise cannot adequately be distinguished from what emanates from it" ("Creatio ex Nihilo Recovered," *Modern Theology* 29 [2013]: 6-7).

8

DIVINE CREATION
AND MODERN SCIENCE

WHEN THE TOPIC OF CREATION IS MENTIONED in almost any setting, the first thing that comes to mind is the challenge modern natural science poses to the traditional doctrine. In the past two hundred years whole forests of trees have been turned into paper and rivers of ink have been drained in attacking or defending divine creation in relation to modern theories of cosmology or evolutionary biology. One person's list of heroes can serve as another's list of villains. Some would describe this history as religion's long war on science and others as science's battle against God. In reviewing this depressing history one is forced to ask with Matthew Arnold whether

> we are here as on a darkling plain
> swept with confused alarms of struggle and flight,
> where ignorant armies clash by night.[1]

Arnold's words ring all too true of the major episodes and players in this drama. Hence in dealing with these controversies I want to place them in historical perspective and clarify the issues involved.

It's not surprising there should be confusion and controversy where discussions of divine creation are concerned. The first sentence of the Bible speaks of both God and the world: "In the beginning God created the heavens and the earth" (Gen 1:1). It speaks of God, time and space, and origination. If it spoke only of God, it might have generated contro-

---

[1]Matthew Arnold, "Dover Beach" (1867), in *The Poems of Matthew Arnold 1840–1867* (1913; repr., London: Oxford, 1940), p. 402.

versies among religious and metaphysical thinkers. If it spoke only of the world, it would have been a matter for different schools of science to hash out. But it speaks of a relationship between God *and* the world, and therein lies the source of confusion. The doctrine of creation says something about God's act in relation to the world and something about the world in relation to God's act. Claims about God and God's action are clearly theological or metaphysical. But what does the doctrine say about the created world, and what status do such claims have? Are they theological, metaphysical or in some way also scientific? As is now evident, we need clear distinctions between the theological, metaphysical and scientific domains and methods of study. And as the reader might have already anticipated I believe much confusion has been created by confusing these domains and methods. Hence I will focus on clarifying the domains and methods of these areas and only tangentially deal with the material claims of natural science.[2]

ANCIENT SCIENCE

For the sake of simplicity we can divide cosmogonies and cosmologies into three types: mythological, rational-esthetic and empirical-mechanical. A cosmogony is an account of the origin and basic structure of the cosmos, usually a mythological account. It is a reckoning of the structures, components and processes of the physical world in its macro dimensions. Mythological cosmologies are based on analogies from ordinary experience that, with the help of reason and imagination, are used to explain things remote from direct experience. In this way they are not fundamentally different from explanations modern people give. In ordinary experience things come into being when seeds sprout up from the ground (plants), through sexual generation (animals) or through battle and conquests (human societies). Since human beings know their own minds and bodies better than anything else, what better analogy could be found to understand origination, movement and growth within the external universe than the human body? Hence it should not strike us as unintelligent that ancient people thought of the universe as a giant living being with one life force pulsating

---

[2]For a lengthy survey of the theological issues raised by contemporary natural science as well as formal efforts at dialogue among theologians and natural scientists, see Hans Schwarz, *Creation* (Grand Rapids: Eerdmans, 2002), pp. 23-165.

throughout. In experience—even contemporary experience—the forces of nature seem mysterious, powerful and enduring. It should not be surprising that ancient people accounted for the origin, structure and contents of the universe by telling stories of the origin of things from the fundamental divine forces that underlie the world. The world and humankind, in the Babylonian creation myth for example, result from a battle among the gods wherein the defeated god or goddess's dead body is used to construct the world. In societies guided by mythological worldviews the divine is continuous with the rest of the world. Everything above humans in the scale of being, power and life is divine by definition. The forces of nature on earth and the heavenly bodies are self-evidently divine. Hence the divine was directly experienced in the powers of nature on earth and in heaven. The gods were visible and palpable. Unbelief, secularism and naturalism were totally unthinkable. The hidden causes of all events are alive, mysterious and personal, or at least quasi-personal. Clearly the mythological worldview held sway before theological, metaphysical and scientific ways of thinking were distinguished.

We can mark the beginning of the Rational-esthetic worldview with the work of Thales (fl. 585 B.C.) the pre-Socratic philosopher. In contrast to mythology, Thales, Anaximander, Heraclitus, Parmenides and those who followed them pioneered a way of explaining the origin and structure of the world by asking about its basic constituents and how they are ordered to construct to the cosmos. They abandoned the dramatic method wherein the existence and flow of events and things are explained in terms of the motives, conflicts and intentions of characters. Instead, they pioneered explanation of events and states of affairs in terms of impersonal and necessary or contingent causes. Plato, as we have discussed previously, understood the world as composed of two basic things (*archai*), form and matter. The demiurge or god brought the two together to create the world. The world itself is a copy of the ideal living creature and is moved and animated by a world soul, the highest and best part of which is located in the heavens. It moves the stars and planets in their daily revolutions around the spherical but stationary earth. The vivifying power of the world soul is transferred downward to move all things on earth. For Plato, the world was an ordered and beautiful imitation of the perfection of eternity and the Good. Aristotle accepted most of Plato's cosmology but rejected his myth of creation by the demiurge.

For Aristotle, the cosmos has always existed.[3] Above all things stands the highest reality, self-thinking thought, which acts as the unmoved mover of the world. The unmoved mover is pure actuality, but everything below it in the hierarchical order exists in a mixed state of potentiality and actuality. The world soul moves in imitation of the unmoved mover's motionless, eternal perfection and pure actuality. Everything possesses a nature and moves toward its natural end by an internal directionality, which Aristotle calls an entelechy. The nature or form or entelechy of a thing is the spark of actuality that urges the thing forward toward full actuality.

Aristotle's explanatory system consists of the ten categories and the four causes. To know what a thing is you need to know the answers to ten questions: What are the thing's substance, quantity, quality, relation, place, time, position, state, action and affection?[4] And to know how a thing came to be, you need to know all its causes: material causality (its medium), formal causality (its blueprint), efficient causality (the power to combine blueprint and medium) and final causality (the end it serves).[5] Aristotle's science assumes that every kind of thing possesses a distinct nature and its behavior must be explained as arising from its nature, which is determinative of its behavior. All living things possess souls. Aristotle understands the world in teleological terms. Natures tend toward their ends. Behaviors, such as a body falling to earth or a gas floating upward, are explained by the inward tendency of things to move toward their natural place in the cosmos. Aristotle's science differs from modern science in many ways, in its methods and in its conclusions. But it seems to me that the decisive difference is his attribution of causal efficacy within the physical world to non-material entities: natures, forms, entelechies, individual souls, the world soul and the unmoved mover.[6] Modern natural science, especially physics and chemistry, rejects a role for such "occult" (hidden) forces.[7]

---

[3]For a brief treatment of Aristotle's science, see David C. Lindberg, *The Beginnings of Western Science: The European Scientific Tradition in Philosophical, Religious, and Institutional Context, 600 B.C. to A.D. 1450* (Chicago: University of Chicago Press, 1992).

[4]Aristotle, *Categories* 4.

[5]Aristotle, *Physics* 2.3.

[6]R. G. Collingwood, *Essay on Metaphysics* (Chicago: Regnery, 1972), p. 250, sees the difference this way: Aristotle's was a science of quality whereas Galileo (and modern science) is a science of quantity.

[7]The important but neglected story of the gradual evolution of science in the middle ages and Renaissance is too complicated to tell here. For my purpose, I need only to present idealized types. For a study of this history relevant to the issues of creation, see Nicholas H. Steneck, *Science and Creation in the Middle Ages: Henry of Langenstein (d. 1397), on Genesis* (Notre Dame, IN: University of Notre Dame Press, 1976).

EARLY MODERN SCIENCE

The empirical-mechanical worldview developed in self-conscious rejection of Aristotle's science and metaphysics. As a comprehensive worldview, it considers the "real" and the empirical synonymous, and since natural science came to mean the study of the empirical, science became—in theory anyway—its exclusive way to knowledge.[8] On any reading Galileo was a central figure in the development of the modern understanding of the methods and scope of natural science. As impressive as his astronomical discoveries and his study of the mechanics of motion are, his thoughts on the scientific method are of equal and perhaps greater lasting influence. He narrowed the study of the motion of physical objects to those aspects that could be comprehended in mathematical terms. The nature of the object or its qualities or any other of Aristotle's ten categories—except quantity— makes not the slightest difference to the physical motion of the object. According to Galileo, only "primary qualities," that is, those that can be stated mathematically, such as a length, weight, density and so on, can be studied in mechanics. Secondary qualities such as color, texture, smell should be left out of account. They are not objectively real but only real in the animal psyche.[9] There is no world soul or inner teleological tendency within the world.[10] In Galileo's own words:

> To excite in us tastes, odors, and sounds I believe that nothing is required in external bodies except shapes, numbers, and slow or rapid movements. I think that if ears, tongues, and noses were removed, shapes and numbers and motions would remain, but not odors or tastes or sounds. The latter, I believe, are nothing more than names when separated from living beings, just as tickling and titillation are nothing but names in the absence of such things as noses and armpits.[11]

---

[8]I am speaking here of a comprehensive worldview, which is "scientistic." I need not address its other components: ethics, atheism, positivism, etc. I am interested only in the nature and scope of modern natural science.

[9]Many of Galileo's thoughts on these subjects are found in his letter to Casarini to which he gave the title "The Assayer." See Galileo, *Discoveries and Opinions of Galileo,* trans. Stillman Drake (New York: Doubleday, 1957), pp. 231-280.

[10]For discussion of Galileo's view of motion, which views gravity not as an internal quality or form but an external force, see Amos Funkenstein, *Theology and the Scientific Imagination: From the Middle Ages to The Seventeenth Century* (Princeton, NJ: Princeton University Press, 1986), pp. 174-79.

[11]"The Assayer," in Galileo, *Discoveries,* pp. 276-77.

Note how for Galileo metaphysical and scientific ideas interact. Aristotelian science, the system in opposition to which Galileo develops his philosophy of science, seeks knowledge of the natures, forms and ends of things as rational explanations for their appearance and behaviors. Practicing Aristotelian science involves acquiring knowledge of what many contemporary thinkers consider metaphysical aspects of things and hence combines the empirical and the nonempirical in one system of knowledge. The birth of modern empirical science requires a clear demarcation between the metaphysical and the empirical. There are two paths open to this goal. One could without prejudice divide natural philosophy into two distinct disciplines, with metaphysics taking on the nonempirical aspects of things and natural science addressing only questions that can be answered with empirical study. Natural science would in effect limit itself to the task of rationalizing the interrelationships among the phenomena of perception. Following this path, however, leads natural science either (1) to give up all claims to knowledge of reality or (2) to claim that only empirical phenomena are real. Galileo and most natural scientists in his wake, that is, physicists and chemists,[12] took an eclectic path by denying or ignoring the objective reality of things not measurable in mathematical terms while affirming the reality of primary qualities. Though most working scientists are not philosophically reflective, they instinctively consider the knowledge they produce to be about the objectively real world. Hence modern natural science finds itself in an ambiguous position. As human beings scientists want to believe that their work leads them closer to understanding the objectively real, but their methods lead only to greater and greater clarity about the interconnections among empirical perceptions. Only with the help of extra-scientific human motivations and metaphysical and epistemological assumptions can natural science do more. Furthermore, not even physics or chemistry can reduce *things* to numbers or all relationships to mathematical relationships. The equations of physics and chemistry are not purely mathematical, for they refer to named "things." Even energy, time and space—not to mention fields and particles, or more particularly electrons, protons, photons and the whole

---

[12]Biology, because it studies living organisms, needs also the category of "function." Function is a relative of teleology. An organ plays a part in maintaining the whole organism, which is the "end" of the organ. But the category of function has relevance only within an organism and not to the ecosystem or to the world.

pantheon of "fundamental" particles and forces—are names. And names refer to things themselves, not the measure of things. All numbers used in physics and chemistry are measurements of *things*.[13] The famous equation $e = mc^2$ says in words: "the measure of energy (in joules) equals the measure of mass (in kilograms) times the velocity of light (in meters per second) multiplied by itself." *E* and *m* and *c* refer to things as well as to the measurements of things. Finally, there can be no scientific knowledge, however theoretical and abstract, unless it can be explanatory or predictive of some empirical experience, which can only be attained through the senses. Even data generated by such sophisticated machines as the giant particle accelerator in Geneva must finally be read off a screen with human eyes.

The Competence and Limits of Natural Science

These limitations of natural science can be seen by considering one of its central tasks, scientific explanation. In the mid-twentieth century, Carl Hempel developed the deductive-nomological (DN) model of scientific explanation. Hempel's work marks a new beginning of reflection on the subject.[14] In the DN model there are two major components, a statement of a phenomenon to be explained (the *explanandum*) and the statement of the facts and laws that account for the phenomenon (the *explanans*). The logical relationship between the *explanans* and the *explanandum* is deductive; that is, if the *explanans* is true the *explanandum* must be true. According to Hempel, the *explanans* must contain true propositions and at least one scientific law. The DN model, in Hempel's words, "shows that, given the particular circumstances and the laws in question, the occurrence of the phenomenon *was to be expected*; and it is in this sense that the explanation enables us to *understand why* the phenomenon occurred."[15] It has been pointed out that the DN model does not take into account causation, for a scientific law is a generalization that does not necessarily appeal to causation. But without discovering that the cause of *e* was *c*, wouldn't explaining *e*'s

---

[13]It is even doubtful that mathematics itself can escape reference to the world of things. After all, mathematicians are themselves not numbers but things!

[14]Carl Hempel, *Philosophy of Science* (Englewood Cliffs, NJ: Prentice-Hall, 1966). For a readable summary of the DN model, see Peter Kosso, *Reading the Book of Nature: An Introduction to the Philosophy of Science* (Cambridge: Cambridge University Press, 1992).

[15]Carl Hempel, *Aspects of Scientific Explanation and Other Essays in the Philosophy of Science* (New York: Free Press, 1965), p. 337.

existence by c be rather unenlightening?[16] Hence philosophers of science have proposed various modifications and alternatives to the DN model.[17] Ernan McMullin traces the history of scientific explanation from Aristotle's *Posterior Analytics* to Charles Peirce (1839–1914). He concludes that modern science as it actually takes place in the activity of working scientists possesses three aspects, abduction, induction and retroduction. The scientific inference moves from data that need explaining to a hypothesis. Induction involves methodical examination of the data through experiment and measurement. Retroduction concludes to a causal theory that explains the phenomena by reference to unobservable and underlying causes. The whole process is very complicated and unfolds only in the history of science.[18]

For my purpose the most important thing to notice is that explanation, whether in the pure DN model or in augmented models, always seeks to make understandable the relationship of one set of empirical phenomena to another set of empirical phenomena by means of law-like generalization(s) or postulated causal relations or some other theoretical mediation. All of these mediating principles may be reduced to patterns of empirical phenomena. Or, if the theory refers to unobservable entities, these entities are still physical and manifest themselves in observable phenomena. This is the nature and the limit of natural science, whether physics, chemistry, biology, geology or paleontology. Natural science studies the relationships within the created world among empirical phenomena, that is, the perceptions or sense data received through the five senses. The decisive difference between Aristotle's explanations of natural phenomena and those of modern natural

---

[16]Kosso, *Reading the Book of Nature*, pp. 58-65. For detailed discussions of this criticism, see James Woodward, "Scientific Explanation," in *Stanford Encyclopedia of Philosophy* (Winter 2014 ed.), ed. Edward N. Zalta, http://plato.stanford.edu/entries/scientific-explanation/; Kosso, *Reading the Book of Nature*, pp. 63-68; and Wesley Salmon, *Explanation and the Causal Structure of the World* (Princeton, NJ: Princeton University Press, 1984).

[17]Wesley Salmon, "Statistical Explanation," in *Statistical Explanation and Statistical Relevance*, ed. Wesley Salmon (Pittsburgh: University of Pittsburgh Press, 1971), pp. 29-87, proposed a model that defined scientific explanation in terms of statistical relevance (SR). As Woodward explains it, "The intuition underlying the SR model is that statistically relevant properties (or information about statistically relevant relationships) are explanatory and statistically irrelevant properties are not. In other words, the notion of a property making a difference for an *explanandum* is unpacked in terms of statistical relevance relationships" (Woodward, "Scientific Explanation"). Other models have been proposed: the causal mechanical model, which attempts to revive the idea of causality within an empiricist framework, and the unification model, which contends that a good explanation is the one that explains a wide range of empirical data in a unified way.

[18]Ernan McMullin, *The Inference That Makes Science* (Milwaukee: Marquette University Press, 1992).

science is not found in their logical aspects or in accuracy of observation. The difference lies in the status of the laws and theories that mediate the relationships between the sets of phenomena. Aristotle's laws were not empirical generalizations but metaphysical principles about "essential natures," not "quantitative relationships."[19] Take for example Aristotle's explanation of why heavy things fall toward the center of the earth and light bodies tend upward. The natures of heavy things possess less resemblance to the absolute perfection at the top of the hierarchical scale of being and light things possess a greater likeness to that divine reality. Hence the one moves further away from and the other closer to that reality, to their proper place in the value-laden order of the cosmos.[20]

Modern natural science is not the science of everything or of every dimension of the world. Science cannot adjudicate the question of whether there is more to the world than the empirical. It is blind to all else. If there is something more, other methods of study will be required to investigate it. Metaphysics is the "science of absolute presuppositions" of all experience.[21] Hermeneutics studies the process of deriving meaning from texts. Christian theology studies the divine revelation believed by the Christian church and deposited in Scripture. And the Christian doctrine of creation seeks to understand the relationship of the triune God to the created world as it is taught in Scripture. The Christian doctrine of creation does not derive from empirical experience of the world or study relationships between sets of empirical phenomena within the world. In theology, the God-relation is everything. Innerworldly empirical relations are of no concern.

Consider again the theological theses on creation we developed in chapter one: (1) The one God is the absolute origin and sovereign ruler over all that is not God. (2) The one God freely established the Creator-creature relation, which is characterized by generosity and freedom and power on the Creator's side and dependence and debt on the creature's side. (3) The creation

---

[19]David C. Lindberg, *The Beginnings of Western Science*, p. 60. See Ernan McMullin, *The Inference That Makes Science*, for a more extensive study of the differences and likenesses between Aristotle's science and modern science.

[20]See David C. Lindberg, *The Beginnings of Western Science*, pp. 58-62, for Aristotle's understanding of motion.

[21]Collingwood, *Metaphysics*, p. 34. Collingwood denies that there can be a science of pure being, for there is no science without presuppositions (pp. 11-16). Presuppositions are either relative or absolute. Relative presuppositions may require support from other propositions. Absolute presuppositions may be accepted or rejected but it makes no sense to demand proof for them.

really exists before God and stands before him as good; that is, as the result of God's act of creation, the creature really is what God intended it to be. (4) The Creator-creature relation established at the beginning, with its characteristic qualities, endures for all time and history. (5) Human beings possess a unique relationship to the Creator characterized by their image and likeness to God and responsibility to him. (6) God acts on and relates to the world from the Father, through the Son and in the Holy Spirit. None of these affirmations says anything about the relation of one set of empirical phenomena to another (natural science in general). None says anything about a particular state of the cosmic order in relation to a previous state or a subsequent state (cosmology). None says anything about how a particular stage of a species or an ecosystem derives from a previous stage or produces a subsequent stage (evolutionary biology). These theses are consistent with any scientific law or cosmic history or history of life. Indeed, there is absolute nothing that is discoverable by the empirical methods of natural science that could falsify or verify even one of these affirmations.

RELATING NATURAL SCIENCE AND CHRISTIAN THEOLOGY

In his Gifford Lectures (1989–1991), Ian Barbour criticized views that distinguish natural science and Christian theology in the way I have distinguished them. He rejects what he calls the "conflict" and "independence" views, appreciates the "dialogue" view in its positive features, but he ultimately favors the "integration" position.[22] According to Barbour, in their efforts to avoid conflict advocates of independence deny that there is a relationship of any kind between science and theology. They must be sealed off from each other in "watertight compartments."[23] Barbour criticizes the independence position (1) for its unrealistic notion that one can divide life into completely "separate compartments," (2) its idea that human beings can give up the quest for a "unified world view" and (3) for its lack of a "theology of nature," which we need to deal with our contemporary environmental concerns.[24] Barbour appreciates the views he calls "dialogue" models because they are interested in relating natural science and theology at the indirect levels of methods and

[22]Ian Barbour, *Religion in an Age of Science: The Gifford Lectures* (San Francisco: Harper & Row, 1990), vol. 1.
[23]Ibid., 1:10.
[24]Ibid., 1:16.

presuppositions even though they decline direct comparison between theological doctrines and scientific theories. But Barbour wants to go further than dialogue can take us, and so he advocates a form of "integration" he designates "theology of nature."[25] A theology of nature takes into account current scientific theories with the goal of reformulating such theological doctrines as "creation, providence, and human nature" so that they are "in harmony with scientific knowledge." He continues to explain how

> Our understanding of the general characteristics of nature will affect our models of God's relation to nature. Nature is today understood to be a dynamic evolutionary process with a long history of emergent novelty, characterized throughout by chance and law. The natural order is ecological, interdependent, and multileveled. These characteristics will modify our representations of the relation of both God and humanity to nonhuman nature.[26]

According to Barbour the best candidate for a theology of nature is process theology, in which God is an aspect of the world, limited in power rather than omnipotent, temporal rather than eternal, interdependent rather than independent, persuasive rather than coercive and able to suffer rather than impassible. Barbour concludes that process theology conforms best to general criteria by which any theoretical claim in science, philosophy or theology must be measured; it must (1) agree with the data, (2) possess internal and external coherence, (3) display a wide scope of explanation and (4) demonstrate fertility in generating new insights.[27] As it turns out Barbour's "theology of nature" is a philosophical synthesis of data and theory from science and religion in which scientific theories are given metaphysical status and then used as criteria by which to revise theological doctrines. More precisely stated, Barbour assumes the truth of current scientific theories, reflects on the metaphysical presuppositions of those scientific theories and then generalizes to a metaphysics to which all theoretical endeavors must conform. It is to such mixing of theology and science that the independence and dialogue positions are directed. This synthesis cannot be accepted as a legitimate Christian doctrine of creation.

The view of the relationship between natural science and Christian theology I am advocating in this study bears some resemblance to Barbour's

---

[25]Ibid., 1:26-28.
[26]Ibid., 1:26.
[27]Ibid., 1:265-67.

independence position and some to his dialogue one. It seems to me that Barbour's criticisms of the independence model come close to caricature. Few advocates of independence would say that science and theology can be separated into absolutely unrelated spheres. Everyone acknowledges that both these endeavors involve human language and thought and will overlap in these areas. The essential point is not *total* independence but *material* independence. It's not how you talk or how you think that differentiates the two but what you speak and think about and on what authority. And if the independence position allows for these methodological commonalities, the distinction between independence and dialogue resolves itself largely into a matter of personal interest. Dialogue is for theologians who are interested in science and scientists who are interested in theology. Independence is for those who simply want to carry out their studies without interdisciplinary involvement and its consequent risk of contamination. Dialogue about possible common presuppositions and methods between particular theological and scientific programs can be conducted by theologians and scientists who hold to independence of material content.[28] Dialogue can produce such salutary results as clarification of the limits and domains of each.[29]

Many have argued that modern natural science takes for granted certain presuppositions that are metaphysical in nature. It is certainly true that a theology of creation makes metaphysical assertions. Here is a place natural science and theology may interact in conflict or agreement; that is, at the

---

[28]I take this to be the position of Ernan McMullin and Thomas F. Torrance. See Ernan McMullin, "Natural Science and Belief in a Creator," in *Physics, Philosophy and Theology: A Common Quest for Understanding*, ed. Robert John Russell, William R. Stoeger and George V. Coyne (Vatican City: Vatican Observatory, 1988), pp. 49-79; and McMullin, "How Should Cosmology Relate to Theology?," in *The Sciences and Theology in the Twentieth Century*, ed. Arthur Peacocke (Notre Dame, IN: University of Notre Dame Press, 1981), pp. 17-57. See also Thomas F. Torrance, *Divine and Contingent Order* (Oxford: Oxford University Press, 1981); and Torrance, "God and the Contingent World," *Zygon* 14 (1979): 329-48.

[29]William R. Stoeger, SJ, of the Vatican Observatory suggests this goal as well as other positive outcomes for the dialogue but rejects the direct integration of the two disciplines ("Contemporary Cosmology and its Implications for the Science-Religion Dialogue," in Russell, Stoeger and Coyne, *Physics, Philosophy and Theology*, pp. 219-47). James R. Pambrun observes, "There is no direct or immediate route between science and theology. . . . The kinds of sources, data, models of investigation each appropriates to fulfill its requirements of being a scientific discipline are quite different. [However] both disciplines appropriate acts of understanding and acquire their status as a discipline by adhering to the criteria of acts of understanding" ("Creatio ex Nihilo and Dual Causality," in *Creation and the God of Abraham*, ed. David Burrell et al. [Cambridge: Cambridge University Press, 2010], p. 193). There is a possibility of dialogue but only through a third medium, philosophical reflection.

metascientific level. Natural science reasons inductively and deductively and works toward a theoretical understanding. So does theology. Since the same human mind works in both, it would not be surprising that there are parallels. But on the material level science and theology have nothing in common. There are no theories or facts that do double duty in both theology and natural science. The relationship is similar to that between two games, for example baseball and golf. The same person may play both and have all the same human motivations for both activities. Both golf and baseball are games and possess rules that must be followed if the game is to be played. And the same ethical rules apply to both. But golf has no place for shortstops or base hits. It would be absurd for a lover of baseball to criticize a golfer for using a golf ball instead of a baseball or for a golfer to object when an outfielder catches the ball. And baseball is not measured by strokes under par and possesses no move called a putt. Hence even though baseball and golf are human activities and both are games defined by sets of rules, they are independent internally or materially.[30]

Big Bang Cosmology and Evolutionary Biology

Let's consider the big bang cosmological theory as our first test case for my material independence view. This theory projects backward in time to previous states of the physical universe from its present state and the processes going on in it now. The physical evidence for this theory is quite strong and much more complicated than I understand or can rehearse here.[31] In 1925 Georges Lemaître, relying on a particular interpretation of Einstein's general theory of relativity, postulated that the universe is expanding. If we project

---

[30]The viewpoint I am advocating, far from being a modern development of neo-orthodoxy, is articulated clearly in Thomas Aquinas. According to Aquinas, scientists and theologians consider the same world, but theologians consider natural objects in relation to God and scientists consider them "as such." Science studies "such things as belong to them by nature . . . the believer [considers] only such things as belong to them according as they are related to God—the fact, for instance, that they are created by God, are subject to Him, and so on." Or again: "For the philosopher takes his argument from the proper causes of things; the believer, from the first cause—for such reasons as that a thing has been handed down in this manner by God, or that this conduces to God's glory, or that God's power is infinite" (SCG 2.4.2-4; Saint Thomas Aquinas, *Summa Contra Gentiles, Book Two: Creation*, trans. James F. Anderson [Notre Dame, IN: University of Notre Dame Press, 1975], pp. 34-35.

[31]For popular summaries of modern cosmology by accomplished physicists, see Stephen Hawking, *A Brief History of Time: From the Big Bang to Black Holes* (New York: Bantam, 1988); and John D. Barrow, *The Origin of the Universe* (New York: Basic Books, 1994).

backward in time, Lemaître reasoned, we must assume that at some finite time in the past the whole universe was smaller than an atom and extremely hot.[32] Indeed, there is no known limit to how "small" the universe can be.[33] The state of infinite smallness and infinite heat is called "the singularity." At that point all the known laws of physics break down, and no experiment can duplicate the conditions at this point. We can know nothing about it. But it is not nothing, for it contains in another form (if "form" is the right word) all that exists in the present state of the universe. Since Lemaître first suggested it, evidence for the big bang has grown and has become more refined. However, it is not necessary for my purpose to trace the theory's history from Lemaître to today or rehearse the scientific evidence for it. Let us assume that big bang cosmology is a highly confirmed theory.

Now it should be clear that the big bang theory conjectures about an earlier state of our physical universe using what is known about the present state of the universe. It tells us nothing about God's relationship to the universe. The image of the whole universe expanding from an infinitely small point may be poetically stimulating and rhetorically inspiring. The infinite smallness of the singularity may evoke the image of God creating the world from nothing. But the conclusion of Paul Copan and William Lane Craig surely goes beyond what we can know when they assert that "the standard Big Bang model . . . describes a universe which is not eternal in the past, but which came into being a finite time ago. . . . The origin it posts is an absolute origin *ex nihilo*."[34] Even their more cautious statement is misplaced: "At a minimum we can say confidently that those who believe in the doctrine of *creatio ex nihilo* will not find themselves contradicted by the empirical evi-

---

[32]Georges Lemaître, *The Primeval Atom: An Essay on Cosmology*, trans. Betty H. Korff and Serge A. Korff (New York: van Nostrand, 1950), pp. 134-63. See Schwarz, *Creation*, pp. 23-26.

[33]I put the word *small* in quotes because size is meaningless when the whole universe is under consideration. As a comparative category, size makes sense only *within* the universe.

[34]Copan and Craig, *Creation Out of Nothing*, p. 223. Pirooz Fatoorchi lists six authors who argue for a "strong interpretation" of the relevance of big bang cosmology to creation from nothing and seven who opt for the "weak" interpretation ("Creation ex Nihilo and the Compatibility Question," in Burrell et al., *Creation and the God of Abraham*, pp. 100-104). Fatoorchi lists Craig in the "strong interpretation" group. In the "weak interpretation" group, Fatoorchi lists William Stoeger. After surveying the science of the big bang, Stoeger concludes, "Thus, we can see that the Big Bang, however we describe it within the framework of cosmology, should not be considered as a beginning either of the universe or of time in any specific or definite sense, much less of creation in the theological sense of that word" (William Stoeger, "Key Developments in Physics Challenging Philosophy and Theology," in *Religion and Science: History, Method, Dialogue*, ed. W. M. Richardson and W. J. Wildman [London: Routledge, 1996], p. 193).

dence of contemporary cosmology but on the contrary will be fully in line with it."[35] Craig's conclusion is trivial, because no set of empirical data can contradict the doctrine of *creatio ex nihilo*. As Janet Soskice puts it, "Creatio ex nihilo is a metaphysical claim, and not an empirical one, and does not dictate a particular cosmology. It is thus not in competition with scientific explanation, nor potentially defeasible by it."[36] Natural science can never demonstrate anything about an absolute beginning.[37] It does not matter how small or how hot, it is still a previous state of the universe that now exists. As William Stoeger says, "Physics as such can specify in great qualitative and quantitative detail how we get from one physical state to another. . . . However, it cannot in principle account ultimately for their existence or for the particular form those structures, regularities and relationships take."[38] I do not believe one can or should deduce any physical law, state or history of the universe from the affirmation that God created it. For all we know God could have created an infinite number of universes, all very different from this one. I see no reason to assume that a universe's possession of some physical characteristics makes its creation more likely than one possessing others.[39] Even the fact that contemporary cosmology provides no answer to the question of an absolute beginning or cannot prove the universe to be eternal should not come as a surprise, uniquely suggestive of divine creation. There are thousands of questions that natural science is incompetent to answer, the importance of which are felt by every existing human being!

As another example consider the theory of biological evolution. In a way similar to the big bang theory of cosmology, the theory of biological evo-

---

[35]Copan and Craig, *Creation Out of Nothing*, p. 248. See also William Lane Craig, "A Criticism of the Cosmological Argument for God's Non-Existence," in *Theism, Atheism and Big Bang Cosmology*, ed. W. L. Craig and Q. Smith (New York: Oxford University Press, 1995), where Craig argues that the standard model of Big Bang cosmology can be used in a strong argument for God's existence.

[36]Janet M. Soskice, "Creatio ex Nihilo: Jewish and Christian Foundations," in Burrell et al., *Creation and the God of Abraham*, pp. 38-39.

[37]William R. Stoeger cautions: "In this regard, even establishing a rough parallel, or consonance between 'the beginning of time' in the Big Bang and 'the beginning of time' in the doctrine of creation . . . is very questionable. It seems highly unlikely that cosmology, or any physical science, will ever be able to reveal a point of *absolute* beginning" ("Contemporary Cosmology," p. 240).

[38]William Stoeger, "The Big Bang, Quantum Cosmology and Creatio ex Nihilo," in Burrell et al., *Creation and the God of Abraham*, pp. 168.

[39]The exception may be a universe that possesses minds. Of course, the laws of physics must be within a certain range for physical beings like us to exist.

lution attempts to account for the present state of a species on earth by showing its derivation from its ancestor species. However, there is one giant difference between evolutionary cosmology and evolutionary biology. Whereas the cosmos is one continuous thing throughout time, living things are very diverse and come into existence and die. We cannot study evolution in one individual organism. We can only study its organic development throughout its life span. Evolutionary biologists are interested in how an earlier species changes over many generations to become a different species. Ultimately they are interested in how all life on earth developed from its assumed simple and single beginning. The current theory assumes that a species' identity is determined by its DNA and that the DNA molecule can sometimes change or be changed (mutation) with beneficial results, which then can be preserved and gain dominance in a population through natural selection.[40] As an outsider looking in, evolutionary biology looks to be in its very beginning stage of scientific explanation compared to cosmology—if completeness would mean accounting for every change in the history of life in terms of biological laws and empirically measurable conditions. In each instance of biological explanation the transformation from a previous state of life to a succeeding one is explained when a set of relevant laws and conditions answers the question of why the transformation took place.

Let me be exceedingly generous for my argument's sake and grant evolutionary biology its goal of complete empirical description and scientific explanation of all changes in the history of life from the first cell to human beings and all other species on earth, living and extinct.[41] What are the implications for the doctrine of creation? Would it disprove it? Or would it reveal God's method of creation? The answer is neither one.[42] If in principle accounting for the history of life scientifically could disprove creation, there is no need to go that far or wait that long. Biological science accounts for

---

[40]For a readable overview of evolutionary theory from an evolutionary-biologist participant in the science and theology dialogue, see Francisco J. Ayala, "The Evolution of Life: An Overview," in *Evolutionary and Molecular Biology: Scientific Perspectives on Divine Action* (Vatican City and Berkeley, CA: Vatican Observatory and Center for Theology and the Natural Sciences, 1998), pp. 21-57.

[41]This achievement is far from being accomplished and may be impossible. Much of the data is lost forever.

[42]In David Fergusson's judgment, "If the engagement with Darwinism has taught theologians one thing it might be this: the sciences must be given their place freely to investigate and hypothesize according to their methods and findings" (*Creation*, Guides to Theology [Grand Rapids: Eerdmans, 2014], pp. 88-89).

your and my existence through sexual generation and the combination of genetic material from our parents. At no point does biological science take into account my relationship to God, as a creature or as a person. But the doctrine of creation does not merely affirm that the first creatures, whether they were atoms or Adams, were created by God. It affirms that all things throughout all time and space, including you and me, are just as much God's creatures as the first creatures were. Evolutionary biology accounts for changes over time in the same way that it accounts for our births, that is, through empirical data and empirically based scientific law. I reject the idea that my identity, my relationship to God or my purpose in life can be reduced to biological categories. And I reject the idea that an evolutionary account of the history of life can prove or disprove creation. God's relationship to his creatures is not an empirical phenomenon and is not subject to investigation by empirical science. It transcends all physical laws; for physical laws are aspects of the physical world and as such are created. The Creator of physical laws is not subject to those laws but is their Lord.

Nor is evolution or the big bang God's *means* of creation. As I argued above, God does not need means or a "how" of creation. This is a rationalist dream. There is no mean between absolutely nothing and the world God created. God is the Creator of everything, time, space and everything in them. The eternal God creates all laws and processes and the entire web of relations internal to creation. Just because b (at t_2) stands in a relationship of causal dependence on a (at t_1) does not mean that God depends on a to bring about b. The whole world process depends on God. I will speak more about this when I discuss the time of creation. But God does not need to relate to the world in creation and providence through the world process in a linear way. God transcends time and space and all that occurs within them. This relationship is beyond the reach of natural science. Science views the world process as an incomplete chain of causes and effects, without beginning and without end. God knows it simultaneously and as a whole.

GOULD, DAWKINS AND DEMBSKI

Of course many would disagree with my conclusion, and some who agree with my separation between natural science and theology do not believe in God and creation. Stephen Jay Gould argues that natural science and theology are nonoverlapping magisteria (NOMA); that is, they depend on different sources for their authority. They speak about different things and use different languages and methods. Gould explains:

> We may, I think, adopt this word [*magisterium*] and concept to express the central point of this essay and the principled resolution of supposed "conflict" or "warfare" between science and religion. No such conflict should exist because each subject has a legitimate magisterium, or domain of teaching authority—and these magisteria do not overlap (the principle that I would like to designate as NOMA, or "nonoverlapping magisteria"). The net of science covers the empirical universe: what is it made of (fact) and why does it work this way (theory). The net of religion extends over questions of moral meaning and value. These two magisteria do not overlap, nor do they encompass all inquiry (consider, for starters, the magisterium of art and the meaning of beauty). To cite the arch clichés, we get the age of rocks, and religion retains the rock of ages; we study how the heavens go, and they determine how to go to heaven.[43]

As is clear, Gould limits natural science in a way similar to what I have argued and gives theology its own domain within which it is sovereign. However, Gould seems to think that the domain of theology is empty, and things excluded from the domain of science have no objective existence. In this essay he refers to himself as "a Jewish agnostic" and says,

> I may, for example, privately suspect that papal insistence on divine infusion of the soul represents a sop to our fears, a device for maintaining a belief in human superiority within an evolutionary world offering no privileged position to any creature. But I also know that souls represent a subject outside the magisterium of science. My world cannot prove or disprove such a notion, and the concept of souls cannot threaten or impact my domain.[44]

For Richard Dawkins, in contrast, evolutionary biology provides an explanation of everything religion used to explain about life. There is no need for divine creation to account for human existence or morality or intelligence or love or religion. Everything human beings are and all the aspects of their experience can be explained by evolutionary principles. The em-

---

[43]Stephen Jay Gould, "Nonoverlapping Magisteria," *Natural History* 106 (March 1997): 16-22. See also his later book *Rocks of the Ages: Science and Religion in the Fullness of Life* (New York: Ballantine, 1999).

[44]Gould, "Nonoverlapping Magisteria," pp. 16-22.

pirical is all there is. God is a delusion.[45] In criticism of NOMA, Dawkins argues that we have no ground to think that the questions science cannot answer can be answered by theology or that they are even meaningful. Why think that the questions theology addresses are any different from the question, "Why are unicorns hollow?"[46] The NOMA doctrine assumes that there is something real for theology to study, but Dawkins objects, "I have yet to see any good reason to suppose that theology (as opposed to biblical history, literature, etc.) is a subject at all."[47] Dawkins asserts that NOMA is popular with some believers "because there is no evidence to favor the God Hypothesis. The moment there was the smallest suggestion of any evidence in favor of religious belief, religious apologists would lose no time in throwing NOMA out the window."[48] Dawkins will not so easily allow meaningful questions to escape the realm of science: "The presence or absence of a creative super-intelligence is unequivocally a scientific question, even if it is not in practice—or not yet—a decided one."[49] Interestingly, some believers agree with Dawkins on this last point.

William A. Dembski argues that the scientist can detect intelligent design within DNA and certain other molecules and biological systems by *empirical* methods. Dembski draws on several well-established sciences that specialize in distinguishing design from chance and law: forensic science, artificial intelligence, information theory, cryptography, archaeology and the search for extraterrestrial intelligence. According to these sciences, the joint presence of contingency, complexity and specificity indicate intelligent design.[50] The presence of contingency rules out automatic and repetitive processes, that is, law-like behaviors. Complexity renders chance improbable and specification ensures the presence of meaningful patterns indicative of intelligence. According to Dembski, studying DNA with the same methods researchers use to seek for signs of extraterrestrial intelligent life compels us to conclude that DNA bears the marks of intelligent design: contingency, complexity and specificity. Dembski distinguishes intelligent design theory

[45]Richard Dawkins, *The God Delusion* (Boston: Houghton Mifflin, 2006).

[46]Ibid., p. 57.

[47]Ibid.

[48]Ibid., p. 59.

[49]Ibid.

[50]William A. Dembski, *Intelligent Design: The Bridge Between Science and Theology* (Downers Grove, IL: InterVarsity Press, 1999). See also his more scholarly book *The Design Inference* (Cambridge: Cambridge University Press, 1998).

from theistic evolution and scientific creationism. Intelligent design theory, according to Dembski, can demonstrate that an object results from an intelligent cause, but it says nothing else about the nature of the designer or the means and the timetable of the designer's activity. The random mutation (chance) and natural selection (law) mechanisms of the Darwinian theory, contends Dembski, cannot account for biological objects that exemplify "specified complexity." Darwinian law and chance can do a good job "conserving, adapting and honing already existing biological structures," but they cannot account for the origin of these information-rich structures.[51] In light of the success of intelligent design in showing the marks of design in nature, Dembski argues that science and theology need not be viewed as totally independent realms of inquiry, much less competing explanations for the same realm in constant conflict. Intelligent design theory shows that science and theology can give each other mutual support. Taken along with the big bang cosmological theory and the fine-tuning of the physical constants of the universe, the intelligence evident in biological structures supports belief in divine creation.[52]

The Bible declares that God created the world and that it proclaims the glory of its Maker. Intuitively and in faith, believers from ancient times to today sense the presence of the divine hand in the vastness, power, beauty and order of creation. Most Christian traditions have affirmed in accord with Romans 1 that God reveals himself in creation. Some theologians even attempt to articulate natural revelation into a system of natural theology or metaphysics.[53] But intelligent design theory does something different. It argues that what is believed by faith and experienced intuitively can be supported *empirically*, that is, by the same methods and data that support contemporary biological or cosmological theories. It argues that information, which weighs nothing, occupies no space, does not reflect light and does not undergo change through time, can be detected by empirical means. It seems to me that Dembski misleads us here. None of the sciences on which intel-

---

[51]Dembski, *Intelligent Design*, p. 180.

[52]I find it interesting that intelligent design seeks the marks of intelligence in molecules and tiny organisms when the existence of intelligence is incontrovertibly proved by the existence of the scientist him- or herself! Why seek it elsewhere?

[53]For my discussion of the distinction between natural revelation, natural theology and natural religion, see *Great Is the Lord: Theology for the Praise of God* (Grand Rapids: Eerdmans, 2008), pp. 11-19.

ligent design depends is strictly empirical: forensic science, artificial intelligence, information theory, cryptography, archaeology and the search for extraterrestrial intelligence. Though these sciences incorporate empirical observations into their methods, each depends decisively on our experience of our own intelligence. Looking for signs of intelligence and information takes a *hermeneutical* (not purely empirical) approach to the perception of meaning and interpretation. So the methods of these sciences are empirical only in a derivative sense. They look for information encoded in empirical media. The information stored in the empirical media is not itself empirical and cannot be detected the way empirical data are detected. Only a mind can detect information or the activity of another mind. A machine can be programmed to detect patterns, but the programmer (a human mind) must program the computer to look for patterns the human programmer considers as marks of mind. And when the computer finds these patterns, it cannot be said even to *recognize* much less understand them. From an empirical point of view, all patterns are qualitatively equal, though they differ in quantity and spatial arrangement.

Dembski's argument would be much more plausible as a philosophical argument. Many years ago while reading Frederick Copleston's *A History of Philosophy* I encountered an argument put forth by Fichte. It impressed me then and continues to impress me as worthy of thought. We have two basic types of experience, that of freedom and that of necessity. Fichte insisted on a strict either/or between them. One or the other but not a mixture must be fundamental. A thinker must choose one or the other as the foundation for a philosophical system and develop the system consistent with the foundation. One becomes convinced of the truth of the basic presupposition or not only in light of the complete system built on it.[54] In this context we translate Fichte's distinction between freedom and necessity into the distinction between the intelligible and the unintelligible. (1) We experience our own minds and their ideas. Some of the content of our minds seems to correspond to an intelligible reality outside our individual minds, most profoundly other human minds. (2) We also experience things that seem merely to resist us and exclude us. They are opaque and unintelligible. We have only external relationships to them, and we assume that they have only external

---

[54]Frederick Copleston, *A History of Philosophy*, vol. 7, *Modern Philosophy From the Post-Kantian Idealists to Marx, Kierkegaard and Nietzsche* (New York: Image, 1994), pp. 37-39.

relationships to each other. The technical term for this reality is matter. Since both of these experiences are primitive on the experiential level, we are faced with a threefold choice.[55] We can rationally reduce one to the other, becoming idealists or materialists, or assert that both are fundamental and eternal, thus becoming dualists. It seems to me that on a purely rational level materialism is by far the least plausible and dualism the most plausible. But both idealism and dualism assert the irreducible reality of mind. Had Dembski argued from the mind's perception of the pervasive presence of intelligible reality within the universe, in itself and other human minds, in biological structures, in cosmological laws and in the micro world of sub-atomic structures to the existence and activity of a cosmic mind, his argument would demand attention. It would be built on a primitive, undeniable and universal human experience. But instead he adopts the appearance of methodological neutrality between the fundamental choices and then attempts to tip the balance with empirical evidence. He attempts to prove what can only be presupposed. As Dembski develops it, intelligent design theory is an elaborate non sequitur and an unfortunate distraction from a truly impressive argument.[56] It makes exactly the same mistake that Richard Dawkins makes, assuming thoughtlessly that a thinker can move from the empirically based theories of natural science to metaphysical conclusions, when in fact the entire system of natural science is founded on a choice among metaphysical presuppositions.[57]

WHY THE CONTROVERSY?

If I am correct that modern natural science by definition studies only empirical relationships within the created order and the doctrine of creation concerns only the world's God-relation and God's relationship to the world,

---

[55]Fichte rejected a middle way as an illogical mixing of opposites, but many thinkers search for it anyway.

[56]Among his criticism's Christopher Doran asks, "Why not criticize metaphysical naturalism philosophically?" See "Intelligent Design: It's Just Too Good to Be True," *Theology and Science* 8 (2010): 229. Doran rightly points out that Dembski "has failed to see the following critical distinction between evolution and an atheistic interpretation of evolution!" (p. 234).

[57]The theories of natural science can "confirm" their metaphysical presuppositions only indirectly by their coherence, explanatory scope, elegance, fertility and other criteria. Hence the work of philosophers of science in articulating the genuine metaphysical presuppositions of natural science plays a very important role in the discussion between science and theology. Do the theories and practices of modern natural science presuppose materialism, idealism, dualism or some other metaphysical commitment? Answering this question is the task of philosophy.

why the conflict or the search for integration or dialogue? The short answer is that both the activities of science and theology take place within a tradition and a culture. Though they are internally and materially independent, they exist within an external cultural context that often obscures their internal independence. We will first consider the wider pressures on science.

Every science, as a system of inquiry designed to acquire knowledge, begins with presuppositions, explicitly stated or hidden. There is no science without presuppositions.[58] Modern Western science makes no sense apart from the history, thought and culture of the Western world. Most working scientists conduct their research as if certain metaphysical and epistemological presuppositions were true: the world really exists as an objective fact; it behaves lawfully; its most basic physical laws apply everywhere in the universe; though it is subject to rational investigation it is largely contingent; that the world is a metaphysically good or at least neutral place; that human sense impressions correspond to some degree to the objective world. Clearly, these presuppositions are culturally conditioned. People living in mythological cultures do not think this way. Gnostics reject the material world as evil. Platonists think empirical experience is unreliable and unworthy. We could continue this list. Note here that these presuppositions are not scientific laws, that is, generalizations of empirical observations. They constitute the foundation or framework to support the normal activities of science. If they are false, the conclusions of science are false or at least not confirmed. Yet no empirical observation or law can confirm presuppositions because the very validity of observation and physical law presupposes them.

At the presupposition level disputed or idiosyncratic metaphysical and epistemological ideas can be inserted surreptitiously or unconsciously without disturbing the science internally. The atheist can insert belief in materialism, which says everything real is material and everything that appears otherwise can be reduced to matter. Or one could assume naive realism as an epistemology, whereby one's empirical observations or generalizations are identified uncritically with objective reality. Or in a third move, one could insert an instrumentalist epistemology, wherein one becomes completely agnostic about the relationship of empirical observations and laws to things in themselves. Two scientists can agree completely on the

---

[58]Such a science would lead to absolute knowledge by transcending the subject/object distinction. The scientist would become like Aristotle's God, self-thinking thought.

internal workings and *empirical* findings of science but hold opposite beliefs about the metaphysical implications of science. A scientist with materialist presuppositions will find those assumptions confirmed, and a scientist who believes in God will find theistic presuppositions confirmed. When atheists contend that empirical science's conclusions confirm atheism, they are merely arguing in a circle, confirming their metaphysical and epistemological presuppositions. And when theists argue that empirical science confirms belief in the existence of God or divine creation, they too are arguing in a circle, finding their presuppositions confirmed in their conclusions. My point is that empirical science considered apart from its superstructure of presuppositions can produce only empirical conclusions. All conclusions that speak of something other than the empirical derive from nonempirical presuppositions that function as foundations for science. But the philosophical-presupposition level is not the only place where natural science interacts with culture.

Modern natural science is greatly valued in our culture, and scientists are held in high esteem. Why? Clearly the main reason for science's social prestige is that science has produced technology that people desire. Human beings want to enjoy health and long life, wealth, exciting entertainment, comfort and leisure, and of course military power. Some people are curious about the world and for that reason are interested in what science discovers. Others mistakenly think science will confirm their metaphysical or religious beliefs. But overwhelmingly science is valued for its material benefits. In their most idealistic moments, scientists may attempt to convince themselves that they pursue science for knowledge alone. Whatever the scientist thinks, however, the culture has another end in mind. There is no other way to account for the vast sums of money governments and businesses spend on research and development and individuals spend on technology. People today do not crave salvation or concern themselves with their God-relation. For many people science has replaced God as the source of well-being and the scientist has replaced the priest as the means of access to the source of good. A kind of mindless worldliness and thoughtless sensuality pervades the consumer culture the scientist serves.

Natural science possesses no natural birthright to the cultural power it holds today. As I indicated, science is held in esteem because people want the things science provides. But science cannot provide everything people need. Science cannot tell you what is right or wrong or make you wise or good. Science cannot endow your life with meaning or make you happy. It cannot give you love or show you how to love. Science cannot forgive your sins or give you hope for eternal life. It cannot give you contentment in life. It cannot give you a genuine identity. It can't tell you whether there is a God or what God thinks of you or what God wants of you. It has no comforting words to prepare you for death. It cannot change the laws of nature or control the future. Science must remain silent or speak foolishly in relation to the existential dimension of humanity. Science is not God. Science is *human* through and through; it derives from the power of our reason to figure out the laws of nature and use them for our ends. It gives the impression of being superhuman for the same reason that governments give that impression: it is a communal undertaking transcending the individual in power and longevity, but it does not transcend humanity as such. Science possesses all the strengths and weaknesses of humanity in an exaggerated form. At the risk of sounding unappreciative of science, it must be said that natural science cannot answer a single one of the top five or ten most important questions we ask or achieve anything of lasting significance. At the end of his *Critique of Pure Reason*, Immanuel Kant gave his list of the three most important questions pressing on human existence: "1. What can I know? 2. What ought I to do? 3. What may I hope?"[59] One could extend that list a long way before one gets to "What is the atomic weight of iron?" Or, as Ludwig Wittgenstein (1889–1951) concluded, "We feel that even when all possible scientific questions have been answered, the problems of life remain completely untouched."[60]

We must consider one more extra scientific factor when evaluating science. The scientist is an existing human being. The scientist is not a machine completely absorbed in the objective world of nature. She or he is a subject, a body, an individual, fallible, mortal and needy, anxious, jealous, hopeful or despairing, optimistic or pessimistic. Science exists only in the minds of scientists. Science can't do scientific research. In addition to being founded on metaphysical and epistemological presuppositions and directed

---

[59]Immanuel Kant, *Critique of Pure Reason*, trans. Norman Kemp Smith (New York: Palgrave, 1933), p. 635.
[60]Ludwig Wittgenstein, *Tractatus Logico-Philosophicus*, trans. D. F. Pears and B. F. McGuinness (London: Routledge & Kegan Paul, 1961), p. 149.

toward social and political ends, science is conditioned by the subjectivity scientists bring to their task. A person can be driven to engage in science by curiosity, love of discovery, love of beauty, desire to serve God, desire to benefit humanity, adherence to a philosophy of nature, desire for wealth or fame, hatred and envy, pride or shame, and many other human motivations. Scientists can be virtuous or vicious, honest or dishonest, caring or cruel. This is true not only because scientists sometimes falsify data or take shortcuts or plagiarize but also because these subjective factors affect what presuppositions they favor and to what ends they direct their research. Even at the levels of observation and interpretation subjective factors play a part, for good or ill.

As these observations make clear, even if the inner workings and the material findings of modern natural science are strictly limited to the empirical, these empirical findings do not stand alone or interpret themselves or put themselves to use. Because science is nestled between epistemological and metaphysical presuppositions and cultural ends and is conducted by subjects, there is plenty of room for conflict and dialogue between what scientists claim as the significance of their empirical findings and other interpretations of reality. But now we must consider theology and theologians.

Augustine or Bellarmine?

I have argued that the theology of creation limits itself to the study of God's relation to creation and creation's relation to God and has nothing to say about creation's inner empirical relationships. This view does not meet with universal approval. As I argued above, none of the six theses on the doctrine of creation gleaned from the Bible assert any *empirical* fact about nature and hence are consistent with any genuine empirical science. But is this also true of the Bible itself? Many believers down through the centuries have used the Scriptures as sources of information about nature, and others have cautioned against such use of the Bible. Augustine of Hippo, explaining the line of reasoning that liberated him from Manichaeism, refers to Mani's presumptuous speculations about the physical cosmos: "He had very much to say about the world, but was convicted of ignorance by those who really understand these things. . . . So when he was found out, saying quite mistaken things about the heaven and stars and the movements of the sun and moon, though these matters have nothing to do with religion, it was very clear that his bold speculations were sacrilegious."[61] Augustine applies this lesson to Christians: "When I hear this or that brother Christian, who is ignorant of these matters and thinks one thing is the case when another is correct, with patience I contemplate the man expressing his opinion." As long as the man does not say something "unworthy" of God, Augustine is willing to let it go. "But it becomes an obstacle if he thinks his view of nature belongs to the very form of orthodox doctrine, and dares obstinately to affirm something he does not understand."[62] Clearly Augustine did not want Christianity to be brought into disrepute because of the doubtful interpretations on matters irrelevant to faith. In the many places where Augustine comments on the first chapters of Genesis he urges caution in interpretation, not wanting to put the interpreter into a position that may be contradicted by the empirical facts.[63] In his *Literal Meaning of Genesis*, he struggles to reconcile the Genesis creation stories with his theological conviction that God must have created the whole universe instantaneously. He finds a possible solution to the problem in the divine command that says let the earth and seas "bring forth" living creatures, plants, fish and land animals (Gen 1:11, 20, 24). God created all the living things in one act as "seeds" that would become actual only when the conditions were right. Ernan McMullin speaks of Augustine's approach as "the other Christian tradition" and traces this tradition from Augustine through Aquinas, who supported it, and Bonaventure, who rejected it, and Cajetan, who defended Aquinas.[64]

But not every theologian took Augustine's advice. After the Protestant Reformation Augustine's reading of Genesis and his caution about using the Bible in ways that could be contradicted by empirical facts was rejected by Protestants and Roman Catholics alike. John Calvin dismissed it and argued for six literal days.[65] Francisco Suarez (1548–1617) wrote a huge work on the six days rejecting Augustine's interpretation in great detail in uncompromising terms.[66] But perhaps Robert Bellarmine presents us with the most

---

[61] Augustine, *Confessions* 5.8 (Saint Augustine, *Confessions*, trans. Henry Chadwick, Oxford World's Classics [Oxford: Oxford University Press, 1991], p. 76).

[62] Augustine, *Confessions* 5.9 (Augustine, *Confessions*, pp. 76-77).

[63] Davis A. Young, "The Contemporary Relevance of Augustine's View of Creation," *Perspectives on Science and Christian Faith* 40 (1988): 42-45.

[64] Ernan McMullan, "Darwin and the Other Christian Tradition," *Zygon* 46 (2011): 291-316.

[65] John Calvin, *Commentary on Genesis* 1.5.

[66] See McMullin, "Darwin and the Other Christian Tradition," pp. 304-6, for his discussion of Suarez.

famous case of ignoring the Augustinian rule. Robert Cardinal Bellarmine (1542–1621) gave his life's work to defending Roman Catholicism against the attacks of Protestant theologians. But he is more remembered today for his role in the 1616 condemnation of the Copernican sun-centered cosmology. The two are related, however. The Council of Trent had declared against Protestantism that no one shall interpret the Scriptures "in matters of faith and customs . . . in opposition to that which has been and is held by Holy Mother Church . . . by giving it meanings contrary to the unanimous consent of the fathers."[67] But the council left undecided the exact scope of "matters of faith and customs." In his 1593 work, *Controversies*, Bellarmine argued against Protestants that "it is certainly a matter of faith that no one is saved without the grace of the Holy Spirit and that Peter, Paul, Stephen and many others had the Holy Spirit and are saved. . . . In the Scriptures not only the opinions expressed but each and every word pertains to the faith. For we believe that not one word in Scripture is useless or not used correctly."[68] The unfortunate consequences of Bellarmine's position become apparent in his letter to Foscarini of April 12, 1515. In this letter he condemns the sun-centered view of Copernicus, Galileo and Foscarini:

> Nor can one reply that this is not a matter of faith, because even if it is not a matter of faith because of the subject matter, it is still a matter of faith because of the speaker. Thus anyone who would say that Abraham did not have two sons and Jacob twelve would be just as much of a heretic as someone who would say that Christ was not born of a virgin, for the Holy Spirit has said both of these things through the mouths of the Prophets and Apostles.[69]

The tradition had interpreted 1 Chronicles 16:30, which says, "Tremble before him, all the earth! The world is firmly established; it cannot be moved," as teaching what at that time seemed obviously true anyway: the earth is stationary. And traditional interpreters had assumed that Ecclesiastes 1:5, which says, "The sun rises and the sun sets, and hurries back to where it rises," spoke the simple cosmological truth: the sun moves around the earth. Consistent with his theological principles, Bellarmine argued that earth-

---

[67]Jaroslav Pelikan and Valerie Hotchkiss, eds., *Creeds and Confessions of Faith in the Christian Tradition* (New Haven: Yale University Press, 2003), p. 2:823.

[68]Robert Bellarmine, *De controversiis* 2.2.12, quoted in Richard J. Blackwell, *Galileo, Bellarmine, and the Bible* (Notre Dame, IN: University of Notre Dame Press, 1991), p. 31.

[69]Blackwell, *Galileo, Bellarmine, and the Bible*, p. 105.

centric cosmology was orthodox and sun-centered cosmology was heretical. On February 23, 1516, the eleven theological experts commissioned by the Congregation of the Index gave their verdict on Copernicanism condemning and censuring two propositions:

(1) The sun is the center of the world, and is completely immobile by local motion. Censure: All agree that this proposition is foolish and absurd in philosophy and is formally heretical, because it explicitly contradicts sentences found in many places in Sacred Scripture according to the proper meaning of the words and according to the common interpretation of the Holy Fathers and of learned theologians.

(2) The earth is not the center of the world and is not immobile, but moves as a whole and also with diurnal motion. Censure: All agree that his proposition received the same censure in philosophy; and in respect to theological truth, it is at least erroneous in faith.[70]

From the perspective of history, Cardinal Bellarmine and these eleven learned theologians look exceedingly foolish, not only because they turned out to be incorrect scientifically, but more so because they assumed that whatever the Bible says about nature, incidentally or illustratively, it also *teaches* as a matter of faith. A much better rule is to assume that the Scriptures do not teach anything about nature as such and that what it says it says according to the common opinion of the day in service of its religious message. Had Bellarmine taken Augustine's advice perhaps the whole Galileo affair could have been avoided.[71]

---

[70]Quoted in ibid., p. 284.

[71]For my take on what the Galileo affair can teach us about the relationship between science and theology, see Ron Highfield, "Galileo, Scientific Creationism, and Biblical Hermeneutics," *Restoration Quarterly* 36 (1994): 279-90.

9

CREATION AND TIME

HOW THEOLOGIANS VIEW GOD'S RELATIONSHIP to time often plays a decisive role in discussions of creation and providence. Did God create the world in time and at a particular time? Or did God create the world *with* time? Is time a creature or a divine attribute? How does an eternal God relate to a temporal creation? The question of God's relationship to time cannot be answered apart from the question of God's relationship to creation as such. And we have already characterized that relationship as God's loving act of giving being to that which had no being. Hence we will begin with a predisposition to see time as an aspect of creation with the same ontological status as the rest of creation, that is, absolute dependence on God.

In the section on the God of creation I addressed the topic of God's eternity. There I focused on God's mode of life as "the simultaneous and complete possession of infinite life" in contrast to the temporal mode of creation. In opposition to those who make God temporal the better to understand his relationship to creation, I argued that for the sake of God's divinity we must maintain his transcendence over time. Jesus Christ, not time, is the mediator between God and creation. For this discussion I will presuppose the view of God's eternity for which I have argued elsewhere.[1] The issue that requires our attention here is the relationship of the eternal God to the time of creation. The nature of this relationship does not depend on a view of time developed in modern physics; nor must we allow our theological understanding to be determined by a metaphysical theory of

---

[1]Ron Highfield, *Great Is the Lord: Theology for the Praise of God* (Grand Rapids: Eerdmans, 2008), pp. 292-311.

time. It must be determined by the way God has in fact related to time in the history of revelation.

THE BIBLE AND TIME

The Bible does not explicitly address the issue of God and time very often. Its most prominent theme when speaking of the subject is the qualitative contrast between God's time and the time of creation. It contrasts God's durability to creation's fleetingness to emphasize God's transcendence or to inspire confidence in his ability to judge, protect and save. Note the contrast in Psalm 90:

> Before the mountains were born
>> or you brought forth the whole world,
>> from everlasting to everlasting you are God.
> You turn people back to dust,
>> saying, "Return to dust, you mortals."
> A thousand years in your sight
>> are like a day that has just gone by,
>> or like a watch in the night. (Ps 90:2-4; cf. Ps. 102: 25-27, and 2 Pet 3:8)

But Scripture does not perceive God's superiority to the creature's time as a problem for God's ability to be present with the creatures. To the contrary, the conviction of God's superiority to time assures us that God can be present for creatures who suffer from the relentless pull of entropy.[2] Those who look to God for help can be assured that

> God is our refuge and strength,
>> an ever-present help in trouble.
> Therefore we will not fear, though the earth give way
>> and the mountains fall into the heart of the sea. (Ps 46:1-2;
>>> cf. Deut. 33:27)

With regard to the act of creation, the phrase "In the beginning" (Gen 1:1) is ambiguous. Taken in the context of the whole Bible it may rule out thinking of creation as everlasting, that is, as possessing an infinite temporal past. But it does not answer all the questions we would like answered. It may give us a hint, however, when we compare it to the creation myths of the

---

[2]In Scripture, the contrast between God's time and ours can be reduced to the contrast between the self-sufficient power of God's life and our dependent and dissipating life.

ancient Near East. As I pointed out in the biblical theology section, the Bible contains no theogony, no story of the birth and struggles among the gods. The act of creation in these myths happens in the middle of time, which is shared by gods and creation. The absence of a theogony in the Bible is significant not only for monotheism and God's absolute sovereignty and transcendence but also for God's relationship to time. This absence places God and creation in different relationships to time.

Four texts in the New Testament speak of a "before" the creation of the world. Jesus possessed glory with the Father "before the world began" (Jn 17:5). The Father loved the Son "before the creation of the world" (Jn 17:24). Believers were chosen in Christ "before the creation of the world" (Eph 1:4), and Christ himself was "chosen before the creation of the world, but was revealed in these last times for your sake" (1 Pet 1:20). One could argue that these affirmations imply that there was time *before* the world was created. But in each case an otherwise contingent historical event is grounded in the eternal God, thus giving the event a universal significance. The significance of the "before" is not chronological but ontological; for before or apart from creation there is only God.

The biblical teaching on God's relationship to time can be summarized under four points:

> (1) *God had no beginning and will have no end.* Unlike creatures, there was no time before God existed, and God will never cease to exist. There was, however, some sort of "before" the creature existed. (2) *God is present to every moment of our time.* God was present and available in the past, is now present, [and] will be present to all future times. (3) *God is the Lord over time.* Time does not wear God out, dissipate his energy, and cause him to grow old and die. (4) *God's eternity is closely associated with his being the source of his own life and hence the source of all other existence and life.* Scripture is interested in everlasting life, not everlasting time. Time is not approached as a container in which life becomes possible but as a mode of life.[3]

AUGUSTINE AND AQUINAS ON TIME

The theme of time and eternity plays such an important role in my argument about the nature of God's relationship to creation that we must spend some

---

[3]Highfield, *Great Is the Lord*, p. 294.

time placing it in the traditional discussion. We do not have the space to write a detailed history of this theme, nor is it necessary for my purpose. We will focus on a few of the most influential theologians.

Augustine of Hippo. Augustine returned again and again to the theme of creation during his long career.[4] As we saw in the sections on mediation and creation from nothing, Jewish and Christian thinkers were driven to compare and contrast the biblical creation narrative to the world-formation theories of the Greeks, especially Plato's *Timaeus*. Following his predecessors,[5] Augustine saw in Genesis 1:1 ("In the beginning God created the heavens and the earth") not a summary statement to be detailed in what follows; rather, this statement refers to the creation of a spiritual or intellectual world (heaven) and unformed matter (earth).[6] Heaven is not coeternal with God and is mutable by nature; nevertheless it participates in God's eternity and thereby remains unchanging and timeless.[7] The formless earth too is changeless and therefore timeless until it is formed by God. For "where there is no form, no order, nothing comes or goes into the past, and where this does not happen, there are obviously no days and nothing of the coming and passing of temporal periods."[8] It is important to note here that, though susceptibility to change is an aspect of creation, actual change is not. God can protect the creature from such change.

For Augustine, God is eternal, immortal and changeless. Hence God's will to create the world is also eternal. But this does not mean that the world itself is eternal. Augustine admits that this is difficult to comprehend. In one sense creatures have *always* existed because time itself is an aspect of creation and time has *always* been; that is, there was no time before time. Still,

---

[4]For a compact survey of Augustine's teaching on creation in his many works, see Simo Knuuttila, "Time and Creation in Augustine," in *The Cambridge Companion to Augustine*, ed. Eleonore Stump and Norman Kretzmann (Cambridge: Cambridge University Press, 2011), pp. 103-15.

[5]For example, see Origen, *On First Principles* 2.1; and Basil of Caesarea, *Hexaemeron* 1.5.

[6]Augustine, *Confessions* 12.9. Clearly, this two-stage creation is an attempt to read the Bible in view of Plato's *Timaeus*. In the first stage, God creates the world of ideas or forms (heaven) and the world of unformed matter (earth). Afterward, God shapes the unformed matter into the ordered world. Unlike Plato, Augustine considers both heaven and earth to be creatures. For a discussion of this aspect of Augustine's view of creation, see Etienne Gilson, *The Christian Philosophy of Saint Augustine,* trans. L. E. M. Lynch (New York: Octagon, 1988), pp. 197-209.

[7]Augustine, *Confessions* 12.9 (Saint Augustine, *Confessions*, trans. Henry Chadwick, Oxford World's Classics [Oxford: Oxford University Press, 1991], p. 250).

[8]Augustine, *Confessions* 12.9 (Augustine, *Confessions*, p. 250). For Augustine, where there is no change there is no time.

the world is not eternal, for eternity is *qualitatively* different from time:

> As, then, we say that time was created, though we also say that it always has
> been, since in all time time has been, so it does not follow that if the angels
> have always been, they were therefore not created. For we say that they have
> always been, because they have been in all time; and we say they have been in
> all time, because time itself could no wise be without them. For where there
> is no creature whose changing movements admit of succession, there cannot
> be time at all. And consequently, even if they have always existed, they were
> created; neither, if they have always existed, are they therefore co-eternal with
> the Creator. For He has always existed in unchangeable eternity; while they
> were created, and are said to have been always, because they have been in all
> time, time being impossible without the creature. But time passing away by
> its changefulness, cannot be co-eternal with changeless eternity.[9]

There was no time *before* creation.[10] For time applies only to the mutable
and changing creation. Speaking to God, Augustine says:

> How would innumerable ages pass, which you yourself had not made? You are
> the originator and Creator of time. What times existed which were not brought
> into being by you? Or how could they pass if they never had existence? Since,
> therefore, you are the cause of all times, if any time existed before you made
> heaven and earth, how can one say that you abstained from working? You have
> made time itself. Time could not elapse before you made time.[11]

But we are speaking about time as if we knew what it was. We experience
the world and ourselves in temporal categories. We remember what happened as past and we anticipate what might happen as future. The past no
longer exists, and the future does not yet exist; for to say that something
exists is to say that it exists *now*. As for the present, it is a fleeting transition
point with no duration. Augustine muses that "perhaps it would be exact to
say: there are three times, a present of things past, a present of things present,
a present of things to come. In the soul there are these three aspects of time,
and I do not see them anywhere else."[12] Were it not for the soul's ability to

---

[9]Augustine, *City of God* 12.15 (*NPNF*[1] 2:236).

[10]Gilson sets Augustine's discussion against the background of Manichaean cosmology, which
    asserted the world's eternity and Stoic cosmology, which solved the paradox in the idea of a
    changing but eternal world by proposing a cyclic view of history in which the world cycles
    eternally between destruction and return (*Christian Philosophy of Saint Augustine*, pp. 190-92).

[11]Augustine, *Confessions* 11.15 (Augustine, *Confessions*, p. 229); See also *City of God* 11.12.

[12]Augustine, *Confessions* 11.26 (Augustine, *Confessions*, p. 235). See Paul Ricoeur, *Time and Nar-*

remember and anticipate we would have no knowledge of time. The sense of duration and of the difference between a long and short time is "distension," that is, a stretching out, of the soul.[13] The memory recalls different states of things at once and measures the "distance" of this series within the soul. Augustine concludes, "That present consciousness is what I am measuring, not the stream of past events which have caused it."[14] Though without change there would be no time, Augustine does not, like Aristotle, identify time with physical change.[15] Time may simply be the soul's way of experiencing a changing world. God, Augustine hurries to remind us, does not know the created world through "distension" between past and future.[16] All times are now in his eternal present.

Thomas Aquinas (1225–1274). Aquinas deals with the question of the relationship between eternity and time in many places, usually under the heading of whether the world has always existed. In *Summa Contra Gentiles*, Aquinas begins his discussion by defending the thesis, "That it is not necessary for creatures to have always existed."[17] In three chapters he records seventeen arguments in favor of the eternity of the world. It seems to me that the force of all these arguments (or at least the best of them) derives from the problem of how the unchanging, eternal God can cause the existence of the world by an unchanging and eternal act, which does not differ from God's own being, without the world itself being eternal. Accordingly, in each of his answers to the seventeen arguments he addresses this problem. In responding to the contention that an eternal cause must create an eternal

---

rative (Chicago: University of Chicago Press, 1990), 1:5-30, for Ricoeur's study of the logic of Augustine's argument about time. Augustine moves first to solve the riddle (or *aporia*) of how time can be thought to exist. He solves this problem with his threefold presence of things past, present and future. Next, according to Ricoeur, Augustine moves to deal with measurement of time, to which the "distension," or stretching out, of the soul is the answer.

[13] Augustine, *Confessions* 11.26 (Augustine, *Confessions*, p. 240).

[14] Augustine, *Confessions* 11.27 (Augustine, *Confessions*, p. 242).

[15] In focusing his reflections about time on the human mind and body, Augustine modifies the central teachings of Plato, Aristotle and Plotinus on time. Indeed, according to Andrea Nightingale, "in introducing this theory of mental distention, Augustine addresses (for the first time in the West) the phenomenon of internal time-consciousness" (*Once Out of Nature: Augustine on Time and the Body* [Chicago: University of Chicago Press, 2011], p. 79). For her study of Augustine's view of time, see chap. 2, "Scattered in Time," pp. 55-104.

[16] Augustine, *Confessions* 11.30 (Augustine, *Confessions*, p. 245).

[17] SCG 2.31-38. This is the thesis statement for chap. 31 (Saint Thomas Aquinas, *Summa Contra Gentiles, Book Two: Creation*, trans. James F. Anderson [Notre Dame, IN: University of Notre Dame Press, 1975], p. 91).

effect, he answers: "Nothing, therefore, prevents our saying that God's action existed from all eternity, whereas its effect was not present from eternity, but existed at the time when, from all eternity, He ordained it."[18] Or again, "Hence, for the will to be a sufficient cause it is not necessary that the effect should exist when the will exists, but at that time when the will has ordained its existence."[19] When we consider the beginning of the *whole* creation, the question takes on a different flavor from when thinking of the beginning of a part within it. The beginning and end of a particular creature can be measured in relation to what came before it or what comes after it. "God, however, brought into being the creature and time together. In this case, therefore, the reason why He produced them now and not before does not have to be considered, but only why He did not produce them always."[20]

Does this mean that time has not *always* existed or that there was a time *before* time? Of course, this way of speaking sounds contradictory, for how could there be a *before* before time or an *always* without time? Aquinas addresses this oddity of language with the idea of imaginary time. He says: "For the *before* that we speak of as preceding time implies nothing temporal in reality, but only in our imagination. Indeed, when we say that time exists *after* not existing, we mean that there was no time at all prior to this designated *now*."[21]Addressing the same question in another place, he speaks of God's duration and defines imaginary time: "God does precede the world in duration, but in eternity, not in time, since God's existence is not measured by time. There was no real time before the world existed, but only imaginary time, in the sense that we, existing now, can imagine that before time began, while eternity existed, unlimited periods of time could have rolled by."[22]

As I indicated above, Aquinas argues against the thesis that creation must have always existed. Apparently, he does this because, although he

---

[18]*SCG* 2.35.3; *Summa Contra Gentiles*, p. 103.

[19]*SCG*, 2.35.4; *Summa Contra Gentiles*, p. 103. See Michael J. Dodds, *Unchanging God of Love: Thomas Aquinas and Contemporary Theology on Divine Immutability,* 2nd ed. (Baltimore: Catholic University Press of America, 2008), pp. 180-81, for his response to William Lane Craig's contention made in *Time and Eternity* (Wheaton, IL: Crossway, 2001) that to create a temporal world the eternal God become temporal.

[20]*SCG* 2.35.6; *Summa Contra Gentiles*, p. 105.

[21]*SCG* 2.36.7; *Summa Contra Gentiles*, p. 108.

[22]Thomas Aquinas, *Quaestiones Disputatate de Potentia* 3.17 (Thomas Aquinas, *Selected Philosophical Writings,* ed. and trans. Timothy McDermott, Oxford World's Classics [New York: Oxford University Press, 1993], p. 270).

believes by faith that creation has not always existed, he does not believe this can be demonstrated by reason. In *Summa Contra Gentiles* 38, Aquinas details and rejects several arguments in which thinkers have attempted to prove that the world is not eternal. Although Aquinas asserts that the Catholic faith teaches that the world has not always existed, he does not believe that there is a cogent objection by way of self-contradiction or lack of divine power that shows why God could not have created the world so that it always existed.[23] The existence of creation depends on God's will,[24] and unless we think God wills the existence of the world necessarily— which Aquinas rejects—we must admit that this issue "cannot be investigated by reason."[25] But even if one supposes, contrary to the faith, that creation always was, "it would not be equal to God in eternity, as Boethius says; for the divine Being is all being simultaneously without succession, but with the world it is otherwise."[26] So even if the world had always existed it would still exist successively and changeably, characteristics opposed to divine eternity.[27]

---

[23]For his arguments for this possibility, see Thomas Aquinas, *On the Eternity of the World*, trans. Robert T. Miller (1997), in *Internet Medieval Sourcebook*, http://www.fordham.edu/halsall/basis/ aquinas-eternity.html. Aquinas's contemporary Bonaventure argued that the idea of an eternally created world is a contradiction in terms. See P. Van Veldhuijsen, "The Question on the Possibility of an Eternally Created World: Bonaventura and Thomas Aquinas," in *The Eternity of the World in the Thought of Thomas Aquinas and His Contemporaries*, ed. J. B. M. Wissink (Leiden: Brill, 1990), pp. 20-38.

[24]For Aquinas, though it is true that creation depends on the divine will, this assertion could be misleading. Aquinas holds that the divine act of creation (*ad extra*) depends on his eternal acts of begetting the Son and causing the Spirit to proceed, each in its own way. See Gilles Emery, OP, "Trinity and Creation," in *The Theology of Saint Thomas*, ed. Rik van Nieuwenhove and Joseph Wawrykow (Notre Dame, IN: University of Notre Dame Press, 2010), pp. 58-76. Emery quotes Aquinas's early work *Scriptum super libros Sententarum* as saying, "The eternal processions of the persons are the cause and the reason [*causa et ratio*] of the production of creatures" and "the temporal going-forth of creatures is derived [*derivatur*] from the eternal going-forth of the persons," and "the going-out of [*exitus*] the persons in the unity of essence is the cause of the going-out of creatures in the diversity of essence." Emery provides extensive evidence to refute those who accuse Aquinas of neglecting the Word's and the Spirit's distinctive roles in mediating creation. Emery argues that the well-known rule of the "common efficacy [in relation to creation] of the three divine persons by virtue of their unique essence" was supplemented in Aquinas by a second rule, "the causality of the Trinitarian processions" (p. 69).

[25]*SumTh* 1.46 2 (Pegis, *Basic Writings*), 1:453. He warns against attempting to prove a matter of faith: "And it is useful to consider this, lest anyone, presuming to demonstrate what is of faith, should bring forward arguments that are not cogent; for this would give unbelievers the occasion to ridicule, thinking that on such grounds we believe the things that are of faith" (p. 453).

[26]*SumTh* 1.46.2 (Pegis, *Basic Writings*), 1:454.

[27]Norman Kretzmann points out that in medieval terminology the word *eternity* is used both for Boethian eternity, which is God's alone, and for the beginninglessness of creation. Even though

Clearly, for Aquinas as for Augustine, God's eternity is qualitatively different from time, though it has a kind of duration in imaginary time. Time and creation belong together, for where there is no change or succession there is no time. Time is God's creature. It seems that Augustine and Aquinas use the word *always* differently but without substantial disagreement. Augustine uses the word *always* to mean throughout all time, which allows him to say that creation has *always* existed. Aquinas uses the word *always* to mean real time plus imaginary time, that is, the time "before" creation, which we imagine as contemporary with God's eternity. This meaning of *always* forces Aquinas to say that creation has *not* always existed. According to Aquinas, the Catholic faith's assertion that creation has not always existed denies that creation is eternal in two different senses: (1) creation does not exist in simultaneous and unending life like God, and (2) creation does not exist beginninglessly throughout what Aquinas calls imaginary time. It is important to note, however, that this faith does not assert that there was a *real* time when creation did not exist; this would disengage time from creation and make it either a divine property or another eternal principle alongside God.

It seems to me that Aquinas's distinction between true eternity, imaginary time and real time demands careful thought. Norman Kretzmann takes Aquinas's critique of the common expression that projects a time before time merely to be saying that we should not take "that convenient way of speaking as the basis of an inference about the nature of reality."[28] It certainly is at least that. But the concept of imaginary time seems to play a positive role in Aquinas's thinking as well. It allows us to imagine eternity as duration that could be accompanied—but is not—by a beginningless, but changing and successive, creation. Correspondingly, it allows us to imagine creation existing for a *finite* temporal duration in relation to eternity imagined as beginninglessly durative. That is to say, it gives us a quasi-quantitative way to relate time and eternity instead of the mere qualitative opposition of the Boethian definition. It is true that Aquinas does not want us to take imaginary time realistically, but he does not say that we should cleanse our lan-

---

theologians knew the difference they nevertheless used the term ambiguously (*The Metaphysics of Creation: Aquinas's Natural Theology in Summa Contra Gentiles II* [New York: Oxford University Press, 1998], p. 143).

[28]Ibid., p. 171.

guage of such expressions. There must be some analogy between begin-ningless succession and eternity, perhaps a kind of unchanging duration. From the perspective of Christian theology, the philosophical necessity of imagining a time before creation in which God's eternity is successionless duration and time is empty and beginningless naturally raises the question of the Trinity. A theology rooted in the economy of salvation, instead of speaking of an imaginary time before creation, speaks of the eternal simul-taneousness of the cause and effect, the "before" and "after" of the begetting of the Son and the proceeding of the Holy Spirit from the Father.

CHRIST AND TIME

Theological reflection on God's relationship to the time of creation must take into account Scripture's twofold theme: (1) God's qualitative superiority to the creature with respect to time and (2) God's absolute presence to the creature in its time. Scripture senses no tension between these two, and the most dramatic proof of this is the incarnation. Hence theology should not define time and eternity as opposites, that is, as time and timelessness. The nature of God's eternity cannot be discerned merely by negating time; other-wise Scripture's untroubled acknowledgment of the eternal God's presence in time would be made into a problem.[29] Let's begin by reflecting on the first theme, keeping in mind the second.

No one doubts that Scripture teaches that God has always existed and will always exist. There was no time when God did not exist. Some argue that all we can conclude from this is that God is everlasting, emphasizing only the quantitative difference between God's time and the creature's time. God's timeline is infinite in both directions, while ours is finite, with a beginning and an end. In my view this minimum conclusion fails to take into account the qualitative distinctions made in Scripture between God's mode of life and creation's mode of existence. Scripture is very clear that time has no power over God. God does not begin or mature or become wiser or grow old; God does not get weak or die. Time cannot change God. Hence God's relationship to time is utterly different from that of creatures. Following

---

[29]Kathryn Tanner warns against thinking we can define God by contrast with creation. She says, "God transcends the application of all ordinary contrastive terms" ("*Creatio ex Nihilo* as Mixed Metaphor," *Modern Theology* 29 [2013]: 139). Tanner develops this position at length in her book *God and Creation in Christian Theology* (Minneapolis: Fortress Press, 2005).

Scripture's qualitative logic, it seems reasonable to conclude that time can impose no limits on God, the kind of limits creatures suffer, that is, becoming, forgetting, worrying and dying. Time however is not wholly negative for creatures. It is a blessing because, for creatures, time and life are inseparable: negating time would mean negating life. Eternity is not time negated but time perfected. And time perfected is time gathered together and made enduring. And that is what Boethius's definition articulates: "the simultaneous and complete possession of infinite life." God's eternal life is not timeless but simultaneous and full. Since God's mode of life is not the opposite of time but the perfection of time, there is no need to consider God's relationship to our imperfect time as a problem. The eternal God has time for us.

Though the entire economy of salvation witnesses to the fact that God is not timeless (as the pure negation of time), it is the revelation of the immanent Trinity through the economic Trinity that provides the clearest and most profound proof of this. The eternal God relates to and acts in creation and time from the Father, through the Son, in the Holy Spirit. The Son and the Spirit were with the Father eternally. Yet the Son is spoken of as "begotten" and the Spirit as "proceeding." Both of these relations are relations of origin, as the church fathers recognized. Although in our experience what is begotten and what proceeds come "after" their origin, in God's life they are simultaneous and eternal in duration. Hence the Father, Son and Spirit are with each other in an unchanging present in which past, present and future are simultaneous. God is God's own time; for the presence, life and simultaneousness they share *is* the divine nature. Likewise, the distinctions of relation between Father, Son and Spirit constitute the divine space in which God lives. God's time and space (simultaneous and coinherent) are the grounds of the possibility of the time and space of creatures.[30] If God were a timeless and spaceless monad rather than the Trinity, our space and time could not exist.

Seen in this light, the doctrines of the incarnation, the resurrection of Christ, the indwelling Spirit, the church as the body of Christ and the resurrection of the dead are astounding in their implications. In the incarnation the eternal Son of God became a temporal creature without ceasing to be

---

[30]Karl Barth makes this point: "True eternity includes this possibility, the potentiality of time" (*CD* II/1, p. 617).

the eternal Son of God. In the words of Karl Barth, "The fact that the Word became flesh undoubtedly means that, without ceasing to be eternity, in its very power as eternity, eternity became time. Yes, it became time."[31] Just as the Son of God becomes present in our world as a human being in real flesh and blood without ceasing to be God, he is present in our time in a temporal way without being enclosed in time. The divine nature, being triune and hence containing relations and otherness, contains within itself the possibility (without the necessity) of other natures and relations, of time and space. Creation is God's making room and time within his space and time for us. God gives life in dependence on his Spirit. In the incarnation, God unites one creature (hence all creation) to himself in the most intimate way possible, so that the time, space and life of Jesus *is* the time, space and life of God without ceasing to be the time, space and life of creation. In his resurrection Jesus' time, space and life become irrevocably united, perfected and filled with the divine eternal life. Jesus Christ is the firstfruits and the beginning of the redemption of all creation. Through the Spirit the resurrected body of Christ, manifest in the world first in the church, grows to indwell and envelop the whole creation, so that ultimately God will become "all in all" (1 Cor 15:28). The whole creation will be filled with God so that there will be no space and time that is not absolutely present and aware and animated by God. Creation looks forward to this end and finds its fulfillment in this telos. And in the broadest sense, the act of creation is complete only at the end.

The reconciliation accomplished in Christ and redemption to be achieved by him includes forgiveness of past sin and liberation from the powers of sin and death, sanctification by the Spirit, union with Christ and adoption by the Father. It also involves the salvation of time. Psychologically and spiritually, the past for us is a time of lost happiness and regret. We are unable to remember the good and powerless to forget the bad. Even if it is true as medieval theologians thought that God cannot change the past, it is certain that God can change the *meaning* of the past. What is divine forgiveness but a kind of healing of the lost and corrupt past by drawing it into the future who is Christ? In forgiveness, Christ becomes our past. For us the future is pure possibility, the possibility of good *and* evil; hence we alternate between

---

[31]Ibid., p. 616.

anticipation and anxiety. However, since in Christ the eternal has been united to time, Christ is our future as well as our past. The ambiguity of pure possibility (i.e., the future) has been resolved into clarity: "If God is for us, who can be against us?" (Rom 8:31). In Christ, past and future have been united into a present reality. From a human point of view the past is the annihilation of everything that once was and the future is the realm of pure possibility, two kinds of nothingness. The present is the fleeting existence of all that is, a fragile skein of being separating the nothingness behind from that ahead. But Jesus Christ is the Creator, reconciler and redeemer, the beginning and the end. He is the past and future held together by God's eternity. In him time is gathered together so that nothing good is lost and nothing evil is remembered.

It is a great mistake to separate (much less isolate) the time of creation from the time of reconciliation and redemption. Jesus Christ is one, his time is one, and his acts are one; hence the times of creation, reconciliation and redemption are one in relation to him. Only in Jesus Christ does creation achieve its end, and since Christ is the Creator, the end was anticipated by the beginning. The meaning of the beginning cannot be known apart from the end. And the end of time is the redemption of time, which can only mean the gathering together of past and future into one simultaneous presence within God's eternity. The "end" is not the end of a causal chain wherein only the last generation or iteration is made eternal and all preceding generations are left to the oblivion of the past. The end is also the whole; that is, the end is retroactive: we will never have not been with the Lord. The search for a beginning of time or a time before creation, in light of the revelation of the end in Christ, is misguided, as misguided as attempting to find the beginning of a circle. For Jesus Christ, who is the end, is also the beginning.

Part Two

✸

DIVINE PROVIDENCE

10

BIBLICAL THEOLOGY
OF DIVINE PROVIDENCE

B ELIEF THAT GOD THE CREATOR of heaven and earth is present and active in every dimension of creation characterizes the Old Testament from beginning to end and within every literary genre. With incomparable wisdom God works in nature, in human institutions and movements, and within human hearts to achieve his plans. What follows can be only a sampling of the texts that express confidence in divine care and foresight in nature and history. At the end of the survey I will draw some general principles that can guide our theological reflections on providence. [1]

PROVIDENCE IN THE OLD TESTAMENT

The Bible begins with the great act of creation (Gen 1:1), and though it speaks of God "resting" on the seventh day, it continues to recount God's interaction with his world. The disobedience of the man and woman sets in motion the history of salvation with its dialectic of disobedience, judgment and grace (Gen 3). Such moral evil follows in the wake of the fall that God intervenes in the dramatic judgment of the flood (Gen 6–9). After the story of the Tower of Babel, where the Lord disperses the nations to the four corners of the earth, the story narrows to Abram and his family (Gen 11). God's call, command, promise, covenant and testing of Abram illustrate the

---

[1]For a recently published study of divine providence in the Bible (527 pages), see the two-volume work by John H. Wright, *Divine Providence in the Bible: Meeting the Living and True God*, 2 vols. (Mahwah, NJ: Paulist Press, 2009).

biblical view of providence. As the Lord commands Abram to leave and go wherever the Lord leads, God promises,

> I will make you into a great nation,
>   and I will bless you;
> I will make your name great. (Gen 12:2)

Many years later the Lord renews the promise in a covenant (Gen 15), and eventually the promised child Isaac is born (Gen 21:1-7). But then occurs one of the most famous incidents in the Bible, the "sacrifice" of Isaac. The Lord tells Abraham to take Isaac and go to Mount Moriah, where the father of the faithful must sacrifice his son as a burnt offering. As Abraham is about to plunge the knife into his son's heart an angel of the Lord stays his hand. As Abraham looks up he sees a ram caught in the bushes. The ram serves as Isaac's substitute. As a result, Abraham calls the place "the Lord Will Provide."

God's knowledge and care over creation continues to be highlighted throughout the patriarchal and exodus narratives. The sin of Joseph's brothers expedites Joseph's passage to Egypt, and the seven-year famine drives Jacob's family to Egypt to secure food. All these events work together to help the family survive and prepare for the greatest event in Israelite history, the exodus. In speaking to his brothers Joseph concludes,

> And now, do not be distressed and do not be angry with yourselves for selling me here, because it was to save lives that God sent me ahead of you. . . .
>   So, then, it was not you who sent me here, but God. (Gen 45:5-8)

The exodus narrative begins with the origin of the Egyptian enslavement of the Israelites and Pharaoh's decree that all newborn Hebrew boys must be killed. The future prophet and liberator Moses was saved as a baby from this fate and adopted by Pharaoh's daughter (Ex 2:1-10). The books of Exodus, Leviticus and Numbers are given to the stories of God's deliverance of the Hebrews from Egypt, the covenant, the giving of the law and the wilderness wanderings. It is a story of divine power and human weakness, of God's faithfulness and the people's unfaithfulness.

John H. Wright concludes that the historical books from Joshua to 2 Chronicles "see all the great events of Hebrew history as due to the active presence of God, blessing, rewarding, punishing, or testing his people."[2]

---

[2]Wright, *Divine Providence in the Bible*, 1:63.

The author of Joshua attributes the miracle at the battle for Jericho and the entire conquest of Canaan to the Lord (Josh 1–12). In the period of the judges Israel's enemies threaten and the Lord delivers the people through chosen warrior judges. The history of this era was seen as a series of cycles of three: on the human side, a time of faithfulness is followed by a season of neglect and unfaithfulness, followed again by a time of repentance and deliverance; on the divine side were times of divine blessing, judgment and restoration. This understanding of the movement of history continues throughout the historical literature until the last king of Judah is taken into Babylonian captivity. The Assyrian destruction of the northern kingdom of Israel and the Babylonian destruction of the southern kingdom of Judah is seen as punishment for disobedience.

The prophetic literature reflects the fundamental view of divine providence we have seen in other parts of the Bible. God is the creator and ruler of heaven and earth. God is wise and just, present and active. Hosea and Amos speak in the name of the Lord, warning Israel of the coming judgment and urging her to repent and seek the Lord (Amos 5:4-6). The book of Isaiah enumerates Israel and Judah's sins and calls on them to repent. More than earlier prophets, Isaiah speaks to God's people within the horizon of universal history. God is the Lord of all nations and all times:

> This is the plan determined for the whole world;
> > this is the hand stretched out over all nations;
> For the LORD Almighty has purposed, and who can thwart him?
> > His hand is stretched out, and who can turn it back? (Is 14:26-27)

The prophet assures the exiles that the gods of Babylon have no power to prevent the Lord from saving them:

> I make known the end from the beginning,
> > from ancient times, what is still to come.
> I say, "My purpose will stand,
> > and I will do all that I please. . . .
> What I have said, that I will bring about;
> > what I have planned, that I will do. (Is 46:10-11)

Jeremiah prophesied in Judah just before and during the exile to Babylon. On the occasion of his calling, the Lord said to him,

> Before I formed you in the womb I knew you,
>> before you were born I set you apart;
> I appointed you as a prophet to the nations. (Jer 1:5)

Jeremiah looks forward to a time when the Lord will make a new covenant with his people, written on their hearts and minds (Jer 31:31-34).

The book of Proverbs contains many sayings that touch on providence.

> Many are the plans in a person's heart,
>> but it is the LORD's purpose that prevails. (Prov 19:21)

> In the LORD's hand the king's heart is a stream of water
>> that he channels toward all who please him. (Prov 21:1)

> The lot is cast into the lap,
>> but its every decision is from the LORD. (Prov 16:33)

And though the books of Job and Ecclesiastes voice skepticism about the easygoing view of Proverbs that God always rewards goodness with good things, they view everything that happens as within God's control. Job complains that he is not being treated fairly, but he never doubts that

> if it is a matter of strength, he is mighty!
>> And if it is a matter of justice, who can challenge him? (Job 9:19)

After the Lord confronts him from the storm the humbled Job exclaims,

> I know that you can do all things;
>> no purpose of yours can be thwarted. (Job 42:2)

The Preacher of Ecclesiastes finds God's providence obscure but does not deny it:

> So I reflected on all this and concluded that the righteous and the wise and what they do are in God's hands, but no one knows whether love or hate awaits them. All share a common destiny—the righteous and the wicked, the good and the bad, the clean and the unclean, those who offer sacrifices and those who do not. (Eccles 9:1-2)

The Psalms, like every other Old Testament body of literature, overflow with confidence that God is universally present and active in nature, history and human hearts. Many psalms celebrate God's creation of heaven and earth and his universal reign over all the earth (Ps 8; 24; 136; 148). For the Psalms, creation is not merely a past event but an ongoing relationship pregnant with meaning for everything that happens. The Creator owns the earth and everything in it (Ps 95:3-5). The glory of God is manifest in creation (Ps 19:1). God's power is unlimited:

> I know that the LORD is great,
>> that our Lord is greater than all gods.
> The LORD does whatever pleases him,
>> in the heavens and on earth,
>> in the seas and all their depths. (Ps 135:5-6)

God rules over all peoples on earth (Ps 22:27-28; 47:3; 82:8).

> The LORD foils the plans of the nations;
>> he thwarts the purposes of the peoples.
> But the plans of the LORD stand firm forever,
>> the purposes of his heart through all generations. (Ps 33:10-11)

The Psalms are filled with prayer and praise referring to divine blessing, guidance, protection and rescue.[3]

Wright observes a threefold pattern to the biblical narrative of providence that holds for the entire Bible: "divine initiative, human response, and divine response to this human response."[4] In his love and grace, God acts in the creation, preservation and reconciliation of creatures. Human beings respond in gratitude and obedience or neglect and disobedience. God responds with blessing or judgment so that in the long run sin and suffering and their destructive effects are incorporated into "God's loving and wise intentions. Even the sinner is made to serve the divine purpose."[5]

PROVIDENCE IN THE NEW TESTAMENT

The New Testament opens with the story of the life and teaching of Jesus. Divine providence is a prominent theme from the beginning. The birth narratives contain visitations from angels, the miracle of the virginal conception, heavenly signs, the divine warning given in a dream and the escape to Egypt (Mt 1–2). Jesus heals the sick and lame; he raises the dead and drives out demons, all of which are signs of the power and presence of God (Lk 11:14-26). Jesus' birth, suffering and death fulfill Old Testament prophecy (e.g., Mt

---

[3]Ibid., 1:180-81.
[4]Ibid., 1:195.
[5]Ibid., 1:196.

1:23; Mk 1:3; Lk 24:44-49). God is at work in all Jesus does and says and in everything that happens to him. In many places Jesus teaches explicitly about divine providence and urges confidence in God's care. There is no need for anxiety about food and clothes because God knows we need these things and will provide them (Mt 6:25-34). God knows when a sparrow falls to the ground, and God knows the number of hairs on our heads, "so don't be afraid; you are worth more than many sparrows" (Mt 10:30-31). In the controversy over the healing of the blind man (Jn 9), Jesus and his disciples agree that the man was born blind for a reason, a divine purpose. The disciples assume the blindness was punishment for sin, either the man's or his parents.' But Jesus declares that "this happened so that the works of God might be displayed in him" (Jn 9:3).

In Acts, the story of the empowerment and spread of the church from Jerusalem to Rome demonstrates divine guidance, direct and indirect, at every step. The Spirit falls on the apostles, and Peter preaches the Pentecost sermon declaring to his audience that Jesus "was handed over to you by God's deliberate plan and foreknowledge [boulē kai prognōsei]; and you, with the help of wicked men, put him to death by nailing him to the cross. But God raised him from the dead" (Acts 2:23-24). Even the persecution that broke out after the martyrdom of Stephen served to spread the gospel (Acts 8:4): Philip evangelized Samaria and converted the Ethiopian official (Acts 8:9-40). The providential significance of the dramatic conversion of Saul of Tarsus is highlighted by its being recounted three different times in Acts (Acts 9; 22; 26; cf. 1 Cor 15:8; Gal 1:15-16). As Saul was praying and fasting after his encounter with Jesus on the road to Damascus, the Lord spoke to Ananias saying: "Go! This man is my chosen instrument to carry my name to the Gentiles and their kings and to the people of Israel. I will show him how much he must suffer for my name" (Acts 9:15-16). The Gentile mission begins with visions to Cornelius and Peter. By sending the Holy Spirit to Cornelius and his household, God demonstrates his approval of Gentile conversion (Acts 10–11). The book ends with Paul in Rome freely preaching the gospel.

Paul's letters contain some of the clearest and strongest affirmations of divine providence. In Romans, after having argued for justification by grace through faith in Jesus Christ, for freedom from the obligation to keep the law by our own strength and for the presence and power of the Holy Spirit in the lives of believers, Paul reminds the Roman Christians that they will nevertheless have to suffer with Christ if they wish to become coheirs with him (Rom 8:17). But even in this suffering God is at work:

> And we know that in all things God works for the good of those who love him, who have been called according to his purpose. For those God foreknew he also predestined to be conformed to the image of his Son, that he might be the firstborn among many brothers and sisters. And those he predestined, he also called; those he called, he also justified; those he justified, he also glorified. (Rom 8:28-30)

In Romans 9–11 Paul deals with the troubling issue of Israel's rejection of Jesus the Messiah. Have God's promises failed? Has God's providence reached its limit? Has his mercy run out? Paul discovers hints of a divine plan even in Israel's unbelief. The failure of Israel provided an occasion for the Gentiles to be admitted to the people of God (Rom 11:11-25). Israel's disobedience brought mercy to the Gentiles, but such disobedience now makes Israel clearly dependent on divine mercy: "For God has bound every-one over to disobedience so that he may have mercy on them all" (Rom 11:31-32). This mysterious and roundabout outcome prompts Paul to break forth in praise of the wisdom and judgment of God: "For from him and through him and for him are all things" (Rom 11:36).

In urging his Corinthian converts not to follow the example of the faithless Israelites in the desert, Paul asserts that "God is faithful; he will not let you be tempted beyond what you can bear. But when you are tempted, he will also provide a way out so that you can endure it" (1 Cor 10:11-13). Paul experienced such danger in the province of Asia that he felt death was im-minent. He interprets this experience as a lesson teaching him to rely not on his own power and wisdom but "on God, who raises the dead" (2 Cor 1:9). In Ephesians 1, Paul praises God for sweeping Christians up into his eternal plan "to bring unity to all things in heaven and on earth under Christ" (Eph 1:10). Our present existence in Christ is rooted in his eternal predesti-nation "according to the plan of him who works out everything in con-formity with the purpose of his will" (Eph 1:11). In a compact statement that brims with the paradox and tension of the biblical doctrines of providence and grace, Paul urges his friends "work out your salvation with fear and trembling, for it is God who works in you to will and to act in order to fulfill his good purpose" (Phil 2:12-13).

Biblical Theology of Providence

It is an often repeated mistake to assume that every question about divine providence that interests us can be solved by one or more biblical quotations or even by thorough critical analysis. And yet Christian theology cannot simply address itself to the nature of God, creation or divine providence independently of the Scriptures. It seems to me best, then, to state a few principles of the biblical teaching on divine providence in a modest way and either be satisfied to leave more speculative questions unanswered or attempt to shed some light on them while remaining within the biblical guidelines. I will follow the latter, bolder but still cautious course. I think we can safely conclude the following principles of the biblical material on divine providence.

(1) *God the creator of all things continues to sustain his creation in being and to act wisely and powerfully in nature and history to realize his plan.* As we saw in our study of creation, God's act of giving being to the creation makes little sense apart from enabling creation to endure in time. But God's relationship in sustaining creation is not impersonal causality. There are no automatic processes in God's life or in God's action on and within creation; for God is free in all he is and does. When we say that God is wise we are saying that God's action toward the world is not merely an arbitrary assertion of will but rational and purposeful. All God's actions are harmonious with one another and are directed to accomplishing God's ultimate purpose for creation. A sound appropriation of biblical theology will not support the idea that divine action might be wasted, misguided or self-contradictory.

(2) *God acts in creation in direct and obvious ways in what are called miracles and in indirect and subtle ways, such as in natural processes and the free actions of human beings.* The Bible highlights God's miraculous actions, though it does not define a miracle. Miracles are recognized by their unusual character and the human reactions they evoke. Most miracles in the Bible are also signs that carry a message about God's nature or about what God requires of human beings. Miracles reveal in dramatic form the nature of God's relationship to creation. They reveal that creation depends on God everywhere and always even when that dependence is hidden in the ordinary flow of nature and history. God is powerful, present and active, and works out his plan in all things and at every time, even when we cannot see it. This is true of the seemingly automatic and repetitive processes in nature and of the thoughts and actions of human beings. Just as God ordinarily

works in natural processes without disrupting them, God works in human hearts, minds and wills in harmony with their ordinary functioning. The Bible offers no explanation of how God does either of these things. As we shall see, our desire to understand how God can work through the free actions of human beings drives a significant contemporary discussion within the theology of divine providence.

(3) *God acts even in or through evil acts and intentions so that God's plan is realized, without God doing evil.* The Bible contains many examples of God using the evil intentions and actions of people to bring about his good purposes: Joseph's brothers, Pharaoh, Assyria, Babylon, Persia and Judas Iscariot. But God never does wrong. God can work through or use these evils without doing evil. Yet those who intend and do evil—even when God uses it to bring about his good will—are nevertheless held responsible. As in God's use of ordinary or virtuous human actions, God's working in and through evil intentions and actions does not disrupt the ordinary functioning of intellect and will. Again, how this is possible has become a point of controversy among philosophers and theologians. *That* such divine working is possible is a basic fact of the biblical narrative and should not be overturned by speculation about *how* it is possible.

(4) *God's providence encompasses "all things" (Rom 8:28), so that everything that is and everything that happens is made to serve the divine purpose.* The Bible narrates many events that appear to go wrong, to depart from the divine plan: Adam falls, Israel becomes unfaithful, one of the Twelve betrays Jesus; and if we are speaking of the immediate context of the errant event we would have to admit that they do. However, on the one hand I don't think one can successfully defend the idea that the Bible really teaches that things could go so wrong that God cannot make them serve his eternal plan. On the other hand, I don't think one can infer from confidence in God's ultimate triumph that everything that happens is good as it stands in isolation from "all things." In the biblical narrative we find give and take, initiative and response both in human-to-human relationships and in God's relationship to human actions. Despite the undeniable presence of divine reaction to human sin and evil in the narrative of salvation history, we must be cautious in drawing lessons about God's eternal being from this divine relativity in the economy. We shall consider this need for caution and the arguments of those who ignore it at length in the pages ahead.

(5) *For those who love God the preceding affirmations about God's providence impart comfort, hope, courage and joy.* In the Bible, divine providence—at least for those who love God and believe in his love and benevolence—is a comforting and joyous theme. It addresses our inability, despite the best human wisdom available, to make things turn out right. Many forces are at work in determining the future, some of them mysterious and some of them hostile, but their sheer number places the future beyond our control. The human condition is such that we can imagine many possible futures; and the human tendency—since we are finite and mortal—is to imagine negative futures more vividly than positive ones. Without a basis for hope that the future will ultimately bring good to us, we are deprived of joy and meaning in the present. Confidence in God's power and will to bring about our salvation and to make all things work together for our good liberates us from the oppressive threat of negative futures and provides energy to continue working for good ends.

(6) *God realizes his aims perfectly. God cannot fail, even in part.* Christian theologians debate the nature and extent of God's plan. Some argue that God plans the history of creation down to the decay of the least subatomic particle, and others contend that God makes only a general plan. But few thinkers would argue that God's final aims for creation could fail. Given the confidence Scripture displays, I think we can safely assert as a principle of biblical theology that whatever God truly intends to accomplish will be accomplished. God knows what he can do.

INTERPRETATION OF THE BIBLE IN THE
THEOLOGY OF PROVIDENCE

Strictly speaking, the subject matter of the theology of divine providence is God's acts and purposes in relation to creation.[6] But in the biblical narratives creatures respond to divine actions by obeying or disobeying divine commands, thus cooperating with or thwarting divine purposes. In those same narratives God in turn responds to obedient human actions with commendation and reward and to disobedient actions with condemnation and punishment. As I observed above, John Wright sees the working of provi-

---

[6]Wright wishes to explain to us "what the Bible says about Providence . . . to give the biblical teaching about God as the Lord of nature and the Lord of history, and about the intersection of divine and human freedom in accomplishing God's purposes in the world" (ibid., 1:7).

dence in the Bible as a dialectical (or dialogical) structure of "divine initiative, human response and divine response to this human response."[7] God's initiatives in creation and his religious and ethical commands to human beings originate from his love and grace and anticipate certain responses, which we can call "proximate" divine purposes. But human beings have it within their power to thwart God's proximate purposes; in Wright's words, "When human beings refuse the will of God they block the divine purpose within the immediate sphere of their action."[8] In the biblical narrative God reacts to human sin with corrective punishment hoping for repentance. If divine punishment achieves its purpose, then

> evil, as sin and as suffering, is taken up into the divine purpose. While human malice and selfishness may block the effectiveness of God's love within the limited situation directly affected by a particular human choice, the further effects of that evil choice are integrated into God's loving and wise intentions. Even the sinner is ultimately made to serve the divine purpose.[9]

In biblical theology the historical dynamism of the divine-human relationship is captured under the rubric of covenant. In Genesis 12, God calls Abram and promises to make him a great nation. In this instance no conditions are stated and no formal ceremony is performed. But in Genesis 15, after some time in which Abram has remained childlessness, the Lord renews his promise to Abram that an heir will come "who is your own flesh and blood" (Gen 15:4). Abram asks the Lord how he can be sure that God's promise will be fulfilled. So the Lord binds himself to his promises by participating in a sacred covenant ceremony: God in the form of a "smoking firepot with a blazing torch" passes between the carcasses of sacrificed animals (Gen 15:17). In Genesis 17, God says to Abram, "Walk before me faithfully and be blameless. Then I will make my covenant between me and you" (Gen 17:1-2). The Lord then adds the covenant requirement to circumcise every male child: "As for you, you must keep my covenant, you and your descendants after you for the generations to come. This is my covenant with you. . . . Every male among you shall be circumcised. . . . Any uncircumcised

---

[7]Ibid., 1:195.
[8]Ibid., 1:196. In my view, we should state this narrative truth more cautiously: human beings can disobey divine commands so that the state of affairs projected by the divine command does not come about. Human beings' moral failures do not constitute God's providential failures.
[9]Ibid., 1:196.

male . . . will be cut off from his people; he has broken my covenant" (Gen 17:9-14). When these three occasions are combined we see all the elements of biblical covenants: divine election, promises and threats, a sacred ceremony and conditions.[10] We see a similar pattern in the renewing of the covenant with the people of Israel at Mount Sinai. The Lord promises to give the people the land as a perpetual possession, but only if they keep the laws given to Moses (Ex 24:3-4; 34:10-28; cf. Deut 28–29). The prophets remind Israel that they broke the covenant God made with their ancestors and have or will receive their deserved divine judgment (Jer 10:11; 22:9; Ezek 16:59). Finally, Jesus instituted a new covenant with humanity sealed with his blood (Mk 14:24; 1 Cor 11:25). The canonical writings of the early church are called the "New Testament," which means new covenant.

In biblical theology the covenant is an explanation of the give and take of God's relationship to creation through time. It explains the dialectic movement of history toward the divine goal. God, who is not bound to human beings by nature or necessity, enters freely into relationship with them for the purpose of blessing them and bringing about God's ultimate goal. God initiates the covenants and takes humanity into partnership in the divine project. Sometimes God's covenants stipulate requirements and sometimes not. Sometimes the covenant is accompanied by a ceremony and sometimes not. But divine faithfulness is always the central theme of every covenant. God is eternally faithful and righteous by nature, but his covenant promises strengthen human beings' wavering faith and help them understand how to relate to God and what to expect from him. God never breaks his promises even if human beings break theirs.[11] Again, from a narrative point of view, God, as it were, makes accomplishing his goals contingent on human action.

There should be no debate that the story of God's relationship with humanity (and specifically with Israel) narrated in the Bible includes divine actions, human reactions and divine reactions to human actions, as Wright claims. The covenant structure of sacred history is pervasive. Within this structure the biblical authors portray God as enduring and enacting the full

---

[10]Not all these elements are present in every covenant.
[11]I am grateful for a conversation with Thomas Olbricht about this section on the covenant. His popular-level chapter on the biblical covenants is well worth reading. See Thomas Olbricht, *He Loves Forever: The Message of the Old Testament* (Joplin, MO: College Press, 2000), pp. 79-91.

range of experiences and emotions. God walks in the garden and asks where Adam is hiding (Gen 3:8-9). When, just before God sent the flood, God saw the extent of human wickedness, and "the LORD regretted that he had made human beings on the earth, and his heart was deeply troubled" (Gen 6:6). God "came down" to see the tower the monolingual humanity was building (Gen 11:5). The Lord becomes angry when faced with disobedience, idolatry and injustice (Num 32:9-11; Deut 6:15; 1 Cor 10:1-13; and many others). In Exodus 32 we find the story of God making an angry decision to destroy the people of Israel in punishment for worshiping the golden calf only to change his mind after being persuaded by Moses (cf. Gen 18:22-33; 2 Kings 20:5). God learns more about the faithful heart of Abraham through a test in which the Lord orders Abraham to sacrifice Isaac (Gen 22:12).[12] Sometimes the Lord experiences surprise and disappointment because he thinks one thing will occur but another happens: "I thought that after she [Israel] had done all this she would return to me but she did not" (Jer 3:7; cf. Jer 32:35). The Lord regretted that he had appointed Saul king over Israel (1 Sam 15:35). The list of characteristics could be expanded further to include features that we associate with embodiment: eyes, hears, hands, fingers, back, wings, arms, nose, mouth, face, voice and breath.[13]

In our efforts to understand the biblical view of divine providence we come up against a tension. On the one hand, as we saw in our survey of the biblical texts that touch directly or indirectly on providence, there is a strong affirmation of the universal lordship and divine sovereignty over nature and history. On the other hand, we noted also texts that pictured God as limited in time and space, possessing bodily characteristics, reacting to human actions and experiencing negative emotions such as distress, anger and jealousy. This tension raises the question of theological interpretation of the Bible. Some contemporary biblical scholars reject listening for the divine voice in the Bible as a goal of scholarship. Biblical scholarship must be ob-

---

[12]The speaker is the "Angel of the Lord," but we can assume that the angel was speaking for God.

[13]From a literary perspective the first set of texts usually take form as follows: (1) retrospective conclusions about the efficacy of God's working as, for example, in Joseph's conclusion about the double intention (divine and human) in his being sold into Egyptian slavery (Gen 45: 5-8); (2) doctrinal or confessional statements (Is 46:10-11; Rom 8:28; Eph 1:11); or doxological statements (Ps 135:5-6; Rom 11:36). Most of the texts that portray God in human form and as exhibiting human emotions occur (1) in narratives where God is a character in the drama interacting with the human world (Gen 6:6; Ex 32); or (2) in doxological or prayer texts where language is being used to request, thank and praise God for his saving deeds or merciful qualities.

jective and purely descriptive.[14] I will not enter into this discussion because it deals with the very possibility of theology and I have dealt with this issue elsewhere. This study presupposes the fact that the Bible speaks theologically to the church.[15] My concern here is to give an account of how I am appropriating the biblical texts relevant to providence and why I reject competing theological uses of these texts.

Modern interpreters adopt certain generally accepted guidelines for intelligent reading of the Bible: one must (1) work from the best available original-language text of the document and make a good translation into the language of the intended audience; (2) take into account the original historical and literary contexts; and (3) seek to understand how the first listeners or readers might have understood the text. It is important to emphasize that these ideal goals cannot be fully attained. Not even the most rigorous application of exegetical methods can tell us precisely how the "first reader" actually understood the text or what the author intended. We do not have access to the information we would need to do this. The author's intent and the reader's understanding can function normatively only as regulative ideals. But theological interpreters cannot be satisfied with even the best exegetical results, naively equating the meaning the first reader might have perceived in the text with the meaning of the text for today. The context within which the text must be read now has expanded far beyond its original setting, and its contemporary readers have very different expectations, worldviews and theological frameworks. For Christians, the Old Testament has been taken up into the Christian canon and must be read in light of Jesus Christ. The life, teaching and work of Jesus and the apostolic witness and reflection on the significance of Jesus was summarized in the church's rule of faith and preserved in the New Testament canon. The church reads Scripture with the expectation that its faith in Jesus Christ will be confirmed

---

[14]Joel B. Green documents both sides of the argument in his book *Practical Theological Interpretation: Engaging Biblical Texts for Faith and Formation* (Grand Rapids: Baker Academic, 2011). Robert Jenson rightly contends that such scholars simply do not belong to the community for which Scripture is canon: "When academic exegetes say that Paul's opinions are too historically conditioned now to be helpful . . . they are simply interpreting Scripture as it now will inevitably be interpreted outside the church. Current academic, political, and publicistic elite communities are indeed alienated historically from the community in which the Bible emerged, and this is the reason and indeed the excuse for their helplessness before this text" (*Systematic Theology*, vol. 2, *The Works of God* [New York: Oxford University Press, 1999], p. 280).

[15]I dealt with the possibility and nature of Christian theology in my earlier work *Great Is the Lord: Theology for the Praise of God* (Grand Rapids: Eerdmans, 2008).

and deepened. Hence the theological significance of biblical texts comes to light only as they are brought into relation with the whole canon, whose central unity is disclosed by the church's confession of the identity and significance of Jesus Christ. The task of Christian theological interpretation of a particular text is to show how it contributes to and coheres with the faith of the church. To do this the interpreter must set the text into dialogical relationship with the entire canon of Scripture, with New Testament proclamation about Jesus, with the scope of Scripture as formulated in the church's rule of faith and with the dogmas it confesses.[16]

Study of the second-century church's struggle with Marcionism and Gnosticism shows the necessity of interpretative interaction between written Scriptures and the church's traditional understanding of the economy of salvation and the gospel, which Ireneaus called "the rule of faith."[17] According to Ireneaus, in its ancient rule of faith, carefully guarded and passed on, the church preserved a true summary and the overarching meaning of the Scriptures. "For Ireneaus and the patristic tradition as a whole, Jesus Christ is the hypothesis. He reveals the logic and architecture by which a total reading of that great diversity and literal reality may be confidently pursued."[18] But the Gnostics choose texts, sayings, parables or words and place them in the utterly different context of their systems, thus completely changing their meaning. Irenaeus's illustration brings out the contrast beautifully:

> Their manner of acting is just as if one, when a beautiful image of a king has been constructed by some skilful artist out of precious jewels, should then take this likeness of the man all to pieces, should rearrange the gems, and so fit them together as to make them into the form of a dog or of a fox, and even that but poorly executed; and should then maintain and declare that this was the beautiful image of the king which the skilful artist constructed, pointing

---

[16]Jenson states this requirement with particular force: "Historical honesty requires the church to interpret Scripture in the light of her dogmas. If the church's dogmatic teaching has become false to Scripture, then there is no church and it does not matter how the group that mistakes itself for the church reads Scripture or anything else. But if there is a church, then her dogma is in the direct continuity of Scripture and is a necessary principle for interpreting Scripture, and vice versa" (*Works of God*, p. 281).

[17]Irenaeus, *Demonstration of the Apostolic Preaching* 3-5 (trans. Armitage Robinson [London: SPCK, 1920], pp. 71-73). For the contemporary significance of the early church's theological interpretation, see William J. Abraham, *Canon and Criterion in Christian Theology: From the Fathers to Feminism* (Oxford: Oxford University Press, 1996).

[18]John J. O'Keefe and R. R. Reno, *Sanctified Vision: An Introduction to Early Christian Interpretation of the Bible* (Baltimore: Johns Hopkins University Press, 2005), p. 41.

to the jewels which had been admirably fitted together by the first artist to form the image of the king, but have been with bad effect transferred by the latter one to the shape of a dog, and by thus exhibiting the jewels, should deceive the ignorant who had no conception what a king's form was like, and persuade them that that miserable likeness of the fox was, in fact, the beautiful image of the king.[19]

Interpretation of Scripture is impossible without some presupposed understanding of its overarching meaning. No neutral interpretation is possible. A Christian theological interpretation must presuppose the church's confession of faith and the continuous identity of the church with that of the apostles.[20] Every new generation of Christians must be taught how to read the Scriptures within this framework; otherwise alien frameworks will be imposed on them.

How shall we appropriate the biblical texts that touch on the topic of divine providence, given the tension mentioned above between those statements that seem to support a strong view of divine sovereignty over nature and history and those that seem to undermine this view? In the first few centuries of the church, Gnostics and Manichees were using the anthropomorphic and anthropopathic language of the Old Testament to argue for the inferiority of its view of God.[21] In defense of the Old Testament, the patristic church insisted that texts that give bodily characteristics and irrational passions to God or that imply that God is limited in knowledge or power should be taken as figurative or spiritual rather than literal. The church had long since understood God to be omnipresent, spiritual, eternal,

---

[19]Ireneaus, *Against Heresies* 1.8.1 (*ANF* 1:326).

[20]Jenson, *Works of God*, p. 280. Green examines five options for relating the Scriptures to the rule of faith: (1) read the rule of faith back into the Scriptures; or (2) treat the rule of faith as the "superstructure" and the Scriptures as the "substructure"; or (3) treat the rule of faith as the "stuff" of Scripture; or (4) allow the rule of faith to guide our understanding of "its overall order or structure"; or (5) allow the rule of faith to question our theological interpretation of Scripture. Green rejects the first three and accepts the last two (*Practical Theological Interpretation*, pp. 77-80).

[21]Augustine relates how in his time as a Manichaean these objections kept him from considering the catholic faith (*Confessions* 5–7). See Jacob Albert van den Berg, *Biblical Argument in Manichean Missionary Practice* (Leiden: Brill, 2010), for a study of a significant Manichaean work *Disputations*, by Adimantus, in which the Old Testament is subjected to criticism based on its inferior view of God. According to van den Berg, rehearsing lists of contradictions between the Old and New Testaments was a standard argument by which Manichaeans pressed their case against Catholic Christians. Until recently Manichaeanism was thought to be of Iranian origin. Recent discoveries provide evidence against this thesis. It seems more likely that it was a variation of Gnostic Christianity (see van den Berg, *Biblical Argument*, pp. 1-3, and the literature cited there).

unchanging and rational. God possesses no defects at all. Given this consensus and the clarity and certainty with which it was held, no text could be interpreted in a way that would overturn these convictions. However, some modern interpreters have revived the literal interpretation of those biblical texts that seem to limit God but with intent opposite to the Gnostics. They want to use Old Testament texts to overturn the patristic view of God, which they argue is inspired by Greek metaphysics instead of the Bible. God really is subject to passion, changeable, bound by time, ignorant of some future events and only partially sovereign over history. My sympathies lie with the ancient church.[22]

Most interpreters of the Bible will admit that even within the horizon of the writers themselves some of the language used to speak about God was understood as figurative. But how does one decide when an expression is being used figuratively and when not? It seems to me that we think an expression must be metaphor when its literal use would contradict something we take to be true in a literal sense. If someone says of an especially obnoxious person, "The ass cannot keep his mouth shut" we know without being told that the word "ass" is being used figuratively. A human being cannot really be a donkey. But there are circumstances when the word "ass" should be taken literally. A sentence can consist of the same words but be located in a different context. Perhaps you take a pack animal into the mountains for an extended excursion and your donkey will not stay quiet when you wish him to do so. A metaphor is an expression that can, and usually does, possess a literal meaning but is being deliberately misused. We use a word literally when we use it the way it is commonly used, in its usual reference, and figuratively when we use it in an extended or unusual sense. Figures of speech please and enlighten us only when we are clued in both to the literal reference and the "misuse" to which it is being put. If I am correct

---

[22]For an extensive study of the divine passion texts, see Jeff B. Pool, *God's Wounds: Hermeneutic of the Christian Symbol of Divine Suffering*, vol. 1, *Divine Vulnerability and Creation* (Eugene, OR: Wipf & Stock, 2009), and vol. 2, *Evil and Divine Suffering* (Eugene, OR: Wipf & Stock, 2010). Old Testament scholar Terence Fretheim has written extensively advocating reading these texts as teaching that God truly experiences these passions. See *God and World in the Old Testament: A Relational Theology of Creation* (Nashville: Abingdon, 2005); Fretheim, *Creation Untamed: The Bible, God, and Natural Disasters* (Grand Rapids: Baker Academic, 2010); and his many studies of Old Testament books. Evangelical open theists also read these texts univocally or realistically. See the works of Gregory Boyd, Clark Pinnock, David Basinger, John Sanders, Richard Rice, William Hasker, Alan Roda and others.

about the need for knowledge gained from outside the text, which the original readers brought to the text, we cannot know the figurative or literal status of an expression from the expression alone. We must bring other information into deliberation. And this insight has significant implications for interpretation both in biblical and systematic theology.

This issue is made more complicated if we distinguish between the horizon of the biblical writer and our own horizon. When we ask whether an expression is literal or figurative, are we asking about the writer's intention, and consequently about the writer's horizon of understanding, or are we asking about our understanding of what can and cannot be taken as fact or truth? Consider Psalm 18:6-15:

> In my distress I called to the LORD;
>   I cried to my God for help.
> From his temple he heard my voice;
>   my cry came before him, into his ears.
> The earth trembled and quaked,
>   and the foundations of the mountains shook;
>   they trembled because he was angry.
> Smoke rose from his nostrils;
>   consuming fire came from his mouth,
>   burning coals blazed out of it.
> He parted the heavens and came down;
>   dark clouds were under his feet.
> He mounted the cherubim and flew;
>   he soared on the wings of the wind.
> He made darkness his covering, his canopy around him—
>   the dark rain clouds of the sky.
> Out of the brightness of his presence clouds advanced,
>   with hailstones and bolts of lightning.
> The LORD thundered from heaven;
>   the voice of the Most High resounded.
> He shot his arrows and scattered the enemy,
>   with great bolts of lightning he routed them.
> The valleys of the sea were exposed
>   and the foundations of the earth laid bare
> at your rebuke, LORD,
>   at the blast of breath from your nostrils.

As modern interpreters and heirs of a long theological tradition, we immediately read these lines as an extended metaphorical description of some sort of personal or communal victory for which the psalmist is thanking God. But we cannot know from the words alone whether the writer understood these expressions figuratively or literally. The history of religion knows of many religious perspectives in which individuals would read these words literally rather than figuratively. How then can we know that the writer uses these expressions figuratively and expects his readers to hear them as metaphors? We can do this only if we accept two premises. First, we would need to establish that within the writer's historical horizon the most informed adherent of the biblical faith would have held views of God that would be contradicted by a literal reading of this psalm and similar literature. It is not my purpose to establish this fact; however, it seems clear that at least some Old Testament writers hold the equivalent of divine omnipresence (Ps 139; Jer 23:23-24) and omniscience (Is 41–48), which would imply a spiritual rather than a physical understanding of the divine nature.[23] Second, we would have to assume that the writer of this psalm shared the generally held spiritual view of God.

But modern Christians hear this psalm from within a larger horizon that includes the New Testament, the rule of faith, church dogma and the history of theological reflection on the doctrine of God and providence. By the time the New Testament was written the spirituality of God and God's freedom from all physical limitations was taken for granted. The modern Christian reader reads Psalm 18 metaphorically without even considering the writer's intentions and historical horizon. Within the historical horizon of the New Testament[24] and beyond, it becomes impossible to think of God as embodied or in any way dependent on a body for life and activity.[25] But is it legitimate to treat Old Testament anthropomorphic texts as metaphorical because a literal interpretation would contradict convictions about the

---

[23]It seems to me that one would have a very difficult time establishing from within the horizon of the Old Testament alone the conclusion that God completely transcends every limitation of space. This conclusion would be necessary to enable us to read every text that pictures God moving in space as metaphorical. If you expand the horizon into the New Testament this becomes possible.

[24]For example: Jn 4:24; Acts 17:24-28; Eph 1:23.

[25]The doctrine of the incarnation does not contradict this truth. The human nature retains its essential qualities in the personal union with the divine Son of God. But the divine nature does not change its essence as Spirit.

divine nature that only became clear in the New Testament era? It seems to me that we are faced with three alternatives: (1) we can assume without being able to establish it that the ancient writers understood but did not say what later came to be expressed explicitly; or (2) although the human writers may not have grasped fully what became clear only later, the Spirit who is the true author of Scripture uses them to speak to us, which legitimates our reading them metaphorically and theologically; or (3) the human writers were expressing their religious experience in the language of their day, and through their figurative language we can attempt to grasp what religious experiences gave rise to their expressions.

The third way of reading the Old Testament possesses no legitimacy in Christian theology because it is not a method of theological appropriation but of humanistic interpretation.[26] Instead, as I have made clear, this study reads the Bible canonically, christocentrically and within the framework of the rule of faith. Hence the second alternative or some variant seems to be the only viable method. Indeed, the New Testament authors read the Old Testament christologically. In light of the actual appearance, teaching and work of the Messiah, the New Testament authors and the patristic authors who followed their example read the Old Testament as speaking of Jesus Christ, even when it is obvious that human authors did not grasp fully Jesus Christ as the referent of their words. By extension I argue that the entire Bible ought to be interpreted in view of the understanding of God that comes to light in the New Testament interpreted according to the church's rule of faith. Hence, given a Christian canonical view, it is legitimate to subject the anthropomorphic and anthropopathic language of the Old Testament to scrutiny in light of principles derived from the New Testament and the rule of faith.[27]

---

[26]Of course, this approach has validity in a social-science-oriented religious studies approach. Mine is a project of Christian theology.

[27]Of the four evangelical views on how to move from the Bible to theology discussed in Gary Meadors, ed., *Four Views on Moving Beyond the Bible to Theology* (Grand Rapids: Zondervan, 2009), my view resembles Daniel M. Doriani's "A Redemptive-Historical Model" and Kevin J. Vanhoozer's "A Drama of Redemption Model" more than it does either Walter C. Kaiser Jr's, "A Principlizing Model" or William J. Webb's, "A Redemptive-Movement Model." Comparison between my views and theirs is somewhat handicapped by their almost total focus on ethics. In my view the history or drama or economy of revelation and salvation leads to Jesus Christ and terminates in him. Everything that came before Jesus Christ points to him, prepares a people for him and possesses no continuing function apart from him; hence the Bible must never be interpreted independently of Christ. Christian faith and doctrine must be derived from this

An examination of the texts in both testaments relevant to providence shows that those texts that are formulated as principles, as universal rules, as doctrine, always support a robust view of divine sovereignty in providence. In other words, the providence texts that take a form most like theology, which clearly attempts to speak of things as they are rather than of the way they appear, support a strong view of divine sovereignty in providence.[28] There are no texts formulated as principles that universalize divine limitations, mistakes or ignorance. Some of the classic statements are as follows: "And we know that in all things God works for the good of those who love him, who have been called according to his purpose" (Rom 8:28); "In him we were also chosen, having been predestined according to the plan of him who works out everything in conformity with the purpose of his will" (Eph 1:11); "For from him and through him and for him are all things" (Rom 11:36); "Now to him who is able to do immeasurably more than all we ask or imagine, according to his power that is at work within us, to him be glory in the church and in Christ Jesus throughout all generations, for ever and ever! Amen" (Eph 3:20-21).

On the other hand, one will search the Scriptures in vain for statements of principles like the following: "God does not know all things" or "God can work in only some things for good" or "God is not able to do everything we can imagine."[29] The tendency of the Bible is the opposite, to heighten the power, knowledge and perfection of God as much as language will allow. This tendency is clear in doctrinal and doxological contexts.[30] In other

---

economy and this center. God is and always was "the Father of our Lord Jesus Christ," Father, Son and Holy Spirit. However, this faith and this doctrine has been preserved, understood and passed on by the church. And therefore theologizing in service of this faith is the task of this community rather than of feral theologians searching promiscuously for truth. Hence individual theologians must theologize in harmony with the rule of faith and in service to the community.

[28]James S. Spiegel, in *The Benefits of Providence: A New Look at Divine Sovereignty* (Wheaton, IL: Crossway, 2005), makes a similar point distinguishing between "didactic" and "historical narrative" passages. He also rightly distinguishes passages that speak phenomenologically and those that speak metaphysically, that is, between how God appears in his action and how God actually works (p. 60). Open theists, Spiegel argues, do not distinguish these two properly, reading the phenomena into the being of God: "They commit the egregious mistake of using biblical phenomenological data as evidence for metaphysical claims about God. Consequently, their doctrine of providence is fundamentally unbiblical, and their portrait of God is woefully incomplete" (p. 61).

[29]Whereas we can find poetic and narrative texts that could be used to support a strong or a weak view of providence, we do not find texts formulated as general principles that support a weak view.

[30]Doctrinal and doxological language tends to be stated either in the language of infinite heightening or total negation. This language emphasizes the qualitative difference between God and creatures. It seems to me that the boundaries of negation or infinite heightening must not be

words, where the intention is to speak doctrinally about God's own being or actions, the language of risk and limits are missing and would be out of place. And where the context indicates the intention to speak of God's involvement and goodwill or judgment in relation to his creatures the authors draw the language of location, change, emotion, risk and limitation commonly used to show human involvement or judgment or goodwill. Complaints about "not taking the Bible seriously" or "rejecting inspiration" simply miss the point. I've never heard anyone, even the most conservative author, deny that the Spirit can use poetry and metaphor as a means of communicating divine truth. Hence when biblical authors describe divine action and judgment in relation to creation in the narrative language of human action or emotion, there is no reason to consider them any less open to God's inspiration than when speaking in propositions. And when theologians such as Ambrose and Augustine understand the anthropomorphic and anthropopathic language as figurative, they are no less submissive to the divine message than those who take it literally. If we set these texts in the larger context of the entire canon and the church's clear perception of God's spirituality and perfection, we are justified in interpreting them in a way that does not undermine this perception. And that is the position I adopt in this study.

---

violated. We should not deny God any of what he is said to possess to an infinite degree, and we must not attribute to God any of what is said to be infinitely absent from him. Within these boundaries some analogies are more refined than others, and they should be used to interpret the less refined analogies or metaphors. One should never allow the metaphorical passages to undermine the negative principles.

11

ALL THINGS
WORK TOGETHER

DEFINING PROVIDENCE

As we begin our systematic analysis of the doctrine of providence and the questions it raises, let us first recall the six theses in which I summarized the biblical teaching on providence:

> (1) *God the creator of all things continues to sustain his creation in being and to act wisely and powerfully in nature and history to realize his plan.* (2) *God acts in creation in direct and obvious ways in what are called miracles and in indirect and subtle ways, such as in natural processes and the free actions of human beings.* (3) *God acts even in or through evil acts and intentions so that God's plan is realized, without God doing evil.* (4) *God's providence encompasses "all things" (Rom 8:28), so that everything that is and everything that happens is made to serve the divine purpose.* (5) *For those who love God the preceding affirmations about God's providence impart comfort, hope, courage and joy.* (6) *God realizes his aims perfectly. God cannot fail, even in part.*

These six theses present a simple and positive affirmation of divine providence that in itself is quite impressive in scope and clarity. Embracing these theses wholeheartedly could indeed found, as the fifth thesis affirms, a life filled with "comfort, hope, courage and joy." But a comprehensive analysis of the teaching of these theses needs to address several questions not answered by them and deal with several problems that could be objected to them.

I shall define providence as *that aspect of the God-creation relationship in which God so orders and directs every event in the history of creation that*

God's eternal purpose for creation is realized perfectly.[1] This simple definition
contains the essential components of the doctrine of providence.
(1) Providence is not a series of totally separate divine acts but an aspect of
the one God-creature relationship. (2) Providence is God's own personal
action, not delegated to angels or left to impersonal causes. (3) God "orders
and directs" the history of creation, not leaving creation to chance or fate or
misguided freedom. (4) Divine providence covers every event in the history
of creation, great and small, good and bad, contingent and necessary.
(5) God's eternal purpose guides God's providential work. God does not
need to adjust his plan or improvise in response to unexpected events.
(6) God realizes his aims perfectly. God cannot fail, even in part.

UNFOLDING THE DOCTRINE OF PROVIDENCE

Before we address the six parts of the definition of providence, we need to
ask about the domain of divine providence. As I argued in the doctrine of
creation, theology reflects on creatures only in relation to God and not as
an independent theme. Theology does not enter the domain of creatures'
natural relationships to other creatures. It leaves this area to natural science.
In the same way, the doctrine of providence concerns God's acts and purposes in relation to creatures and their acts and purposes, but not to creaturely acts and purposes as such or in relation to each other. It leaves these
to psychology, sociology, history and other areas that study the human phenomenon. We cannot prove or disprove divine providence by studying the
history of nature in cosmology, paleontology or evolutionary biology. Un-
raveling the interlocking system of nature in its present state or recon-
structing models of previous states reveals nothing about providence, about
God's relationship to this history. The study of human history understood
as the study of humanity through time, its interaction with nature, its acts
and creations, its passions and sufferings, likewise leaves the question of

---

[1]My definition tracks with Paul's statement on in Rom 8:28. The Greek text reads: *oidamen de hoti
tois agapōsin ton theon panta synergei eis agathon, tois kata prothesin klētois ousin.* The NIV translates as follows: "And we know that in all things God works for the good of those who love him,
who have been called according to his purpose." Charles W. Wood examines the Christian doctrine of providence at a formal or meta level. He first speaks of "the question of providence,"
which says, "How are we to understand what goes on?" He then speaks of "the task of exposition"
of the doctrine as setting "out the principles according to which, in light of the Christian witness,
we are to understand theologically what goes on" (*The Question of Providence* [Louisville: Westminster John Knox, 2008], p. 17).

providence open. The question of divine action and purpose in and through the history of creation is for theology to address; the natural and human sciences have no competence to speak to the issue. But we must also take care from the other direction. We cannot deduce the history of nature or of humanity from a theological or philosophical doctrine of providence; nor can a providential reading of history supplant rigorous analysis of the phenomena of history by the historian.

1. Providence is not a totally separate series of divine acts but an aspect of the one God-creature relationship. As I argued when discussing the act of creation, it makes no sense to think of God's relationship to creation as a series of different types of relations—creation, providence, reconciliation and consummation—with each type containing a series of different acts. This way of thinking pictures God as time-bound and changeable and puts the unity of creation in jeopardy. Many other theological misconceptions and contradictions find their root in this fundamental mistake. I argued that we should think of the God-creature relation as one eternal act with many temporal results.[2] Consistency with this commitment demands that we think of providence as an aspect of the simple God-creature relation.[3] The one, simple God-creature relation is described in Christian theology as the history of creation, providence, reconciliation and consummation. We must use all four of these salvation-historical categories to grasp something that would otherwise remain completely mysterious. None of the four categories by itself could grasp God's relationship to creation adequately. However, since the God-creature relation to which they all point is one and simple in itself, each of the four categories refers to the whole relation. In some sense, then, the entire God-creature relation is creation, is providence, is reconciliation and is consummation. In the case of creation, in whatever way the other three aspects differ from creation they also possess a creative character. For this study I want to emphasize that providence is creative.

As we shall see, many views of providence I find defective are plausible

---

[2] I will address these mistakes as they arise in the debate with alternative views of providence.

[3] The unity of the divine relationship to creation seems to have been the position of post-Reformation Reformed Orthodox theology. According to Johannes Braunius, "In respect of God the same action is creation and providence; God works all things by a single, most unifold will, that they exist, remain in existence and work" (*Doctrina foederum sive systema didacticae et elencticae* [Amsterdam, 1688], quoted in Heinrich Heppe, *Reformed Dogmatics*, rev. and ed. Ernst Bizer, trans. G. T. Thompson [1950; repr., London: Wakeman Great Reprints, n.d.], p. 251).

only because they isolate providence from creation and the other aspects of the God-creature relation, that is, from reconciliation and consummation.[4] Some thinkers separate providence from creation because they want to solve certain problems by speculating about how God works in providence. And seeking the "how" of creation makes no sense as long as one understands creation to be "from nothing." But if the entire God-creature relation possesses a creational character, there can be no thoroughly rational account for "how" God's works in providence. None of God's acts of providence is susceptible of rational explanation in terms of the kind of instrumentality that applies in the creature-creature relation. Nor could one impose on God rules about how God can work in providence that are derived from creature-creature relations; for example, the contradiction between determination and freedom or between foreknowledge and freedom. If God's relation to human acts is an act of creation, such rules do not apply. Now if it makes sense to see providence as creational in this sense, it is a mistake to think of God's act of creation as merely an initial bringing stuff into being whereupon the relationship changes to one of providence. It would be like saying that a builder created the building by dumping the materials in a pile. No, the entire process of bringing the building to completion must be included in the idea of creating the building. So creation is not complete until it has been perfected, brought to its final and definitive state.

Even though divine providence should not be isolated from other aspects of the God-creature relationship, it does turn our attention to something not considered in the doctrine of creation. In creation God gives existence to things that possessed no prior existence. Creatures cannot in any way participate in their own coming into being from nothing. For the creature, being created is passive reception of existence. In giving creatures existence God gives them certain capacities through which they preserve their existence, direct their lives and actualize their potential. The doctrine of providence explores God's continuing action on behalf of creatures. God sustains

---

[4]Wood also laments that providence has been treated in isolation from creation, though his understanding of the unfortunate effects differs from mine. He thinks disconnecting the doctrine of providence from creation results in a preference for the status quo instead of being open to change. Nevertheless I agree with him when he says, "To view creation and providence . . . together as one ongoing work, one eternal act of God being realized throughout time and space, might not only renew the doctrine of providence but also provide a new perspective on every other aspect of the Christian witness" (*Question of Providence*, pp. 71-72).

the existence of his creatures, accompanies and empowers them as they enact their lives, and directs the course of events toward his chosen goal. God does not abandon creatures to their own strength and wisdom, for God has planned an end for his creation that far surpasses creation's power to attain. The word *providence* translates Greek and Latin words that mean "to foresee." Human beings cannot by their limited wisdom foresee the future so well that they can direct their lives to their divinely appointed end. By highlighting God's foreknowledge, the doctrine of providence clues us in that we are now considering God's relation to creaturely time and movement; the doctrine assures us that God can guide history according to his eternal plan to his goal. God possesses the power, wisdom and knowledge to master every event and direct every movement so that all things serve his purpose. Hence the divine relationship to creation in providence involves relating to existing creatures in the exercise of their causal powers, their reason and freedom, and their purposes and acts in ways that bring them to full actualization, realizing God's purpose for creation.

2. Providence is God's own personal action, not delegated to angels or left to impersonal causes. God sustains, orders and guides the world in the same way God created it, from the Father, through the Son and in the Spirit. In this distinction lies the fundamental difference between the Christian view of providence and religious or philosophical views of fate.[5] In early Greek literature the irrational decrees of fate concern only the decisive issues in a human life—birth, wealth, pain, suffering and death—and they do not yet encompass a general plan for the course of all things. In Stoic philosophy the idea of fate (*heimarmenē*) was combined with providence (*pronoia*) and nature to create an all-encompassing system of determinism.[6] Fatalism can

---

[5]Wood argues that in all probably early theologians derived their confidence that God can and will make everything turn out as God intends from Stoicism, which then prevented the doctrine of providence from being reconceived in a trinitarian way (*Question of Providence*, pp. 58-59). Wood's equation is too simple. One need not accept the Stoic concept of fate to believe that God can and will make things turn out as he wills.

[6]Simon Hornblower and Antony Spawforth, eds., *The Oxford Classical Dictionary*, 3rd ed. (New York: Oxford University Press, 1996), s.v. "Fate," pp. 589-90. For the range of possible kinds and levels of determinism in Hellenistic philosophy, see R. J. Hankinson, "Determinism and Indeterminism," in *The Cambridge History of Hellenistic Philosophy*, ed. Keimpe Algra et al. (Cambridge: Cambridge University Press, 2005), pp. 513-41. The Stoics, specifically Chrysippus, aimed for a middle position between "hard determinists," who left no room for responsibility, praise or blame, and indeterminists, who seemed to deny the principle of universal causality. This middle position can be called "soft determinism" (p. 537). Chrysippus argued that although

take many forms, some pessimistic (Arthur Schopenhauer) and others opti-mistic (Friedrich Nietzsche). I understand the essential tenet of fatalism to be that the course of world events is unchangeably determined by impersonal and immanent causes. Even very robust views of providence in which God brings it about that every event happens the exact way it does are not neces-sarily guilty of fatalism. God can work in and through worldly causes, but God also transcends these causes and is directly related to every event as its creator. God's sovereignty over all events does not remove from events their created character—contingent, necessary or freely chosen. Some of the early church fathers opine that God deals personally only with the large-scale matters and assigns the lesser matters to angels.[7] Some contemporary theo-logians hold the view that God assigned particular realms and functions of creation to angels, some of whom rebelled against God. In this way they trace the origin of "natural" evil to misused angelic freedom.[8] But even if God as-signs angels to care for his saints or entire nations (or even whole planets) and prophets to speak to human beings in God's name, this does not put God at a distance or insert room for divine mistakes. God's use of secondary causes and other agents in his providence is not the same as metaphysical mediation, which is the idea that God is so transcendent that God cannot get close to creation and needs the assistance of nondivine mediators.

The personal character of God's providence can be clearly seen in Romans 8:28-39. Paul begins with "And we know that in all things God works for the good of those who love him, who have been called according to his purpose." In the following verses he interweaves references to God's eternal fore-knowledge, election and predestination with God's love for us demonstrated by God's giving his Son for us. God loves us and Christ loves us. How could we think that he will not also "graciously give us all things?" (Rom 8:32). Such a great demonstration of divine love proves that no power "will be able

---

all things are determined it is not true that our fate will be the same whatever we do. Our actions are codetermined with what happens, and since our actions arise from what is within us, we are subject to praise and blame.

[7] Leo Scheffczyk, *Creation and Providence*, trans. Richard Strachan (New York: Herder and Herder, 1970), pp. 61-62. This idea is found in Athenagoras, Justin, Clement of Alexandria and Novatian. According to Scheffczyk, the presence of this notion in these thinkers "is characteristic of a stage in theological thought when Creation and Providence are not closely associated but each is to some extent autonomous" (p. 62).

[8] Gregory Boyd, *God at War: The Bible and Spiritual Conflict* (Downers Grove, IL: InterVarsity Press, 1997); and Boyd, *Satan and The Problem of Evil: Constructing a Trinitarian Warfare Theod-icy* (Downers Grove, IL: InterVarsity Press, 2001).

to separate us from the love of God that is in Christ Jesus our Lord" (Rom 8:39). In Romans 8:28, Paul speaks of God's care for "those who love him," and in the succeeding verses he speaks of the love of God and Christ for us. Clearly God's providence is set in the most intimate personal relationship. God works "in all things . . . for those who love him," not from a distance or through a string of mediators but in the absolute presence and power by which he creates all things and with the intimate presence of the divine love in which Christ gave himself for us and the Holy Spirit indwells us (Rom 8:12-27). The divine agent of providence is the God who is Father, Son and Spirit. When we think of the one in whose care we live, we should think of the generous and all-powerful Creator of heaven and earth, the loving Savior we meet in Jesus Christ and the life-giving Spirit who will bring all things to perfection.[9]

3. God "orders and directs" the history of creation, not leaving creation to chance or fate or misguided freedom. I stated this thesis a bit more narrowly than usual. Theologians from many theological traditions divide God's providential action into sustenance, concurrence and governance.[10] My definition leaves sustenance and concurrence out of the definition of what God does in providence. The reason for this is that I consider sustenance and concurrence to fit better in the creation aspect of the God-creature relationship. Creation would be meaningless apart from the duration of creatures. What would it mean to create something that instantaneously ceases to be? Sustenance is God's continuous act of creation, or to put it another

---

[9]Charles M. Wood criticizes the traditional doctrine of providence for being effectively monotheistic rather than trinitarian. In his trinitarian revisioning of the doctrine, Wood assigns one of the three traditional aspects of providence to the appropriate persons of the Trinity. The Father sustains the creature, the Spirit concurs with the creature's acts and the Son directs all things (*Question of Providence*, p. 80). The traditional doctrine's exclusive focus on the unified action of God *ad extra* tends to picture divine action exclusively as the exercise of sheer power while obscuring other ways of divine operation: the gentle persuasion of the Spirit and the self-emptying way of the Son's directing (ibid., pp. 89-91). I agree with Wood up to a point. However, I reject the notion that, because the Spirit acts in subtle ways and the Son acts in the apparent weakness of the cross, these forms of action cannot achieve God's will perfectly. This does not follow. But I suspect this is where Wood is headed. It seems to me that the most important benefit of reconceiving the doctrine of providence in a trinitarian way is that it brings the goal of providence clearly into focus, that is, it directs us toward the union of creation to God though the Son and in the Spirit.

[10]For Lutheran Orthodoxy's use of these three terms, see Heinrich Schmid, *Doctrinal Theology of the Evangelical Lutheran Church*, trans. Charles A. Hay and Henry Jacobs (1899; repr., Minneapolis: Augsburg, 1961), p. 175. For the usage of the three terms in post-Reformation Reformed dogmatics, see Heppe, *Reformed Dogmatics*, pp. 256-64.

way, the Word through which God lets things be continues its effectiveness. Sustenance concerns the continued existence of creatures. But, in thought at least, we can distinguish the existence of a creature from its action. Of course, in reality it is impossible for a creature to exist without moving or changing or acting. And just as a creature cannot sustain its own existence by an act of will, it cannot act or sustain the effect of its act solely by its own will. Hence the concept of concurrence or cooperation is needed to describe fully the idea of divine creation. Concurrence is God's continued act of creation as it empowers creatures throughout their action and sustains the effect of their acts. Traditional theologians have pointed out that acts bring things into being and the event of coming into being requires divine creative action. Concurrence can also be understood as God's continued sustenance of creatures as they act. Creatures, their powers and the material on which they act require divine sustenance. Without God's action of sustenance creatures could do nothing because, of course, they would not exist.

It must be pointed out that the concepts of sustenance and concurrence do not include the concept of governance, or what I called "ordering and directing." We can imagine something being sustained in being and enabled to act without the direction of the action being determined or channeled to a particular end beyond the immediate intention of the agent. This possibility is why I focus on "ordering and directing" or governance as the heart of the work of divine providence. Just as the world could not bring itself into being from nothing or sustain itself in existence by its own power, it cannot order and direct itself. Creation apart from order and direction is just as inconceivable as creation apart from sustenance. To create is to create something in particular, and order and direction are the difference between real objects and mere chaos. Chaos and nothingness have within them no principles of existence and order. Hence the act of creating must include ordering and directing. Indeed, in our experience creating is exclusively ordering and directing. We cannot create the materials out of which we create. In the special case of divine creation, everything about the creature depends on God—material, order and direction.

God's ordering of the world is conceptually distinguishable from God's directing activity but, in fact, the two cannot be separated. In ordering the world God sets things into intelligible relationships with each other according to a plan. Trees, frogs and stars exist partly because of their orderly internal and external relationships. All physical things are "things" (or unities or wholes) because of the harmonious and stable relationships of their component parts. But internal order must be accompanied by commodious external order. Trees and frogs cannot exist apart from a supportive environment, and stars cannot exist apart from their galactic environment. The idea of matter does not include the idea of order as such, much less a particular order. Even adding a particular set of physical laws to matter, which of course are not necessary to the idea of matter, only provides the possibility for some kind of order and does not determine which order will arise. Hence belief in God as the providential orderer of the world cannot be limited to creating matter and giving it a particular set of laws. This divine act would give the initial components of the world a stable order and thus would provide a foundation on which higher-level orders could be built. But it would not determine which possible order would actually come to be.

That God wills to create a particular order seems to entail that God also wills to accompany creation through time and continue his ordering care, making sure that order arises.[11] The necessity of continued ordering leads to the idea of providential directing. Directing shepherds an act or a process to a particular end when, apart from such direction, the act or process would be indeterminate with respect to the end the director has in mind. If God wills a certain order to arise, the coming to be of this order can be thought of as an end to be achieved. That is, God's ordering of x at time t may be the result of God's directing processes a, b, c and so on at $(t\text{-}n)$ to produce x at t.

That God "orders and directs" the history of creation means that God brings it about that the created world is and remains the world God intended it to be and that in all worldly events, processes and free acts God brings it about that his will is achieved. I will do what I can in later sections to defend this view of providence from critics. But here I want summarize the argument implicit in the paragraphs above. I have tried to show that when the Bible affirms God as the creator, it does not mean that God created matter and left it to form a universe by pure chance. Nor does it mean that God created matter *and* the laws of physics and left them to form a universe by a combination of chance and necessity. It does not mean that God created matter, the laws of physics and an initial order and let them explore their

---

[11]This is true unless one insists that God created all order in the first instant of time, a view I hold as unbiblical as it is implausible.

more constrained but still infinite possibilities by chance. No, when the
Bible affirms that God is the creator of heaven and earth it means that God
created the order we now experience, the ones that came before and those
that will follow until God has created the definitive order in realization of
God's eternal plan. God was, is and will be the creator of heaven and earth.
Hence a robust view of divine creation and a robust view of divine providence stand or fall together.

   *4. Divine providence covers every event in the history of creation, great
and small, good and bad, contingent and necessary.* Just as God creates and
sustains everything that exists and concurs with everything that happens,
nothing can escape God's ordering and directing governance. As I argued
above, ordering and directing are necessary aspects of God's creative activity; creation as the Scriptures present it is inconceivable apart from order
and direction. Hence God's ordering and directing activity is coextensive
with his creative act, and his creative act applies to everything that exists. We
must keep in mind that the categories great and small, good and bad, contingent and necessary apply only in this world. Measured in human terms,
events are considered great or small according to how many lives they affect,
at what level and for how long. But we are limited in our capacity to see the
true effects of a contemporary event. Even ten thousand years may not
suffice to distinguish great from small. Perhaps even genuinely small
events—again from the human perspective—will contribute something indispensable to the total history of the world. We cannot judge. And if something is indispensable to the total picture, how can we judge it small? How
can anything God creates be thought insignificant? To think this way is to
impose a human perspective on God.

   Good and bad in the senses I am using them are relative to life in this
world.[12] Something is good insofar as it aids living a vigorous, happy, successful and long life. Something is experienced as "bad" when it detracts or
seems to detract from the good life. But wisdom teaches that not everything
that looks good really is good and not everything that looks bad really is bad,
even in the relative sense. In some instances, a pain though seemingly bad
may turn out to be for the good, and likewise, a pleasure can seem good but
really be harmful in the long term. So we need to distinguish between two

---

[12]I am not using the words *good* or *evil* in their metaphysical or moral senses.

senses of the terms *good* and *bad* even in relation to this life. To judge something "good" can refer to a subjective and momentary feeling of delight or be used in relation to an ideal of what human life ought to be. The word *bad* also carries this double meaning, an immediate subjective response or a considered judgment about what retards human life in the long term.

God orders and directs both good and bad. But God directs good and bad to his ends, not to ours. God does not order and direct good and bad to our immediate pleasure or even to our flourishing in this life in the long term. There is nothing in the doctrine of providence that promises that God will order and direct events so that we experience health, wealth, respect and long life. Nor are we promised exemption from suffering, disease and early death. The comfort in the doctrine of providence is that "in all things God works for the good of those who love him" (Rom 8:28). The "good" in this verse is neither of the two "goods" mentioned above. Paul has our ultimate good in mind; he does not judge from an immediate or even a whole-life perspective. This good can be measured only from the perspective of eternity. Even though I said earlier that God orders and directs all things to his ends and not ours, the end for which God works will turn out to be for our good in eternal perspective. Our true good is that which contributes to our attaining eternal life. The truly bad is that which prevents us from attaining eternal life.

Likewise, I am using the terms *contingent* and *necessary* in senses that apply only within the physical world. By *contingent* I mean an effect that is underdetermined by its causes, and by *necessary* I mean an effect that is wholly determined by its causes. Effects are necessary if, given the presence of specific causes, those precise effects must follow. *Contingent* does not mean uncaused but that given the same set of causes more than one effect is possible.[13] Some philosophers think God is bound by certain logical or causal necessities and that certain contingencies cannot be directed without removing their nonnecessity and hence their contingency. The conclusion follows that God cannot order and direct necessities because, being necessary, God's governance would be superfluous, and that God cannot order

---

[13]For brief discussions of necessity and contingency, see Robert Audi, ed., *The Cambridge Dictionary of Philosophy*, 2nd ed. (Cambridge: Cambridge University Press, 1999), s.v. "Necessity" and "Contingency." As I said above, I am discussing only physical necessity and contingency and not metaphysical necessity or contingency.

and direct contingencies without acting as a necessitating cause, which would deprive them of their contingency. I am not pursuing a solution to this problem in this section.

Physical necessities and contingencies are modes of the coming to be of worldly events; they are creation-creation relationships and, hence, are subjects for scientific or psychological study. As I have emphasized throughout this study, the God-creature relationship should not be confused with creature-creature relationships. God's ordering and directing of physically contingent or necessary events fall under the law of the God-creature relationship; hence rules derived from creature-creature relationships do not apply. The necessary or contingent character of a physical event describes the type of relationship it has with its immediate causes. To think of God as directing and ordering the world by becoming an additional physical cause is to confuse the God-creature with creature-creature relationship. The archetypical model for all aspects of God's relationship to creation is the formula *creatio ex nihilo*. There is never any physical cause prior to God's relationship to creation. God's relationship to the whole creation and every event in the history of creation is eternal. God's eternal "Let it be" is directed to the whole creation, which only exists as its parts and events are set in creature-creature relationships. Those relationships possess different characters. Some of them are necessary, some are contingent and some possess the special character of being freely chosen. God is equally the creator of them all.

Perhaps an illustration will help. Who is Ron Highfield? Is he the baby born in June 1951 or the teenager who came of age in the 1960s? Am I the professor of religion at Pepperdine University in December 2014 who writes these lines? I believe all these answers are correct but incomplete. Will Ron Highfield ever possess a definitive identity so that there is a final and unchanging answer? Yes. The eternal God knows who I am definitively, which for me is still future. Now suppose God wishes to create Ron Highfield. Which one does God create? Surely the answer must be that God creates the total history whose name is Ron Highfield; for God's relation to me is eternal, even though I am temporal and have come to be in the middle of time. So "when" God creates me, God creates my entire history, everything that happens to me and everything I have done. But every event that touches me and every act I perform possesses a character appropriate to its nature—necessity, contingency or freedom—in relation to the causes that brought it into being.

I believe we can extend this illustration to the entire history of creation. What world does God want to create? Is it the one that existed long ago, that exists now or far in the future? All of these are God's worlds, but will the world ever possess a definitive identity? God knows the world in all its phases and in its totality. From an eternal perspective God creates the whole history of the world instantaneously. If we understand creation in this way, God's providential "ordering and directing" can be understood as God's eternal act of creating the world (with its order and direction) viewed from within time. Let's briefly anticipate a problem I will deal with in depth later. Does not thinking of God's act of creation as eternal and applying to the entire history of creation mean that our future and the future of all things is "already" settled, indeed "already" real? But if everything is "already" settled and real, are contingency and freedom excluded from the world? In answer to this question note first that I put the word "already" in scare quotes, indicating that the word is being used improperly. In speaking this way we use the word "already" to refer to the created results of God's eternal act of creation as if eternity were really endless time. In doing this we are mixing the creaturely temporal frame of reference with the eternal frame of reference. We begin to think of the entire history of time as already existing somewhere and our movement in time as a moment-by-moment coming on what was already there. But this fails to take seriously God's eternity and the eternity of God's act of creation. God is eternal (the simultaneous possession of endless life), and there was no time when God was not the Creator. From the eternal perspective God creates the entire history of creation in one eternal act. But creation itself is temporal; it is not static but constantly changing. The past is gone and the future does not exist. We are not traveling along a path that exists already before we get there. The future, that is the next present, comes directly from eternity, out of nothing. From a temporal perspective the world is being created, ordered and directed at this moment. Providence is happening now because God's eternal act of creating, ordering and directing is contemporary with every moment.

5. God's eternal purpose guides God's providential work. God does not need to adjust his plan or improvise in response to unexpected events. No one who believes in divine creation doubts that God creates the kind of world God wants and gives it the order he wishes. And surely God's acts correspond to God's intentions. And so when we say that God orders and directs all things we are safe to add the qualifier "according to his eternal purpose" (Eph 3:11) or "in conformity with the purpose of his will" (Eph 1:11). The Christian doctrine of providence assures us that as we experience God's providential ordering and directing in time, God's present action, however mysterious it might appear, is the manifestation of his eternal wisdom and is the outworking of a plan for bringing creation to its glorious fulfillment. It is as important for the theme of providence as it is for doctrine of creation to distinguish divine acts from human acts. To speak of a genuine human act of creation, the agent must have in mind a goal and an idea of what is to be created. Otherwise we can speak only of an accident. Human beings, in contrast to God, need help from other powers in order to create: we need ideas, material and tools. Because we are not the sole cause of our creations we do not control every aspect of our creative acts. What we actually do may differ greatly from the "plan" we had in mind; so we have to adjust to unexpected outcomes and can at best achieve only an approximation to our plans.

The God who is Father, Son and Spirit needs nothing outside God. Everything other than God is God's creature, created from nothing. God's Word is the eternal archetype for creation, and the Spirit is God's power for its coming into being. Nothing limits God from enacting his intentions perfectly. But God's plan for creation is not merely a blueprint for a static order. The eternal God creates a dynamic creation, and hence the plan for creation includes temporal and causal relations as well as spatial ones; that is, the eternal plan covers God's acts of ordering and directing as well as the acts of giving existence, sustenance and concurrence. As I explained above, the God-creature relation is eternal and encompasses the whole history of creation. Events that are truly contingent and free within time and for us surprising and unexpected are in relation to God simply the order that God creates. Events catch us off guard and require us to adjust our plans and change our actions. But God does not live a temporal life in which God needs to wait until something happens in time to know what to do in response. In an eternal act God creates the world God wills to create. The act of creation and the plan for creation are both eternal, and they correspond perfectly. There is no time differential or nondivine medium between them.

Understanding present events as existential enactments of God's eternal plan contributes significantly to the pastoral and theological power of the Christian message. Theologically, New Testament writers understand the events of the life, death and resurrection of Jesus Christ; the outpouring of the Holy Spirit; and the admission of the Gentiles to the people of God as revelations of the eternal plan of salvation. On the one hand these events reveal the actual trajectory of eternal plan as never before, but the events themselves are given significance far greater than they could achieve on their own by the prior belief in an eternal divine plan. In reading the New Testament one senses that the first Christians felt themselves overwhelmed by events in which heaven opened and all history became clear.

Pastorally, reference to the eternal divine plan addresses the human condition in a powerful way. We need meaning and order, and we are terrified of chaos. The confusion, pain and suffering we experience in the present and anxiety about the possibility of continued or new confusion, pain and suffering in the future can easily rob us of peace, meaning and joy in the present. When the New Testament writers address this inner turmoil pastorally they appeal either to God's present knowledge and care or to the eternal divine plan. In the Sermon on the Mount, Jesus reminds his anxious audience that God is their "heavenly Father" who knows and cares about them. God takes care of the birds and the grass. God will surely take care of you. Paul, in Romans 8:28-39, combines the two perspectives into a grand vision of providence. God foreknew, predestined, called and glorified you. God sent his Son to die for you, so God will do whatever it takes to bring you to glory. Nothing can come between you and God. The biblical writers find great encouragement in accepting things good and bad, great and small, necessary and contingent that befall us as part of an eternal divine plan. It reinforces Jesus' teaching about God's present care, it gives meaning and significance to daily events, and it assures us that nothing can befall us that will not be worked for our eternal good.

6. God realizes his aims perfectly. God cannot fail, even in part. Few Christian theologians would argue that God's eternal plan could fail, that the purpose for which God created the world might not be achieved or might be achieved only in part. For surely God knows what he can and cannot do, and being wise and good would not attempt to do the impossible or the improbable. We cannot believe that God would act on a plan if the full realization of that plan were anything but certain. Most Christian thinkers would reject as unworthy of consideration theories that envision God as capable of making mistakes, not understanding his limits or not anticipating other factors that help determine the outcome of the history of creation. Such theories do not deserve to be treated as serious candidates for a Christian doctrine of God or of divine providence. The central question at issue among Christian theologians lies elsewhere. Even theologians who agree fully with proposition 6, which asserts that God's plan will most certainly be realized, define the divine plan in different ways. Some reason that, since God cannot be certain of accomplishing a plan that depends on angels and human beings making all the right choices, God's plan must be very general and open-ended. Additionally God, being holy and loving, would not choose to create a world so plagued by evil as is ours, if better worlds were possible and feasible.[14] God cannot plan on making his ideal world and must settle for the best feasible world. And God would not include in his plan the unconditional damnation of particular people to hell. So God's plan must accommodate the possibility (but not certainty) that some people might finally reject God's offer of salvation. In all these theories, God's plan is conditioned on other factors, such as human free choices. But would the New Testament authors' appeals to God's eternal purpose be as reassuring as they purport to be if the eternal purpose were conditioned on anything other than God's good will and faithfulness?

I believe my understanding of the relationship of creation and providence sheds some light on this topic. Few people doubt that God could create any creature or any world God desires exactly according to his specifications. Since God creates everything about the creature, his act is conditioned by nothing other than God's own will.[15] In this case, God's plan for the creature would be perfectly realized in the actual creature. But if we include in that world free beings that make a series of choices, some thinkers object that God cannot make world history conform to an eternal plan. However, this line of reasoning makes the mistake to which I referred above. It applies the rules of creature-creature relationships to the God-creature relationship. If we keep clearly in mind that God creates the world with all its internal relationships, the problem of tension between human freedom and the certainty

---

[14]These technical terms are used by contemporary Molinists like William Lane Craig and Thomas Flint. I will address the Molinist theory in a later section.

[15]I will leave discussion of the question of logical restrictions to another occasion.

of fulfilling the eternal plan is resolved. God creates the world he wants, a world that includes our free acts whereby we, as free subjects, become causes of new states of affairs. The problem of mutual exclusivity of divine and human freedom arises only when one thinks of human freedom as absolute, as exempt from all metaphysical laws and from dependence on God.

12

MODELS OF
PERFECT PROVIDENCE

Foreknowledge

IN THE PRECEDING CHAPTERS I have developed a robust doctrine of providence whose central features would be recognized and accepted by theologians from many Christian traditions throughout most of Christian history. There is a consensus that God knows the future exhaustively and that God works in all things to achieve his will. The debates within this tradition center not on whether God exercises comprehensive providence but on *how* God does this without negating human freedom and responsibility and becoming the author of sin. As we begin to think about how God might be understood to exercise the type of providence I have described, recall my discussion of the question of how God created the world in part one. I examined the concepts of act, cause, relation and means and argued that none of these concepts can do justice to the Christian doctrine of creation. The Christian answer to the how of creation is that the Father creates through the Son and in the Spirit. God does not need nondivine means to create. Any attempt to speculate about how God creates the world in terms that would make that act comprehensible will inevitably place God in a dependent relationship with some nondivine means through which the act of creation becomes understandable.

As should be obvious from my understanding of providence I do not believe the how of providence can be understood in human terms any more than the how of creation from nothing can be so understood. How does God exercise providence? God needs no nondivine means to "order and direct" the history of creation to its appointed end. Ordering and directing is a trinitarian act and is just as much a mystery as the trinitarian act of creating from nothing. Hence proposals for understanding how God orders and directs creation should be judged in part by whether they make God dependent on nondivine means.

Though they vary widely in certain respects, it seems to me that models that attempt to shed light on how God exercises perfect or comprehensive providence fall into two classes.[1] One focuses on God's nondetermining foreknowledge, and the other places God's omnipotence at the center. In foreknowledge models God can make his eternal plan according to his eternal foreknowledge of what could or will or would happen.[2] Since God's plan is based on perfect wisdom, power and foreknowledge, God can enact it perfectly down to the smallest detail. It is important to point out that a central feature of foreknowledge models is their distinction between knowledge and causality. Divine foreknowledge does not cause its object to exist or to act as it does. In this way God can know all future events, including free acts, without determining them. Of course, God knows everything God causes, but God does not cause everything God knows. Foreknowledge becomes in this way a means of making a plan that can be enacted without destroying creaturely freedom. Without foreknowledge God could not devise a plan that involved free creaturely acts with certainty that it would be accomplished perfectly.

The second set of models argues that God makes his eternal plan on the basis of his goodwill alone and accomplishes it through his power alone. God's foreknowledge is not passive but active. That is to say, God's knowledge of the future, including future free acts, is not caused by the object of knowledge; rather, God knows the future by knowing what God can do and what he will do. Most thinkers in this group, nevertheless, wish to make room for the efficacy of secondary causes, to preserve human freedom and

---

[1]A view of perfect providence that would be accepted by all parties in the discussion would be the conditions under which the eternal divine plan and purpose for creation map in a one-to-one correspondence to what actually comes to be and happens in the entire history of the world. For a study of eleven different models of divine providence, see Terrance Tiessen, *Providence and Prayer: How Does God Work in the World* (Downers Grove, IL: InterVarsity Press, 2000). Tiessen examines views of providence in order of how much control over the course of events they give to God, beginning from the least to the greatest.

[2]Note the modal qualifiers. I am trying to include all of the foreknowledge models, some of which use modal distinctions in their arguments.

to exempt God from all culpability for sin. To do this they employ a variety of strategies. The divine causality by which God orders and directs the world, as the universal cause of the world's being, operates on a different level from particular causes within the world. Freedom must be defined in a way that is compatible with God's determination, and sinful acts must be defined in a way that does not require divine causality in the same sense that good acts require. God causes sinful acts as acts but does not cause their sinfulness, which is a defect.

THE FOREKNOWLEDGE MODELS

To my knowledge the most powerful and ambitious foreknowledge model of providence[3] is Molinism, or the middle knowledge model.[4] Modern versions of this model hale back to Spanish Jesuit theologian Luis de Molina (1535–1600). The title of Molina's work explains his goal: *A Reconciliation of Free Choice with the Gifts of Grace, Divine Foreknowledge, Providence, Predestination, and Reprobation.*[5] Before Molina, theologians in their specula-

---

[3]There are many variations on the foreknowledge model. David Hunt defends a "simple" foreknowledge model over against "augmented" models like middle knowledge approaches. Simple foreknowledge models assert that we need not explain "how" God attains complete foreknowledge to show its compatibility with human freedom and divine agency and its usefulness in God's providential action. See David Hunt, "The Simple Foreknowledge View," in *Divine Foreknowledge: Four Views*, ed. James K. Beilby and Paul R. Eddy (Downers Grove, IL: InterVarsity Press, 2001), pp. 65-103; and Hunt, "Divine Providence and Simple Foreknowledge," *Faith and Philosophy* (July 1993): 394-416. William Hasker calls into question the usefulness of simple foreknowledge even if God possesses it. Any attempt to change what will happen based on foreknowledge would falsify the original foreknowledge and any attempt on the basis of foreknowledge to bring about what is foreknown would be redundant (*God, Time and Knowledge* [Ithaca, NY: Cornell University Press, 1989], pp. 53-74). See also John Sanders, "Why Simple Foreknowledge Offers No More Providential Control Than the Openness of God," *Faith and Philosophy* 14 (1997): 26-40.

[4]Among the contemporary advocates of Molinism are Alfred J. Freddoso, introduction to Luis de Molina, *On Divine Foreknowledge (Part IV of the "Concordia")*, trans. Alfred J. Freddoso (Ithaca, NY: Cornell University Press, 1988); Alvin Plantinga, *The Nature of Necessity* (Oxford: Clarendon, 1974); Thomas P. Flint, *Divine Providence: The Molinist Account* (Ithaca, NY: Cornell University Press, 1998); Edward Wierenga, *The Nature of God: An Inquiry into Divine Attributes* (Ithaca, NY: Cornell University Press, 1989); William Lane Craig, "God Directs All Things: On Behalf of a Molinist View of Providence," in *Four Views on Divine Providence*, ed. Dennis W. Jowers (Grand Rapids: Zondervan, 2011), pp. 79-100; Craig, *The Problem of Divine Foreknowledge and Human Freedom from Aristotle to Suárez* (Leiden: Brill, 1980); Craig, *The Only Wise God: The Compatibility of Divine Foreknowledge and Human Freedom* (Eugene, OR: Wipf & Stock, 2000); Craig, "The Middle Knowledge View," in Beilby and Eddy, *Divine Foreknowledge*, pp. 119-43.

[5]*Liberi arbitrii cum gratiae donis, divina praescientia, providentia, praedestinatione et reprobatione Concordia.* For a translation of the section on divine foreknowledge, see Molina, *On Divine Foreknowledge.*

tions about the nature of God's knowledge surmised that, even though all God's knowledge of creation is eternal, two aspects or logical moments can be distinguished: natural knowledge and free knowledge. God's natural knowledge is God's grasp of all things possible, and free knowledge is knowledge of everything God wills to create.[6] Both of these forms of knowledge can be stated as divine self-knowledge. God knows the extent of his power or what he could do, and God knows his will or what he will do. All possible or actual objects are covered by these two aspects of divine knowledge. Since all God's knowledge is divine self-knowledge, there is no need to postulate the independent existence or causal efficacy of any object of God's knowledge. There are only two types of objects to be known: those God could cause and those God in fact causes. Hence God is the cause of everything that ever comes to exist.

Molina proposed a theory designed to avoid the determinism to which the traditional model of divine knowledge seems to lead. Molina agreed with traditional theology that God knows eternally what he *could* do (natural knowledge) and God knows eternally what he *will* do (free knowledge); but as a bridge between these two God knows what any free being *would* freely do under any set of circumstances. This move, according to Molina, saves human freedom and responsibility from being destroyed by vicious determinism. In the traditional view, God knows everything you and I will freely do because God causes it.[7] In middle knowledge, God knows everything you and I would do in every possible set of circumstances in an act of causally neutral supercomprehension. This knowledge enables God to select from among the possible sets of circumstances (natural knowledge) the world God wishes to create. Once God decides on which world to create, God possesses free knowledge (free knowledge) of everything that will ever come to be and everything that will happen. In God's omnipotence God can

---

[6]These two moments of knowledge correspond to the two powers of God, absolute and ordinate, a distinction that became important in late medieval debates.
[7]Among the possibilities God knows in his natural knowledge are counterfactuals of creaturely actions; that is, God knows what he could cause me freely to do but determines not to cause. This divine knowledge of counterfactuals does not possess the same status as that of middle knowledge, because God knows these counterfactuals by knowing what he could do. In middle knowledge, God's knowledge of counterfactuals of creaturely freedom is caused by eternal truths grounded in what free agents would do were they faced with particular decisions that will never obtain. When God creates the world, God knows not only what he is doing but also what he is not doing.

create, sustain, concur, order and direct the history of creation exactly according to his eternal plan and precisely to his eternal goal.

We will examine the work of William Lane Craig as a modern representative of a Molinist doctrine of providence. In a recent work, Craig sets out a compact version of the Molinist theory of providence.[8] First, Craig reviews the historical background of the Molinist theory and presents its basic components, which I have set out above. Craig claims that the Molinist theory

> affords God a means of choosing which world of free creatures to create. For by knowing how persons would freely choose in whatever circumstances they might be, God can, by decreeing to place just those persons in just those circumstances, bring about his ultimate purpose *through* free creaturely decisions. . . . [Hence] God can plan a world down to the last detail and yet do so without annihilating creaturely freedom.[9]

Craig speaks of the infinite set of different possible circumstances and free decisions logically possible in those circumstances as "possible worlds."[10] So in God's natural knowledge God knows the full range of worlds logically possible. In God's middle knowledge, God knows what every free creature would actually do if placed in each and every possible circumstance. Craig designates this set of different complexes of circumstances and their corresponding free acts as "feasible worlds." They are called feasible because God could create them without destroying freedom. It is important to note that the set of feasible worlds, though still virtually infinite, is smaller than the set of logically possible worlds. In other words, there are some logically possible worlds that God knows—in his middle knowledge—that he cannot create because free creatures would not actually make the necessary choices for those worlds to exist. So having middle knowledge of the full range of feasible worlds God could create, God can select one of them to actualize. After God decides which feasible world to create— which decision can now be called the decree or the eternal plan for the history of creation—God knows not only which worlds God can create

---

[8]Craig, "God Directs All Things," pp. 79-100. This book also contains Craig's three responses to the other views presented in the debate.

[9]Craig, "God Directs All Things," p. 82.

[10]The term "possibility" refers to logical and not physical possibility. Something that is logically possible may not fall within the scope of a particular being's powers. Traditionally, God's power was understood to extend to everything logically possible.

without destroying human freedom but also everything that will ever happen in the actual world he creates. According to Craig, Molinism provides a way to conceive a perfectly provident God without sacrificing genuine freedom.

Craig also thinks that Molinism provides the best answer to the problem of evil. Why is there so much sin and suffering in the world? Why did not God create a better world? To prepare for Craig's answer we need to understand the ontological status he gives to the truths God knows in his middle knowledge. The power of middle knowledge gives God access to all members of the set of what are called "true counterfactuals of creaturely freedom." Here is an example: "If Highfield gets hungry after 6:00 p.m., on July 9, 2012, he will eat leftovers rather than take the time to fix a new dish." For human beings the truth value (true or false) of this hypothetical statement cannot be known before Highfield actually does or does not do, at the time specified, what would reveal the truth value of the statement. But God knows eternally the truth value of this counterfactual. True counterfactuals of creaturely freedom are eternal just like $2 + 2 = 4$; they cannot be changed even by God. They are just there, presumably in God's mind, and God's knowledge of them is passive. Why then did not God create a perfect world or one much better than our world? According to Craig, God did not have the option to create such a world, "the wrong counterfactuals being true."[11] There was no feasible world that contained a set of true counterfactuals of creaturely freedom in which every free being always makes the right choices. We can only assume on the basis of God's goodness that God chose the *best* feasible world in his decision to create our world.[12] By choosing the *best* feasible world God ruled out "pointless suffering. . . . Thus, we can rest assured that God has morally sufficient reasons for permitting the evils in the world."[13]

---

[11]Ibid., p. 88.

[12]Gregory Boyd misrepresents Craig and Molinism when he says in response to Craig's theodicy argument: "The Molinist must accept that, however nonideal this world may be, it nevertheless is the best of all possible worlds, down to the last detail" ("Response to William Lane Craig," in *Four Views on Divine Providence*, ed. Dennis W. Jowers [Grand Rapids: Zondervan, 2011], p. 138). As I stated above, Craig need affirm only that our world needs to be the best of all *feasible* worlds.

[13]Craig, "God Directs All Things," p. 88. What makes the best feasible world the best? I assume that the best feasible world is the closest to perfection in God's judgment.

The Problem with Molinism (And Other Foreknowledge Models)

We must acknowledge that Craig's Molinist model of providence affirms that God is perfectly provident. In harmony with my five-point summary of biblical theology, consistent with my definition of providence and largely in sync with my six-point exposition of providence, Craig argues that God's eternal plan and purpose will be accomplished in the history of creation in every detail. In my view, his objective of defending perfect divine providence makes Craig's theology of providence qualitatively superior to any theory, such as open theism or process theology, that gives it up.[14] But Molinism consists not only of the thesis it defends but also of a set of suppositions and arguments about the nature of reality, the way God knows and the way God decides what world to create. And some of Craig's Molinist suppositions and arguments are open to serious objections. I want to develop three lines of criticism, which force me to reject Molinism as a viable option for explaining how God can order and direct the world.[15]

1. The passivity in divine knowledge. Craig describes divine natural knowledge as God's knowledge of all logically possible worlds and God's free knowledge as God's knowledge of whatever God has chosen to create. Craig describes his understanding of how God knows as "conceptualist," that is, God knows things by means of concepts; and concepts can be stated in propositions.[16] But as I pointed out above, natural knowledge can also be understood as God's direct knowledge of his being and power. By knowing himself God knows everything that God can do, that is, everything that is possible. God's free knowledge can also be understood as God's

---

[14]I will treat open theism and process theology below.

[15]I developed these three arguments first in my response to Craig's Molinism: "Response to William Lane Craig," pp. 114-22. I added another argument in "'From Faith to Faith' (*ek pisteos eis pistin*): In Defense of the Autonomy of Theology Against Colonization by Analytic Philosophy" (paper delivered at the annual meeting of the Evangelical Theological Society, San Francisco, 2011). I will not enter the debate over the "grounding objection," which is one of the most serious philosophical objections to the idea of middle knowledge. How does God know what a libertarianly free being would do in a set of circumstances that will never obtain? The advocate of simple foreknowledge and the theological determinist can explain how God knows the future because either God sees eternally what will take place or God causes what takes place. But there is no comparable explanation for middle knowledge. For critiques of middle knowledge along these lines, see Robert M. Adams, "Middle Knowledge and the Problem of Evil," *American Philosophical Quarterly* 14 (1977): 109-17; and Stephen B. Cowan, "The Grounding Objection to Middle Knowledge Revisited," *Religious Studies* 39 (2003): 93-102.

[16]Craig, "God Directs All Things," p. 85.

knowledge of his will, that is, everything he will do. But middle knowledge cannot be articulated as divine self-knowledge; rather middle knowledge is divine knowledge of something else—truths, propositions, concepts— that is neither a creature nor God. In middle knowledge God knows the truth value of all counterfactuals of creaturely freedom without deter- mining them. These true counterfactuals exist eternally as truths, pre- sumably in the divine mind, "like a mind's knowledge of innate ideas."[17] God knows them to be true because they are true; they are not true because God knows them. This conceptualist distinction is essential to the theory, for it separates divine foreknowledge from divine causality, which is the key to preserving human freedom.

Consider now the price paid for separating divine knowing from divine being and acting. We must accept the existence of a class of unchangeable and eternal truths that determines both what God can know and what God can do. True counterfactuals of creaturely freedom are not laws of logic, which are assumedly necessarily true; rather, their being true depends on a free act of a creature. More than this, their truth value depends not on what a creature actually does, because they do not yet exist and may never exist. It depends on what a creature *would* do, given certain circumstances. I will ignore the famous "grounding objection," which asks how a counterfactual can be true or false when the condition that would make it true or false is both contingent and nonexistent. I am concerned about the threat that these eternal truths pose to God's deity, God's freedom and independence. The conviction that all things other than God are God's creatures has been axi- omatic for Christian theology since *creatio ex nihilo* came to explicit articu- lation in the second century. There are and can be no nondivine beings that are not God's creatures. But Craig says that true counterfactuals of creaturely freedom exist eternally, are true, exercise causal power on the divine mind and constrain the divine will. In his effort to escape the idea that divine knowledge determines its object, Craig affirms that the object determines divine knowledge. The conclusion is inescapable: the object of middle knowledge is middle being. This theory may be judged brilliant philosophy by some standards, but it leaves out of consideration the central Christian revelation of God. In place of the Word of God through whom God created

---

[17]Ibid.

the world and who became incarnate in Jesus Christ it substitutes another "means" and archetype for the world. Hence the philosophical problem of mediation, which I discussed in part one, shows up again in Molinism. And for the same reason: the human intellect just cannot comprehend how God could create directly the world we experience without not-quite-divine mediators. In the Molinist universe God is not the creator of all things, and as I will show in the next objection the existence of these middle beings prevents God from executing his perfect will for creation.

2. God's options eternally limited. According to Craig, middle knowledge provides God a means of knowing which possible worlds are also feasible because in middle knowledge God knows the truth value of all counterfactuals of creaturely freedom. So God could decide to create any one of them with complete foreknowledge of the free decisions within it. In any of them God's will would be done in every detail. Middle knowledge puts God in a position of being able to choose which feasible world to create, but God is limited to the set of feasible worlds available. Keep clearly in mind that knowing a world is logically possible does not give God the knowledge that God can actually create it.[18] The character and content of each of these feasible worlds depends on the truth value of its counterfactuals of creaturely freedom and is beyond God's control. Imagine that God, being perfectly omniscient and good, possesses a vision of the creaturely world that he would like to create, a world that fittingly corresponds to his goodness and holiness. This ideal world is of course logically possible; but as God surveys the range of feasible worlds, God may or may not see among them God's first choice of worlds to create. Whether God sees his ideal world among feasible worlds does not depend

---

[18]According to Molinism, God does not possess middle knowledge of the truth value of divine counterfactuals of freedom. God does not know what he would do in the same way he knows what creatures would do. Hence God would have to choose which feasible world to create without knowing how he would respond to creatures' free decisions in those innumerable worlds. Only in deciding which feasible world to create could he decide simultaneously how he would act in that world. But this raises a problem: how can God's middle knowledge of feasible worlds be exhaustive, apart from knowledge of divine counterfactuals within that world? God's decision to create would increase his knowledge of this world by the addition of divine counterfactuals. Might not the world God creates differ from the corresponding feasible world known in middle knowledge? And if even one change is made to the circumstances in this world, would this not again place free creatures in a different situation and possibly cause them to change their decisions? See Flint, *Divine Providence*, pp. 55-71, for a discussion of the Molinist view of counterfactuals of divine freedom.

on God. God must pick a world off the lot or go home empty-handed. God cannot simply do what he would prefer.

Let us explore God's predicament a bit further. From the point of view of natural knowledge of logically possible worlds, the character of feasible worlds cannot be predicted. Perhaps God's ideal world will not be found among the feasible worlds. Or perhaps the world closest to God's ideal is still a pretty lousy world, very far from the ideal. It is possible that every feasible world is so bad that it would be better not to create at all than create even the best one. Again, God has no control over the quality of feasible worlds from which God must choose. Clearly the problem of evil is beginning to come into view, and I will deal with it in the next objection. Here I want to underline the eternal limitation that the middle-knowledge scheme places on God. God cannot simply create the world God wants. Craig's own words will show that I am not exaggerating: "It is correct that Molinism holds that God cannot create whatever he wants or accomplish whatever he chooses, since some possible worlds are infeasible for God. So he is not free to create just any world he pleases."[19] Another reality limits God's choices. If there is a world worth creating among the given set of feasible worlds, God has nothing to do with it. Call it luck or fate or chance; but for God it is a brute fact, anything but divine sovereignty. To conclude this objection I quote from my published response to Craig on this subject:

> The irony in this situation is stunning. The theory of middle knowledge was supposed to rid the world of fate and chance while preserving human freedom. To accomplish this task, however, it limits *God's* freedom and subjects *him* to a kind of fate worse than the one from which it supposedly liberates human beings. If God determines our "fate," at least we can believe our destiny is determined by a free, loving, and just God. But if God is fated to be constrained eternally by impersonal laws and truths, we can take no such consolation. Indeed it was against precisely this type of impersonal fate that the patristic theologians fought so hard![20]

3. Divine excuses. Why is there so much sin, suffering and evil in the world? According to Craig the answer to this question given by the Augustine-Calvinist theory of providence "seems, in effect, to turn God into

---

[19]William Lane Craig, "Response to Ron Highfield" in Jowers, *Four Views on Divine Providence*, p. 171.
[20]Highfield, "Response to William Lane Craig," pp. 120-21.

the Devil."[21] If God determines everything to happen as it does, then God causes sin and every evil. Even though God could have easily caused everyone to do justice and never to experience pain and distress, God did not. According to Craig, Molinism avoids this problem, maintains God's goodness and absolves God of causing sin. Apparently in his middle knowledge, God found no feasible world free of sin and suffering. God did not have the option to create a world *with* freedom but *without* sin and suffering. Hence we cannot say that God chose our world *because* it contained evil and sin along with stars and planets and oceans. Drawing on our faith in God's goodness we can affirm that God would not have created our world unless God had morally sufficient reasons.[22] We can believe that God did the best he could.

But does Molinism really present a satisfactory answer to the problem of evil? As we noted above, the effectiveness of Craig's theodicy depends on showing the unavailability of a feasible world better than ours. God simply cannot create his ideal world. Above I quoted the first half the following admission by Craig. Now I add the claim that follows:

> It is correct that Molinism holds that God cannot create whatever he wants or accomplish whatever he chooses, since some possible worlds are infeasible for God. So he is not free to create just any world he pleases. *But I deny that that places any nonlogical limit on his power, any more than does his inability to create a married bachelor.*[23]

Craig does not deny that those infeasible worlds are logically possible. So how can it be logically *impossible* for God to create a logically *possible* world? How can a world be both logically possible and logically impossible at the same time? To prepare for the answer to this question imagine one of those logically possible but infeasible worlds. Imagine World Alpha (W_α) in which the angels who sinned in our world do not sin and every other libertarianly free being remains completely faithful. Clearly there is no logical contradiction in such a state of affairs. Nevertheless, according to Craig, God can no more create this world than God can create a married bachelor or make a square circle. Why? Because, although possible, W_α is not feasible. And

---

[21]Craig, "God Directs All Things," p. 91.

[22]Ibid., p. 88. What makes the best feasible world the best? I assume that the best feasible world is the closest to perfection in God's judgment.

[23]Craig, "Response to Ron Highfield," p. 171 (emphasis added).

why is it not feasible? Because there are no worlds (like W_α) in which libertarianly free angels and human beings remain completely faithful. How does God know this? God knows it through his middle knowledge!

But how do *we* know that God knows that W_α is infeasible? As far as I can tell, the only means we have of knowing this is our experience of the actual world in which we live. Our world is not W_α or anything close. In our world many angels and all human beings—save one—became and will become sinners. And we know that if God could have created W_α instead of our world, God would have done so because not to have done so would call into question his goodness. In this case God would not have done the best God could.

But how do we know that God would have preferred to create W_α and innumerable other worlds than to settle for our world? Only because of our negative experience of this world! We conclude that no perfectly good God would choose to actualize *our* world if God could have created whatever he wanted. We judge that it would be morally reprehensible, and possibly blasphemous, to assert that when God chose to actualize this world, God was doing exactly what he wanted. But isn't this judgment too bold? How does a finite and sinful mortal gain the perspective to make such a judgment about the totality of time and space and history? Shouldn't such judgment be left to God? Shouldn't we harbor some suspicion that a human attempt to justify the ways of God to humanity might really be a subtle way of justifying the ways of humanity to God?[24] Ultimately, then, the need to excuse God for making this world is the unjustified belief that we can imagine a better world than the one God created. God is excused, however, because our imaginary better worlds are infeasible.

Where, then, is the logical contradiction in creating the infeasible W_α, even though W_α is logically possible? The contradiction, according to Craig, rests on the principle that God cannot know what is true to be false or what is false to be true. The eternal and immutable counterfactuals of creaturely freedom possess the truth value they do, and not even God can change them.[25] Since the counterfactuals in W_α do not have the truth value they

---

[24]To bring God before the bar of human judgment, whether to defend or accuse, is itself to stand under his judgment. God is always the judge and never the defendant. This system's approach to the problem of evil arises not only from an agonized question but from a moral judgment about God's justice in creating our world.

[25]They are made true by what free agents "would" do in every possible situation, a problem I addressed in my response to Craig.

possess because of logical necessity, W_α is logically possible. But they possess a kind of necessity similar to that of past events, which were open before they happened but now are determined. The counterfactuals in W_α could have been different considered logically but are in fact determined by what free beings would do.

But would actualizing W_α (or any other so-called infeasible world) really involve logical contradiction, as Craig argues? If you do not conclude the infeasibility of W_α *simply because W_α is not the actual world,* you can ask the following question: "Is there a logically possible way for God to guarantee that possible W_α is feasible?" Now, since Molinists insist on speaking of the incompatibilist brand of libertarian freedom, determinism and every other means that excludes alternative choice are ruled out. But what about persuasion? Do Molinists believe that persuading someone to believe in God or to buy a Ford violates libertarian freedom? Is there any *logical* problem with the Holy Spirit persuading a human being to believe in God or to buy a Ford? Are there any logical limits to the extent of the Holy Spirit's success in matters of persuasion? Finally, is there any logical contradiction in saying that God can always persuade libertarianly free agents to choose the good every time? I cannot see any.[26] I deny, then, the Molinist's contention that infeasible worlds are infeasible because it is logically impossible to create them. So, if the Molinist continues to insist on their infeasibility this can be only because they think God is limited in power and wisdom. God is not really omnipotent in the traditional sense because there are things logically

---

[26]Duns Scotus, in his commentary on Peter Lombard's *Sentences,* proposes a type of infallible persuasion that God exerts in his grace. God moves without fail those whom he chooses to will his good will. According to Reginald Garrigou-LaGrange, Thomists limit infallible persuasion to the vision of God. I agree that only the vision of God would necessitate willing the good, but the will need not be moved by necessity to be moved infallibly. There could be conditions other than the vision that are sufficient for God's purposes. For further discussion, see Reginald Garrigou-LaGrange, OP, *Predestination: The Meaning of Predestination in Scripture and the Church* (Rockford, IL: Tan, 1998), p. 113-14. The early post-Reformation Reformed theologian Girolamo Zanchi (1516–1590) argues that God's grace never acts by coercion on the human will. Instead, God leads the will persuasively: "For even when he [God] changes it and makes it good out of bad, and willing out of not-willing, he does not exert power onto it but persuasively leads it so that it, being led, spontaneously even moves itself immediately." The editors in commenting on this text point out that only after the Remonstrants began to employ the concept of "moral persuasion" did the Reformed theologians cease using the idea of persuasion to explain God's action in conversion (*The Fall of the First Man, Sin and the Law of God* 1.6, translated and cited in *Reformed Thought on Freedom: The Concept of Free Choice in Early Modern Reformed Theology,* ed. Willem J. van Asselt, J. Martin Bac and Roelf T. te Velde [Grand Rapids: Baker Academic, 2010], pp. 74, 92).

possible but naturally impossible for God. Specifically, although it is logically possible for God to persuade all libertarianly free beings to make the right choice all the time, God cannot do it. This conclusion places the Molinist theodicist in a dilemma. Either God is not good because God does not actualize W_α (or any other so-called infeasible world) when he could or God's inability to actualize W_α (or any other so-called infeasible world) is natural rather than logical, that is, due to God's inability to persuade free agents to choose their own highest good.

In sum, I do not believe foreknowledge models succeed in showing how God can exercise perfect providence. Indeed, Craig's effort to find this "how" illustrates the fatal flaw built into all such projects. Every theory that proposes a *means* for God to exercise perfect providence assumes that God *needs* a means to exercise perfect providence. But I question whether the Molinist understanding of "perfect providence" really should be called such. According to Craig, God's providence is perfect in that once God has chosen a feasible world God can execute that plan perfectly. But how can perfect providence be limited to executing a plan perfectly, if the plan itself is not perfect? I would think that perfect providence would include the perfect execution of the perfect plan. Molinists do not believe God can do this.

13

MODELS OF
PERFECT PROVIDENCE

Omnipotence Models

THE SECOND FAMILY OF MODELS for explaining and defending God's perfect providence places God's power and universal activity at the center. God's eternal plan and purpose perfectly reflects his goodwill, and in his infinite wisdom and omnipotence, God enacts his plan and achieves his purpose down to the smallest detail. Omnipotence models reject the separation between divine knowledge and divine power. There is no class of objects that God knows merely passively. God is the creator of everything that is not God, and God knows what he does. Whereas foreknowledge models insist on the passivity of God's knowledge of free acts to preserve human freedom and absolve God from the taint of sinful acts, omnipotence models reject passivity in divine knowledge for the sake of divine freedom and sovereignty. And just as foreknowledge models find themselves charged with undermining divine freedom, omnipotence models must defend themselves against accusations of destroying human freedom and making God the author of sinful acts.

The central elements of this model were adopted by nearly all Western Christian theologians from the fifth to the seventeenth centuries. Augustine of Hippo is often credited with being the founding architect of the omnipotence model of providence. But the list of advocates includes such illustrious names as Anselm, Bernard of Clairvaux, Thomas Aquinas, Luther, Zwingli, Calvin, Jonathan Edwards and Karl Barth. This view is advocated by the seventeenth- and eighteenth-century post-Reformation Reformed Orthodox, the Dominican order in the Roman Catholic Church and many contemporary theologians. Rather than attempt the impossible task of presenting a history of the omnipotence model, I shall select a few of its most articulate and influential representatives.

THOMAS AQUINAS

Thomas Aquinas asserts that God can do not only anything God wills but "everything possible in itself."[1] In his providence God orders all things toward his goodness, neither arbitrarily nor by necessity; for God is beyond this dichotomy. As Aquinas says, "So the coming to be of creatures, though it finds its first reason in God's goodness, nevertheless, depends on a simple act of God's will."[2] As the creator and governor of all things, God causes all things: "But the causality of God, Who is the first agent, extends to all things not only as to the constituent principles of the species, but also as to the individualizing principles. . . . Hence all things that exist in whatsoever manner are necessarily directed by God toward the end."[3] But we cannot conclude from this that all things happen by necessity. In God's universal causality God causes some things (with respect to their immediate causes) to happen necessarily, some contingently and some freely. Aquinas explains:

> A similar difference must be noted in regard to God's will; for God's will is to be thought of as existing outside the realm of existents, as a cause from which pours forth everything that exists in all its variant forms. Now *what can be and what must be* are variants of being, so that it is from God's will itself that things derive whether they must or may or may not be and the distinction of the two according to the nature of their immediate causes. For he prepares causes that must cause for those effects that he wills must be, and causes that might cause but might fail to cause for those effects that he wills might or might not be . . . although all depend on God's will as primary cause, a cause which transcends this distinction between *must* and *might not*.[4]

---

[1]Thomas Aquinas, *Disputations on the Power of God* 1.3.7 (Aquinas, *Selected Philosophical Writings*, ed. and trans. Timothy McDermott, Oxford World's Classics [Oxford: Oxford University Press, 1993], p. 249).

[2]*SCG* 3.97 (Aquinas, *Selected Philosophical Writings*, p. 274).

[3]*SumTh* 1.22.2 (Aquinas, *Basic Writings of Saint Thomas Aquinas*, ed. Anton C. Pegis, [New York: Random House, 1945], p. 232).

[4]Thomas Aquinas, *Commentary on Aristotle's* De Interpretatione (Aquinas, *Selected Philosophical Writings*, p. 283).

As indicated in the above quote, Aquinas asserts the compatibility between divine causality and human freedom. However, he rejects psychological determinism and affirms a freedom of indifference. The indifference of the will, we must note, is only to finite alternatives presented to the will. Aquinas reasons that the human will must necessarily will happiness, "for there is a sort of natural compulsion on every one to want to be happy. . . . If then there were goods which were the *sine qua nons* of happiness, these too would compel desire."[5] But in our decisions in ordinary life we cannot know whether choosing this or that is connected to happiness as a sine qua non. Hence our wills are not determined wholly by the objects of our judgment. In this way Aquinas defends the freedom of indifference against psychological determinists.

If God causes all things, does God cause our sins? Sin is not a substance that requires a cause or that can cause other things. Only good things, real things, have causes and can become causes of other things. According to Aquinas, evil is "the privation of what is connatural and due to anyone. . . . Now privation is not an essence, but is the non-existence of something in a substance."[6] Sin is a defect in a free act. "And our free choice causes sin by a lack of response to God. God then causes the free choice but not the sin."[7] And further:

> What we must say then is that God is the first source of all movements, and that some things are so moved by him that they also move themselves, having free choice . . . but if they are not properly ordered then there will result disordered or sinful action, in which what there is of action can be traced back to God's causality but what there is of disorder and deformity does not have God as cause but our free choice. For this reason then we say that God is responsible for sinful actions but not for sins.[8]

POST-REFORMATION REFORMED ORTHODOX THEOLOGIANS

The omnipotence model of divine providence was adopted and defended by

---

[5]Thomas Aquinas, *Commentary on Aristotle's* De Interpretatione (Aquinas, *Selected Philosophical Writings*, p. 284). See also *SumTh* 1.83.1: "And just as by moving natural causes He does not prevent their actions from being natural, so by moving voluntary causes He does not deprive their actions of being voluntary; but rather is He the cause of this very thing in them, for He operates in each thing according to its own nature" (Pegis, *Basic Writings of Saint Thomas Aquinas*, 1:786).
[6]*SCG* 3.8 (Pegis, *Basic Writings of Saint Thomas Aquinas*, 2:13).
[7]Thomas Aquinas, *Disputations on Evil* 3.1 (Aquinas, *Selected Philosophical Writings*, p. 293).
[8]Thomas Aquinas, *Disputations on Evil* 3.2 (Aquinas, *Selected Philosophical Writings*, p. 296).

the chief architects of the early sixteenth-century Protestant Reformation, Luther, Zwingli and Calvin. Later sixteenth- and seventeenth-century Reformed Orthodox theologians maintained and developed the Reformers view of providence. Heinrich Heppe in his chapter on the Reformed doctrine of providence describes the orthodox view in thirty-three points. I will summarize those points under a few general headings.

1. *The definition of providence and its relationship to creation.* Reformed dogmatics of this era emphasized the unity of creation and providence.[9] Johannes Braunius (1628–1708) explains: "In respect of God the same action is creation and providence; God works all things by a single, most unifold will, that they exist and remain in existence and work."[10] Johannes Heidegger (1633–1698) asserts: "Preservation is not an act distinct from creation but is continued creation."[11] Emphasis was also given to the connection between the eternal decree and its actual execution in history; so providence in some thinkers was divided into these two general aspects. The Reformed theologian Amandus Polanus (1561–1610) defines providence in this way: "God's providence is God's transeunt [i.e., outside the divine mind] action, by which He cares for and administers the world created by Him and all things that are and are made in it according to His own will for His own glory and the salvation of the elect."[12]

2. *The threefold execution of the divine plan.* Providence, though one divine act in relation to God, can be considered threefold in relation to creatures: preservation, concurrence or cooperation and government. Preservation is the act "by which God maintains and perpetuates the things made by Him as regards their existence, essence and natural faculties."[13] Since God created creatures from nothing, they do not possess the power of self-preservation. They are forever dependent on their Creator for their continued existence and activity. Concurrence is the divine act by which God enables creatures to begin, continue and finish

---

[9]For the following exposition I shall be relying on Heinrich Heppe, *Reformed Dogmatics*, rev. and ed. Ernst Bizer, trans. G. T. Thompson (1950; repr., London: Wakeman Great Reprints, n.d.).

[10]Johannes Braunius, *Doctrina foedorum, sive systema theologiae didacticae and elencticae* 1.2.12.1 (Amsterdam, 1688), quoted in Heppe, *Reformed Dogmatics*, p. 251.

[11]Johannes Heidegger, *Corpus theologiae Christianae* 7.22, quoted in Heppe, *Reformed Dogmatics*, p. 251.

[12]Amandus Polanus, *Syntagma Theologiae Christianae* 6.1 (Hanover, 1624), quoted in Heppe, *Reformed Dogmatics*, p. 253.

[13]Heidegger, quoted in Heppe, *Reformed Dogmatics*, p. 257.

their acts. In concurrence God acts with creatures "so as to urge or move them to action and to operate along with them in a manner suitable to a first cause and adjusted to the nature of second causes."[14] As these words witness, the Reformed Orthodox were keen to emphasize that God's action on and in creatures does not replace or destroy the genuine efficacy of creaturely acts. In fact, God's act, as truly creative of something other than God, establishes their efficacy. The Reformed view of concurrence must be distinguished from other views that use the term but differ in their understanding. Divine concurrence cannot be limited to mere preservation of existence while the creature acts independently or to an indeterminate energy given to the creature—like wind in the sails or the sun's energy on the green plant—which the creature then determines to this or that end of its choosing. In concurrence God acts on and in our acts so that we begin, do and finish that which God wills to be done by us; but this divine determination in no way vitiates our freedom. Like Aquinas, the Reformed insisted that God moves creatures according to their natures and operations: if the connection between causes and effects is necessary, he moves them to their necessary effects, if contingent, then contingently, and if voluntary, then freely.[15]

Governance is the third aspect of providence. God directs all things to their ends and especially to their final end, which is the glory of God. The *Leiden Synopsis* (1624) defines governance this way: "We also subject to the providence of God the ordering of things to an end, especially the final one. For there not only belongs to it the ordering of means to an end, but also the achievement of the end."[16] Within the matrix of created causes, things are ordered and directed not only to their individual ends—the acorn becomes the oak—but also to the final end. Hence it is beyond the power and wisdom of individuals, apart from divine governance, to achieve their divinely appointed ends. Things without freedom God governs by law implanted in them, whereas things possessing the power of voluntary movement God governs by "a peculiar method suited to their freedom, which is entirely moral and consists in Him laying down what is to be done and forbidding what is to be left undone . . . and in Him personally fulfilling or not fulfilling

---

[14]Heidegger, quoted in ibid., p. 258.
[15]Ibid., pp. 259-61.
[16]Quoted in ibid., p. 262.

in them by His movement, aid and grace according to his pleasure, what the law has ordered them to do or to omit."[17]

3. Contingent and free acts are subject to universal divine determination. According to the Reformed Orthodox, contingent and free acts are subject to divine determination and causality just as are events that happen by necessity.[18] According to Zacharias Ursinius (1534–1583), contingency is "the order between a cause and a changeable effect."[19] An event can be contingent with respect to its immediate creaturely causes and necessary with respect to God's eternal will. Ursinius continues to explain this distinction: "In respect of God the order between cause and effect is unchangeable, but in respect of creatures the order is changeable between the cause and the same effect."[20] Petrus Van Mastricht (1630-1706) puts this relationship this way: "The Reformed are of the opinion that all effects whether they be contingent or necessary, happen surely and undeviatingly, provided their causes have been aroused and predetermined by the divine influx."[21] Hence we must confess that human free acts are under complete divine control even though reason cannot comprehend how this is possible. Johannes Braunius (1628-1708) affirms both truths without resolving the tension between them: "We admit that we cannot clearly and distinctly conceive, how the freedom of our will consists with the operation of God. Yet it is certain (1) that in all our volitions our minds depend on God. It is also certain (2) that our will always acts most freely; no one has ever experienced compulsion of his will; so that if freedom were removed, will could be done away with too."[22]

4. Rejection of mere foreknowledge and other errors. The Reformed reject theories that rely on passive foreknowledge or an indifferent and

---

[17]Heidegger, quoted in ibid., p. 263.

[18]In this era the term *determination* did not have its modern meaning of all-determining fate. "Rather, determination means that a cause gets directed to one effect. A natural cause is determined by its nature to the act; a free cause determines itself by freedom to one of its possible acts. Hence, determination refers to the state of a cause: being undetermined means that the (free) cause has not yet determined itself to a certain effect. A determined cause will produced its determined effect, but still the effect can be either contingent (determined by a free act) or necessary (determined by a natural act)" (Willem J. van Asselt, J. Martin Bac, and Roelf T. te Velde, eds., *Reformed Thought on Freedom: The Concept of Free Choice in Early Modern Reformed Theology* [Grand Rapids: Baker Academic, 2010], p. 31).

[19]Heinrich Heppe, *Reformed Dogmatics*, p. 266.

[20]Ibid.

[21]Ibid.

[22]Ibid., p. 272.

general providence to resolve the paradox of divine and human freedom. The *Leiden Synopsis* affirms that "there is in creatures no freedom of will not derived from participation in the highest increate freedom which is the first, proper and inmost cause of all created liberty and of all free actions, insofar as they are such."[23] The orthodox also reject the idea that God moves the free will by solely fallible moral persuasion; God moves the will *physically*. By "physically" they do not mean by a natural cause or by coercion but rather an act in which God operates directly on the will to communicate the effect God wills. The difference can be grasped by noting the difference between the direct, inner illumination of the Holy Spirit and the indirect and external appeal of divine promises and threats revealed in the law.

   5. In providence God permits and causes sinful acts without authoring or excusing sin. God also works through sinful acts, and in a certain sense causes them, though not in the same way God causes good acts. God actively causes good and permissively allows sin. Polanus describes divine permission as "the act of the divine will by which God, in whose power it is to inhibit the actions of others, if He willed, does not inhibit them, but according to His eternal and righteous decree allows them to be done by the rational creature."[24] Apart from the divine action of grace, fallen and sinful human beings will sin. God sometimes allows sin to take its course by not acting to stop it. God does this not because God approves of sin, but to punish prior sin or in some other way to manifest his righteousness and glory. But the Reformed insist that we must not take the act of permission as mere withdrawal; instead, it is a positive act. God wills for his own reasons that this particular person do this particular sinful act at this time and place. If we are careful to distinguish various aspects of the sinful act we can say that God is the *deficient cause* of the sin. Many Reformed Orthodox theologians make a threefold distinction when speaking of sinful acts: the material of the act, the defect or unlawfulness of the act and God's ordering of the material act to a good end.[25] God causes the material act actively and uses this act according to his purposes. But the unlawfulness in the act, being a defect and hence ontologically empty, cannot have an active cause. It exists only as the material act exists; hence the *divine* cause of the material act is

---

[23]Ibid., p. 269.
[24]Ibid., p. 274.
[25]Ibid., p. 276.

the cause of the defect only as a defective cause. However, the *human* actor performs the material act in opposition to divine law and so is the author of the sin in the act. God never wills sin as sin but only as punishment for sin. "In fact God punishes sins with sins. Hence He wills sin not as sin but as a punishment or act of justice, because all punishment of sin is just and thereby good" (Gulielmus [William] Bucanus [d. 1603]).[26]

Karl Barth

As my example of a contemporary theology of providence that falls within the omnipotence family I have chosen its most influential advocate, Swiss Reformed theologian Karl Barth. Barth treats divine providence in the third part of his four-part doctrine of creation and outlines his doctrine of providence roughly along the lines of the Reformed Orthodox studied above. In the end Barth defends a view of providence very much like those of Aquinas and post-Reformation Reformed Orthodoxy. It is worthy of note even before we begin our exposition, however, that Barth criticizes his medieval and his Reformed predecessors for developing their doctrines of providence in too great isolation from the Christocentric and trinitarian pattern of Scripture. I shall summarize Barth's 531-page volume under a few headings.

1. The Christian doctrine of providence must be defined as the execution of the decree of the election of Jesus Christ to be Lord of all. Barth follows the Reformed Orthodox in treating the doctrine of providence as an aspect of the doctrine of creation rather than as Aquinas, who divides it between the sections on God and creation. Unlike some of his Reformed predecessors, however, Barth rejects the idea that providence is a continuation of creation. Though the acts of creation and providence are "one in their divine origin," they are genuinely distinct.[27] God's relationship to the creature in the act of creation is direct, while his relationship to the creature in providence is indirect. Providence presupposes the existence of creatures; creation does not. In creation creatures are completely passive, whereas in providence they are also active.

Barth insists that we cannot understand God's providence in a Christian way without understanding it in relation to the character and purposes of the God revealed in Jesus Christ. According to Barth, we must not limit

---

[26]Ibid., p. 277.
[27]CD III/3, p. 7.

providence to the story line of the Bible on the one hand or "to the common denominator of a doctrine of general world occurrence" on the other.[28] The theme of the Christian doctrine of providence is that the general history of creation is "really ordered in relation to the history of the covenant." In the end, God "will finally be manifest Himself in this One [Jesus Christ] as the Lord of the whole."[29] Barth explains:

> The faith awakened at the one point by the revelation of God . . . is necessarily faith in His lordship even at points where there is no such revelation, where to all appearances we have to do only with creaturely occurrence, where the orders and contingencies of natures, the works of caprice and the cleverness or folly, the goodness or badness of man seems to be the only reality.

While noting the inadequacy of each metaphor, Barth describes creation variously as an instrument in God's hands, a theater for displaying God's glory, a servant and a mirror to reflect his glory. The common factor among these metaphors is conviction that world history serves an end beyond itself toward which it is being guided. The fulfillment of the eternal covenant made in Jesus Christ is the goal and meaning of world history. God alone can *preserve, empower* and *direct* creation to its divinely appointed end. And this threefold divine act is what the doctrine of providence treats.

2. "God fulfils his fatherly lordship over His creature by preserving, accompanying and ruling the whole course of its earthly existence."[30] Barth here follows the traditional threefold division of the doctrine of providence: preservation, concurrence and government. First, God *preserves* the creation. It is not enough to speak of God as a supreme being who sustains dependent beings by the same power through which God created them. Instead of such categories as causality or participated being, Barth uses the personal language of the Bible. God preserves his creatures in the eternal love, grace and faithfulness manifested in Jesus Christ. God elected humanity in and for his Son, and his preservation of creation is his faithfulness to that choice. In keeping with his distinction between creation and providence Barth emphasizes the indirectness of God's preservation. God preserves creatures in the context of creation and within the mutual relationships of dependence of creature on creature. To put this in terms of

[28]Ibid., p. 37.
[29]Ibid., p. 41.
[30]Ibid., p. 58.

traditional theology, the first cause, God, maintains the efficacy of the worldly network of second causes. Barth says, "It is God alone who does everything according to his own free good-pleasure. But He does it by maintaining this relationship and therefore by maintaining the creature by means of the creature."[31] Barth's final distinction under preservation is that the "nothing" from which God preserves human beings should not be understood in metaphysical terms alone—as mere nonexistence—but in a theological sense as "nothingness." Nothingness is more threatening than mere nonexistence. It is the whole realm of that to which God says, "No," a realm of chaos, imaginary possibilities, destruction, evil and sin. God preserves us from falling into nothingness, only one aspect of which is nonexistence.[32]

Second, Barth treats the divine *accompanying*. God not only gives creatures existence and preserves them from nothing but also continually gives them their "own autonomous activity."[33] God does not create creatures merely to let them go their own way but accompanies them all the way to the end. God's activity surrounds the creature's activity "in that which borders upon it, in its environment, in the nexus of being in which it has its duration."[34] Barth can even say, with due qualification, that God "co-operates with the creature, meaning that as He Himself works He allows the creature to work."[35] Instead of limiting the freedom of the creature, God's accompanying and overruling the activity ensures its genuine freedom. There is no competition between divine and human freedom. Quite the opposite is true; for God's free activity opens room for human freedom. Without God we can do nothing.

Barth addresses at length the suitability of the familiar concept of causality (primary and secondary causality) for reconciling perfect divine providence with human freedom. He concludes that there is no reason to prohibit using the concept of causality provided certain precautions are maintained. (1) The "cause" should not be taken as something that works automatically. We are speaking of free causes. (2) Neither God nor human beings is a "thing" or substance that causes effects. Things are open to comprehension and mastery, but God and human beings are open to being

---

[31]Ibid., p. 65.
[32]We will deal in the fourth point with Barth's understanding of evil.
[33]Ibid., p. 91.
[34]Ibid.
[35]Ibid., p. 92.

known only in self-revelation. (3) The idea of cause cannot be taken as a univocal concept that applies in the same way to God and human beings. It does not describe the activity of two subjects but two utterly different "active subjects."[36] Barth makes clear the difference: "The great unlikeness of the work of God in face of that of the creature consists in the fact that as the work of the Creator in the preservation and overruling of the creature the work of God takes the form of absolute positing, a form which can never be the proper work of the creature."[37] Barth suspects that the concepts of primary cause and secondary cause violate this rule, and suggests, rather, that we should speak of "*causa divina* or *creatrix* and *causa non divina* or *creata*."[38] This change would make clear that divine action does not operate the same way and on the same level as human action and so cannot displace human freedom. (4) If the preceding three cautions are observed the fourth should also follow: We must not allow a philosophical system to gain the upper hand. Theology must maintain its autonomy, sticking to its source and following its methods tenaciously. (5) The concept of causality may be used only if we emphasize that the "first cause" is "the operation of the Father of Jesus Christ in relation to that of the creature."[39] The Father of Jesus Christ "takes it to Himself as such and in general in such sort that He co-operates with it, preceding, accompanying and following all its being and activity, so that all the activity of the creature is primarily and simultaneously and subsequently His own activity, and therefore part of the actualization of His own will revealed and triumphant in Jesus Christ."[40] God "bends" the activity of every creature to serve his grand and gracious purpose that will be fulfilled in Jesus Christ.

Barth makes it unequivocally clear that the Creator's freedom in action is completely superior to the creature's action. The creature's action in no way limits or conditions God. Barth rejects the "Jesuit and Arminian" disjunction between omnipotence and foreknowledge. "The foreknowledge of God is a movement of His omnipotence. . . . What God knows He wills, and what He wills He does. Not only does He know all in all but he also works all in all.

---

[36]Ibid., p. 102.
[37]Ibid.
[38]Ibid., p. 104
[39]Ibid., p. 105.
[40]Ibid.

And He does so as the eternal God."[41] Every activity of the creature and every event in the history of the universe is also the activity of God. God's activity is "unconditional and irresistible."[42] Only if we think of God as an unknown supreme being or an impersonal law or an all-determining force will we find God's Lordship anxiety-provoking and threatening. But if we keep clearly in mind that the God of providence is the gracious and loving Father who in the incarnate Son has been revealed as eternal love, we cannot feel threatened or grow anxious for our freedom and dignity. Barth insists that we gain our understanding of the identity, character and purpose of the God of providence from the grace revealed in Jesus Christ. Hence God's providential activity arises from his eternal love and is directed toward our freedom and salvation. We cannot ask for any greater freedom and dignity than "to be ruled absolutely by the divine activity of grace."[43] It is as far from the truth as east is from west to think that we might be better off to put some distance between us and God's gracious action. Barth warns: "And if anyone thinks it necessary to diminish the sovereignty of the activity of God or to set a limit to His omnipotence, let him consider well what he is doing. For if that is the direction in which his thoughts and utterances run, then he is contending for the greatest possible evil that could ever befall the creature as such."[44]

Barth appreciates the modesty of theologians who say that the *modus* or "how" of God's action in relation to creation is a mystery incomprehensible to human reason. However, he insists that the Scriptures actually give an answer to this question. God acts in creation through his Word and his Spirit. And although the work of the Word and Spirit are still mysterious, we have in the Bible many instances of God's action from which to learn something about this "how." God creates through the Word and gives life in the Spirit. The prophets and apostles spoke the Word in the power of the Spirit. In the power of the Spirit the Son of God became incarnate. In every instance, creatures are enlivened, expanded, illuminated, empowered and liberated by divine action. Only by looking away from the biblical record of God's revelation do we become anxious that God's "unconditional and irresistible" lordly action might diminish or limit us.[45]

---

[41]Ibid., p. 120.
[42]Ibid., p. 131.
[43]Ibid., p. 149.
[44] Ibid.
[45]Ibid., p. 131.

Divine *ruling* is the third aspect of providence. The power God exercises in creatures is not a blind and aimless force but intelligent guidance toward the goal God has set from all eternity. And the goal is God himself, his glory and the salvation of creation. God alone rules creation. God works in and through but always above "the cosmic antithesis of necessity and freedom."[46] And why does God will to control all things?

> The answer is that God controls all things because in and with and by and for all things He wills and actually accomplishes one thing—His own glory as Creator, and in it the justification, deliverance, salvation, and ultimately the glorification of the creature as it realizes its particular existence as a means of glorifying the Creator.[47]

3. Nothingness (das Nichtige) and the problem of sin and evil. Barth's eighty-page section "God and Nothingness" has been quoted often, nearly as often dismissed as nonsense, but only rarely understood. Barth does not choose the word evil (German *Böse*) as his main descriptor. Though he does not explain why the word *evil* is not acceptable, I think it is because this word is too closely associated with what Barth calls "the shadow side" of creation; that is, since the creation is not God it is plagued with vulnerability, changeability and impotence. Barth needs a more metaphysical concept that gets at the existential or ontological aspects of the problem of sin and evil. Traditional terms that served this purpose were *nothing* or *defect* or *privation*. These concepts can be understood in their opposition to those of being and perfection. Christian theologians chose these concepts to give an adequate account of evil without falling into the errors of those philosophies—like Gnosticism and Manichaeism—that make evil into a metaphysical substance in eternal opposition to God and goodness. The Manichees were attempting to avoid the choice between understanding evil as divinely created or as nothing at all. Obviously evil is real and something to be resisted; hence it must be an eternal godlike enemy of light and goodness against which God struggles in cosmic war. The Manichaean religion was composed of teachings, rituals and ascetic practices designed to join with God in the war against this evil substance.

Orthodox Christianity rejected this way of thinking. For God is perfectly

---

[46]Ibid., p. 164.
[47]Ibid., p. 168.

good, and everything God created is also good; yet God is also omnipresent, omnipotent and perfectly provident, so cannot be locked in a struggle with a power with any real potency against God. It cannot be that in his struggle against evil God sometimes wins and sometimes loses, sometimes is present and sometimes is pushed out by evil. Combining these two convictions the church fathers concluded that evil cannot be either a creature of God or an eternal substance that exists on its own. Hence they used the category of defect or privation to describe the ontological status of evil; it is something that possesses a sort of reality but not the kind that can exist on its own. A broken chain, a lame limb and a torn page are examples of things with defects that damage their integrity and inhibit them from fulfilling their purposes. Disorder is a kind of defect, an absence of the right order that enables things to work properly. Defects also characterize immoral actions and wrong choices. Sin is absence of right order in human willing. In sin human beings will a created good out of order. Defects and disorder cannot exist on their own; yet insofar as they are present in a created thing they "cause" damage and malfunction. By themselves they are "nothings."

Why did Barth not adopt the traditional terms such as *nothing, privation* and *defect*? Barth agrees with the tradition that evil is nothing in itself, a privation and a defect. He rejects in the strongest terms the idea that evil is a substance against which God is locked in eternal struggle. But Barth thinks that the biblical description of evil cannot be captured by the traditional terms. They give the impression that evil is innocuous or nothing at all; and they fail to take into account evil's enmity toward God and his creation. So Barth chooses the term *nothingness*. Adding the *-ness* ending gives the word *nothing* a substantial feeling, a feeling that evil possesses a sort of reality. In this move Barth attempts to take into account the Bible's imagery of war, struggle, suffering, death and sacrifice in relation to sin and evil.

What is nothingness? Barth lists some bad things as indicators of the realm of nothingness: "the whole complex of sin, guilt and punishment, the whole reality of calamity, suffering and death in the world-process."[48] But this list only indicates nothingness and does not define it. First, we must be careful not to confuse the negative side of creation with nothingness. Creation is not God. Unlike God, creation is temporal and spatially limited,

---

[48]Ibid., p. 291.

changeable and vulnerable, mortal and limited in knowledge and power. It is not a manifestation of nothingness that we have bodies that need care and that feel pain, or that our time is limited. Barth rejects the conclusion that pain, early death and sickness can be unequivocally identified with nothingness or evil. To think of nothingness as the negative side of creation shows that we have chosen the wrong source for our knowledge of nothingness. We mistakenly conclude that whatever we experience as negative, undesirable and fearful must be evil and hence rejected by God.

How then do we recognize nothingness or genuine evil? Real nothingness shows itself only in the revelation of God's relationship to it. "The objective ground of our knowledge of nothingness is really Jesus Christ himself."[49] What God rejects, counts as his enemy and deals with decisively is exposed in the death and resurrection of Jesus Christ. In Jesus Christ, God did not reject the weaknesses and limitations of creation; for the Son of God became a weak and limited and mortal creature. According to the New Testament, Jesus came to save sinners, to condemn sin and to bring righteousness. In Jesus Christ, then, "the concrete form in which nothingness is active and revealed is the sin of man as his personal act and guilt, his aberration from the grace of God and its command, his refusal of the gratitude he owes to God."[50] But nothingness is more than sin. Sin sets in motion a train of consequences that also manifest nothingness. "Contrary to his will and expectation the sin of man is not beneficial to him but detrimental."[51] Sin brings real suffering and death. By death Barth does not mean merely the termination of life, and by suffering he does not mean the pain that attends all sentient life. Suffering that manifests nothingness is "the suffering of evil as something wholly anomalous which threatens and imperils this existence and is no less inconsistent with it than sin itself, as the preliminary experience of an absolutely alien factor which is radically opposed to the sense and purpose of creation and therefore of the Creator Himself."[52] And by death Barth means "the intolerable, life-destroying thing to which all suffering hastens as its goal, as the ultimate irruption and triumph of that alien power which annihilates creaturely existence and thus

---

[49]Ibid., p. 306.
[50]Ibid., p. 305.
[51]Ibid., p. 310.
[52]Ibid.

discredits and disclaims the Creator."[53] In sum, nothingness is "the comprehensive negation of the creature and its nature."[54] And by negating the creature and its nature it negates the command of God by which the creature exists and lives.

How does nothingness come to possess its strange existence? If nothingness is not an eternal substance, a natural possibility of creaturely life and freedom, or a divine creation, then why does it exist? Barth's answer to this question is sure to strike many readers as strange. Nothingness exists only as and because God actively wills it not to be. God graciously creates the world and directs it to the goal God has chosen. In the same act God also rejects all other "possible" worlds and "possible" goals for creatures. Barth says, "God elects, and therefore rejects what He does not elect. God wills, and therefore opposes what He does not will. He says Yes, and therefore says No to that to which He has not said Yes."[55] Or further:

> That which God renounces and abandons in virtue of His decision is not merely nothing. It is nothingness, and has as such its own being, albeit malignant and perverse. A real dimension is disclosed, and existence and form are given to a reality *sui generis*, in the fact that God is wholly and utterly not the Creator in this respect. Nothingness is that which God does not will. It lives only by the fact that it is that which God does not will. But it does live by this fact.[56]

Nothingness exists in its own strange way only because God graciously willed to create. Strictly speaking, in relation to God nothingness is nothing, and it has no power against God directly. But because creatures have become threatened and have actually fallen prey to it, nothingness challenges God's goodness and lordship. Only because creatures exist and only through the creature can nothingness become an event in the world. In his gracious affirmation of creation God also says no to everything that would negate and destroy the creature. Yet through God's commands—both positive and negative—nothingness enters the human imagination either as a negation of what God has affirmed or an affirmation of what God has negated. Nothingness, as that which God has rejected, enters the world and exists in a

---

[53]Ibid.
[54]Ibid.
[55]Ibid., p. 351.
[56]Ibid., p. 352.

parasitic way in the human imagination as a seductive image of an impossible world that invites human beings to exert themselves futilely to make the impossible world a reality. And in this parasitic way nothingness takes on a sort of personal and active character whose trajectory is destruction. Apart from the creature nothingness is nothing at all, but in the creature it lives as it were a demonic existence by possessing God's creatures and using their created powers against the Creator and creation. Nothingness exercises power over human beings and moves them to move themselves away from God and toward destruction, toward real suffering and death.

Finally Barth assures us that nothingness will not forever trouble creation. Barth does not allow the dialectical relationship of affirmation and negation to continue endlessly. God's goodness does not need an opposite to be itself. And God can free the creature from the "impossible possibility" of denying what God has affirmed and affirming what God has denied. In Jesus Christ, God takes up the creature's cause and as a human being wins the victory over nothingness. In Christ, nothingness has no place and no power. He is the human being for whom the "Yes" to God is no longer accompanied by a "No" even in the imagination. If nothingness still troubles us, it is only because God permits it. In this form God makes it "an instrument of His will and action. He thinks it good that we should exist 'as if' He has not yet mastered it for us."[57]

---

[57]Ibid., p. 367.

14

THE OPEN THEIST
MODEL OF PROVIDENCE

Foreknowledge and omnipotence models of providence defend God's perfect providence. From the point at which God decrees his plan and purpose for creation both models assert that God enacts his plan down to the smallest detail. Open theism abandons this thesis and makes some significant changes in the traditional doctrine of providence.

OPEN THEISM'S REVISIONS TO THE TRADITIONAL DOCTRINE OF PROVIDENCE

The central alterations are (1) the abandonment of divine foreknowledge of contingent events such as human free decisions. Such foreknowledge is logically impossible. For a contingent event to be known as a certainty, it must have already occurred or be bound by necessity to occur.[1] (2) Open theism discards the Boethian understanding of divine eternity—"Eternity is the simultaneous and complete possession of infinite life"[2]—and replaces it with temporal everlastingness, that is, God has always existed, exists now and will always exist temporally. As a corollary to everlastingness, open theism asserts that God experiences only the present as real because

---

[1]This abridgment of divine knowledge is also found in such philosophers as J. R. Lucas, *Freedom and Grace* (London: SPCK, 1976); Richard Swinburne, *The Providence of God*, 2nd ed. (Oxford: Oxford University Press, 1993); and in other open theists such as John Sanders, *The God Who Risks: A Theology of Divine Providence*, 2nd ed. (Downers Grove, IL: IVP Academic, 2007); David Basinger, *The Case for Freewill Theism: A Philosophical Assessment* (Downers Grove, IL: Inter-Varsity Press, 1996); and William Hasker, *God, Time and Knowledge* (Ithaca, NY: Cornell University Press, 1989).
[2]Boethius, *The Consolation of Philosophy*, trans. W. V. Cooper (London: Dent, 1902), pp. 160-61.

as a metaphysical necessity only the present is real. The past no longer exists and the future does not yet exist. God does not experience all time simultaneously but like us experiences his life and all things moment by moment. (3) Consistent with the previous two revisions, open theism abandons the idea that God possesses an eternal plan or decree that includes every event in the history of creation. In place of a detailed plan it substitutes an open-ended outline of what God wishes to happen with inbuilt flexibility that allows God to respond and adjust to the free decisions of creatures. The telos, or goal, of creation must be open-ended and general enough to be achievable even on a worst-case scenario. (4) God cannot determine all things to happen as God wishes if God also wants to allow human beings to exercise their free will. Since God *has* chosen to allow human beings genuine freedom, God cannot make sure that all things turn out exactly according to a detailed plan. However, God is infinitely resourceful and works toward achieving the maximum good possible in the long run even though human beings are able to frustrate God's will in the short run. (5) Nevertheless, some evils must be considered gratuitous, that is, they cannot be justified as means to any conceivable higher good. Hence the idea that God permits evil for the sake of a greater good must be abandoned. God does not permit evil directly; God permits freedom. God permits evil only indirectly as a possibility or risk inherent in creaturely freedom.

GREGORY BOYD ON DIVINE PROVIDENCE

As my example of an open theist doctrine of providence I will examine the work of Gregory Boyd, focusing primarily on his essay "God Limits His Control."[3] This essay summarizes a view of providence elaborated in much greater detail in several of Boyd's earlier books.[4] And as the occasion warrants I will refer to those other books.

Four Christocentric criteria. Boyd begins by asserting that the doctrine of providence should be developed and must be evaluated from a Christo-

---

[3]Gregory Boyd, "God Limits His Control," in *Four Views on Divine Providence*, ed. Dennis Jowers (Grand Rapids: Zondervan, 2011), pp. 183-208.

[4]Boyd's other works relevant to this topic are *God at War: The Bible and Spiritual Conflict* (Downers Grove, IL: InterVarsity Press, 1997); Boyd, *God of the Possible* (Grand Rapids: Baker Books, 2000); and Boyd, *Satan and the Problem of Evil: Constructing a Trinitarian Warfare Theodicy* (Downers Grove, IL: InterVarsity Press, 2001).

centric perspective. To that end, he lays down "four Christocentric criteria" by which to judge viewpoints on providence:[5]

1. God wages spiritual warfare. Jesus engaged in conflict with Satan and his demons in anticipation of the coming kingdom of God. These spiritual, malevolent beings "genuinely resist" the coming kingdom. In his earlier work *God at War*, Boyd argues Satan's exercise of freedom and power prevents God from simply "controlling evil" and makes it necessary for God to war against it.[6] In this war God faces "formidable" enemies that put up "genuine" resistance.[7] According to Boyd, "running the cosmos . . . is no easy matter, even for the Creator."[8] God works toward his goals "in the face of genuine, powerful, opposing forces."[9] This criterion demands that a doctrine of providence do justice to "the need for God to do battle" against the forces of evil.[10] In other words, a doctrine of providence must show why God *cannot* simply prevent all evil or rid the world of evil by force.

2. God relies on power and wisdom. The Bible shows that the Creator is powerful, and any view of providence must acknowledge the level of power attributed to God in Scripture. But the way Jesus worked to defeat malevolent spiritual powers and to redeem humanity shows that God also relies on "superior wisdom." God cannot win every battle by force; sometimes "God needs only to *outsmart* his opponents."[11] Boyd concludes that a doctrine of providence "viable for followers of Jesus must be able to render intelligible why God often relies on his superior wisdom to defeat his foes and accomplish his purposes."[12]

3. God relies on other-oriented love. God not only "outsmarts" his enemies but also attempts to win them with his self-sacrificial love. According to Boyd, in deciding to love human beings God gives them power to influence God, to affect his emotions and to cause him to change his mind. So any doctrine of providence must account for "the loving, communal way God

---

[5]Boyd, "God Limits His Control," p. 184.
[6]Boyd, *God at War*, p. 291.
[7]Ibid., p. 93.
[8]Ibid., p. 148.
[9]Ibid., p. 98. References to Satan's power and God's struggle against him can be found throughout *God at War*. The war is "absolutely real for God" (p. 87). God "must battle" (p. 89) against "real opposition" (p. 92). The battles God must fight are "genuine" (pp. 98, 162, 283), for Satan is a force that "Yahweh must reckon with" (p. 144).
[10]Boyd, "God Limits His Control," p. 185.
[11]Ibid., p. 203.
[12]Ibid., p. 186.

operates in the world as well as God's willingness to be affected and influenced by humans."[13]

4. God wins by bringing good out of evil. Any model of providence must show how God can bring good out of genuine evil given the constraints mentioned above.

The rest of Boyd's essay is devoted to laying out the essential elements of the open theist vision of divine providence and showing that it measures up to his four christological criteria better than other approaches. Central to the open theist doctrine of providence is the thesis that God created the world out of love and for the purpose of establishing a community of human beings who love God in return. But this goal cannot be achieved by fiat or coercion. Love cannot be forced but must be freely given. If God wants a world in which exists a community of beings who love him and each other, God must give them freedom. In Boyd's words, "If love is the goal of creation, however, then the creation must include free agents."[14] Boyd illustrates his assertion with a story. A scientist invents a microchip that can be implanted in a person's brain and, once in place, can cause a person to experience emotions and express words that perfectly mimic love toward the scientist. Would the chip-enabled person really love the manipulative scientist? Of course the answer is no. Love must be free. To be free, a loving act must be chosen, and the person choosing must have the power to withhold as well as to give love. Hence if God wants us to love him genuinely, God must give us the power and opportunity to refuse as well as return his love. According to Boyd, then, Scripture, logic and intuition "teach us that love requires freedom."[15]

But if love requires freedom and freedom involves the power of alternative choice, then to create a love-conducive world God cannot exercise "unilateral control" of his creation.[16] By the very choice to create our world, God "limits the exercise of his power."[17] According to Boyd, this does not mean that God ceases to be omnipotent, since God's limits are not natural but logical. Just as God cannot create "a round triangle or a married bachelor"

---

[13]Ibid., p. 187.
[14]Ibid., p. 189.
[15]Ibid., p. 190.
[16]Ibid.
[17]Ibid., p. 191.

God cannot determine the free acts of his creatures.[18] Libertarian freedom "is the capacity to choose to go this way or that way."[19] Both alternatives must remain open until the agent decides one way or another. A choice cannot both be open and not open at the same time and under the same conditions. A sentence that affirms both of them (*a* and ~*a*) is a logical contradiction, and God cannot actualize logical contradictions.

God's lack of foreknowledge and the issue of omniscience. The logical flow of Boyd's argument so far leads to what he calls the most controversial thesis of open theism: God does not possess foreknowledge of everything that *will* happen in the future; specifically God cannot know which alternatives a free agent will choose in a genuinely free act in advance of the actual choice. Nevertheless, Boyd argues that open theism does not deny God's omniscience. He seems to understand an omniscient being as one that knows perfectly everything about what is now real.[20] Everyone who affirms omniscience believes this, and so do open theists. So the decisive issue, argues Boyd, concerns how we understand the real. To quote Boyd: "Open theists hold that if God is omniscient and thus knows reality exactly as it is, and if the future is in fact partly comprised of ontological possibilities, then God must know the future as partly comprised of such possibilities."[21] This statement requires careful analysis. Note first that Boyd speaks of the "future" in terms of its status in being, that is, whether it actually exists or exists only as a possibility within the present state of things. Strictly speaking, by the "future" Boyd means a state of affairs that arises out of present reality. Given Boyd's assumption that, in genuinely free acts, human beings alone determine the future state that results from their acts, it makes sense that prior to those acts the "future" exists only as indeterminate possibilities within present reality. This indeterminate state of possibility is all that exists of what we call the "future," and only what exists now can be known because only what exists now is real. One more nuance in Boyd's argument: the indeterminate "ontological possibility" for multiple futures exists not only as "*logical or epistemological*" possibilities. Boyd insists that these possibilities "reflect a state of being."[22] This point is easily missed. Here Boyd is distin-

---

[18]Ibid.
[19]Ibid., pp. 191-92.
[20]Open theists also hold that God remembers the past perfectly. But the past is not now real.
[21]Ibid., p. 195.
[22]Ibid., p. 195 n. 24.

guishing between a mere logical possibility (which is purely about the relationship among propositions that exist in the mind), an epistemic possibility
(which is about what we know or can know or any observer can know) and
the ontological possibilities latent in an actual state of affairs. Clearly here
the "the state of being" refers to the actual existence of the free agent and its
attendant and supporting circumstances. This distinction is important for
Boyd; it affirms that in knowing the "future" as an "ontological possibility"
God knows a present, actual state of affairs rather than knowing merely a
range of logical possibilities resident in the divine mind.[23] God has access
to a particular "future" state of the world only by knowing the present state
of things wherein this future is a mere ontological possibility. God never
knows a particular future as in itself actual. Again, only what is real can be
known and only the present is real. The present actually exists, but the
present actual state of the world contains possibilities for other actual (and
"future") states. God knows reality perfectly. God knows the actually existing
world, and, in knowing the actually existing world, God knows its indeterminate, ontological possibilities for other and future actually existing worlds.
Hence in Boyd's estimation open theism affirms divine omniscience.

   In a second form of his argument Boyd shifts from speaking about futures in terms of their ontological status to speaking of the modal status of
propositions we use to speak about the future. In Boyd's view, classical theology speaks of God's knowledge of the future as God knowing the truth
value of all propositions that assert what *will* happen: "In January 2013
Barack Obama *will* be inaugurated for a second term as president of the
United States." In the classical way of thinking, this statement is either true
or false as I write this (July 2012), and God knows which. Boyd complains
that critics accuse open theism of denying God's omniscience because it
denies that God knows the truth value of this statement. But this criticism
misses the point because, according to Boyd, this statement about Barack
Obama refers to no actual state of affairs. Propositions that refer to states of
affairs as actually existing when they exist only in the mode of possibility

---

[23]The range of possible futures may be considerably narrowed by limiting the number of possible
futures to the number of possibilities in an actually existing state of affairs as opposed to all the
logically possible states of affairs. Since God knows human thoughts and motives, the actual
state of affairs just before a free decision may narrow down to an alternative between just two
futures, yes or no, this or that. But the range of logical possibilities in this situation would remain
much larger.

cannot be either true or false. The future does not exist now as a determinate state of affairs but only as an indeterminate ontological possibility for multiple futures. To speak meaningfully about future states of affairs we must use modal qualifiers that reflect the mode in which the future exists now, that is, as a range of indeterminate ontological possibilities. So we must use a modal qualifier such as *might* or *might not* rather than *will* or *will not*. Now let's return to the question of Barack Obama's presidency. If we reframe the proposition to read, "In January 2013 Barack Obama *might* be inaugurated for a second term as president of the United States," it refers to an actual state of affairs and its ontological possibilities. Hence this statement possesses a truth value of true or false, and God knows the truth value of this "might" proposition. To affirm God's omniscience is to assert that God knows the truth value of all meaningful *is* and *was* and *will be* propositions and of all meaningful *might* propositions.[24] This range covers everything that was, is and will come about by necessity, and those present ontological possibilities latent within what actually exists now.[25]

How do Boyd's views on the nature of human freedom, the ontological status of the future and God's knowledge of the future as ontological possibility affect his view of how God works to achieve his purposes in the world? As I pointed out at the beginning of this section, Boyd rejects the traditional belief that from all eternity God has decreed in every detail, to what end and by what means God will bring about the world God has chosen. According to Boyd, this course of action is impossible for God. Instead God works step by step toward the more general goal of creating a loving and holy people who achieve eternal life in union with the Father, Son and Spirit. Boyd asks us to think of the history of creation as a story with many subordinate authors under the overall authorship of God. God has set the "overall structure" of the plot and has anticipated every alternative event and all the possible endings of the story. God knows how to respond to every choice. But the actual flow of events and the final end will not be determined

---

[24]The meaningfulness of "will" propositions depends on whether or not the states of affairs they describe are already determined in the present. God, of course, knows what he has unconditionally determined to do under every possible set of future circumstances and God knows what is determined by present causal forces that necessitate one and only one future state of affairs. For Boyd's more popular discussion of this issue, see Gregory Boyd, *God of the Possible*, pp. 120-130.

[25]God knows nothing that does not exist now. God's knowledge of the past is his memory and God's knowledge of the future is either of latent possibilities or of his determinations to bring about unconditionally a definite future.

by God alone but also by the decisions of every free agent within the story. Boyd assures us that God is infinitely wise in anticipating and responding to human decisions and is intimately involved in every decision, attempting to persuade free agents to make right choices. We can be sure that God will achieve his eternal purpose.

Boyd's self-assessment. Now Boyd is ready to submit his view of providence to evaluation by his four christological criteria. (1) According to Boyd, the open view better fits with the biblical teaching that God wages spiritual warfare. Understanding that God cannot determine or foreknow the outcome of a free action relieves the theologian of the obligation to explain why God caused (the omnipotence view) or selected (the foreknowledge view) a specific horrible evil event to become actual. God knew beforehand only that such events *might* happen, and with this knowledge God determined to do everything possible to persuade free agents to avoid evil. We need only justify why God would take this risk. It is easier to make a plausible case for God risking the occurrence of evil in view of the goal of eternal life for those who choose it than justify God for choosing every evil and occasion of suffering because of the good they bring about. Additionally, the open view allows one to deny that "there is a specific reason for each specific evil."[26] (2) For Boyd, the open view better explains why in the Bible God relies on wisdom instead of coercive power to bring about his will. God's purposes are such that they cannot be achieved by power alone. Free agents may be influenced and persuaded in various ways without destroying their freedom, but they cannot be dragged into a loving relationship with God. (3) The open view comports better with the biblical picture of God as "other-oriented love," which we see lived out in Jesus Christ. According to Boyd, open theism better conforms to the concept of love. Love not only requires freedom but also demands that the one who loves become vulnerable and open to change in the relationship. And in open theism God gives human beings power over how the story turns out and even some power over the part God plays in the story. (4) Finally, Boyd argues that open theism can affirm just as strongly as other models God's ability to bring good out of evil. Since God is infinitely wise and knows eternally every possible alternative that free agents could choose,

---

[26]Boyd, "God Limits His Control," p. 202.

God can plan the best response to each possibility. Whenever states of affairs become actual, God is perfectly ready to continue working for good. Returning to his metaphor of the master author of a story and the subordinate authors, Boyd asserts: "We may thus rest assured that if there were possible story lines that could not result in God's bringing good out of evil, let alone story lines that threaten God's objectives for creation as a whole, the Author of the adventure of creation would simply exclude them from the adventure."[27]

CRITICAL OBSERVATIONS ON THE OPEN THEIST MODEL

It is impossible with the limits of this section to assess every dimension of the open theist model of divine providence. Even Boyd's brief summary of the open theist doctrine of providence raises more issues that I can deal with in this subsection.[28] I will limit my comments to three issues where critics of open theism have directed their heaviest fire. Boyd writes his essay in awareness of these allegations and appears eager to respond to three charges: (1) The open theist view of providence calls into question God's greatness and perfection by denying some and modifying other of the traditional divine attributes; (2) open theism denies God's omniscience; and (3) the openness view of providence undermines Christian confidence in God's care and hope in God's triumph over all evil.

Divine perfection? Does Boyd's view of providence undermine our confidence in God's greatness and perfection? The traditional divine attributes were developed by theologians to articulate in the philosophical language of being and perfection the nature and identity of the God revealed to Israel and in Jesus Christ. The Scriptures extol God's knowledge, love and power as beyond all rivals. Indeed the Scriptures give us warrant to attribute to God all the powers and perfections we can imagine to the infinite maximum. Anselm of Canterbury speaks for all traditional theologians when he praises God as "that than which nothing greater can be conceived."[29] Traditional theologians articulated the concepts of simplicity, eternity, omnipotence, immutability, impassibility, omniscience and others to express God's infinite

---

[27]Ibid., p. 207.
[28]For more extensive but still not exhaustive comments on Boyd's chapter, see Ron Highfield, "Response to Gregory A. Boyd," in *Four Views on Divine Providence*, ed. Dennis Jowers (Grand Rapids: Zondervan, 2011), pp. 231-42.
[29]Saint Anselm, *Basic Writings*, trans. S. N. Deane, 2nd ed. (La Salle, IL: Open Court, 1968), p. 7.

perfection and freedom from every defect. Open theists deny outright divine simplicity, eternity, immutability and impassibility, and then attempt to modify omnipotence and omniscience without giving up their names. Here I will say only a sentence or two about the others but reflect at greater length on eternity. I will deal with Boyd's claim to adhere to omniscience in the next section.

The doctrine of divine simplicity denies all composition in the being of God. Traditional theologians considered any being composed of parts as imperfect. Whatever must be put together can come apart and die. But God cannot die. Whatever exists because of the relationships among its parts cannot be the ground of its own existence. It needs a creator. But God is the ground of his own existence. Hence God is simple. When open theists deny divine simplicity they become vulnerable to the charge of denying divine perfection. Divine immutability, too, is a negative concept. It denies that God is subject to change. Only imperfect and unstable beings change. Only immature beings grow and only ignorant beings learn. But God is fully actual and perfect. To attribute change to God would imply that God grows or declines or learns or forgets. Open theism insists that God changes and thereby calls into question God's greatness and perfection. Divine impassibility also is a negative concept; it denies that God possesses evil passions or emotions or that God is ever driven or changed by passion or emotion, good or evil. This doctrine does not deny that God loves us and has mercy on us. It affirms, rather, that God's stance toward us is eternally perfect for all our changing circumstances. God does not change and does not need to change. But open theism denies the doctrine of impassibility and places itself under obligation to show why this denial does not insert evil and imperfection into the concept of God.

As I pointed out in the introduction to open theism, these thinkers deny the traditional teaching that God is eternal in the sense of Boethius: "the simultaneous and complete possession of infinite life."[30] Open theism substitutes the concept of everlastingness for divine eternity. As everlasting, God has always existed and always will exist; but God exists only in the present moment. Open theists make this change for a variety of reasons, some philosophical and some theological. If God knows the whole creation

---

[30]Boethius, *Consolation of Philosophy* 5 (*Consolation of Philosophy*, pp. 160-61).

from beginning to end in one simultaneous moment, how can human decisions be genuinely free or our moment-by-moment lives possess any significance for the outcome of the whole? Everything is eternally settled. But free choices cannot be determined or known in advance, so God cannot be eternal. Open theists also fear that an eternal God could not relate to a temporal creation or be genuinely responsive to the flow of events in our lives. But in the Bible God interacts with and responds to creatures in a temporally sequenced way. Hence God cannot be eternal. I believe these concerns can be answered adequately within the traditional framework, but I will not take the space to do this here.[31] I want to focus, rather, on the negative implications of giving up eternity for everlastingness.

The Boethian definition of eternity like others of the traditional attributes is designed to remove imperfections from our concept of God. It asserts that the entirety of God's life is his possession "now." God is already with his total being. God does not need to remember what God was or imagine what he will be. God's "now" never fades or changes. We temporal beings are by definition imperfect, incomplete: we change, becoming what we were not and ceasing to be what we were. We live in a series of fleeting "nows," never experiencing our whole lives simultaneously. We are open to many futures because our present state possesses the potential for many different but mutually exclusive future states. As temporal we can never actualize all our potential. But God is fully God "now." God has no unrealized potential, no nostalgia for what once was or anticipation and anxiety for what might be. God is eternal.

Sustaining its substitution of everlastingness for divine eternity requires open theism to show why making God temporal does not insert potentiality, change, multiplicity and becoming into the life of God. And if open theism admits that these things are true of God's life and being, what then? How can God's freedom and absolute independence be maintained? Potentiality is a kind of deadness that must be awakened by something living. It is a sort of a not-yet-being, awaiting a forming or activating cause. Traditional theology contended that we must understand God to be "pure act," for otherwise God is not the Lord of his own being. Time implies multiplicity, different phases and moments succeeding one another; one passes away as the

---

[31]See my treatment of these objections in *Great Is the Lord: Theology for the Praise of God* (Grand Rapids: Eerdmans, 2008), pp. 292-311.

other comes into being. Are these different phases of God's life to be considered different gods, finitely actualizing part of the divine potential? If not, what makes God self-identical through time? What holds all these phases together? Is it something other than God? But if God himself is the ground of his own unity and self-identity, we have returned to the essential aspect of the concept of eternity. Only eternity transcends and spans everlastingness so as to be able to unify it. So if God is the cause of his own unity throughout everlasting time, God is eternal! But if God is not this cause, he is not God.

Omniscience? As we saw in our exposition of his understanding of divine knowledge, Boyd denies that God knows the future exhaustively. God cannot know genuinely free decisions in advance of their performance. God knows only the present, for only the present is real. God knows the future only in his own present decisions about what God will do or as a set of ontological possibilities latent in the present state of the world.[32] God's determinations can be stated in *will* propositions, but the present's ontological possibilities must be stated in *might* propositions. Open theism's critics object that in thus limiting God's knowledge it denies divine omniscience. As we noted, Boyd defines omniscience as exhaustive knowledge of everything real and asserts that his understanding of divine knowledge qualifies as omniscience. Can a being possess any greater knowledge than knowing everything about everything real? How shall we assess Boyd's claim?

By defining omniscience as knowing everything about everything real, Boyd deflects the charge that open theists do not believe in divine omniscience—a damning conclusion if true—by attempting to change the focus of the debate from whether his view of God's knowledge measures up to omniscience to the issue of what is real. If successful, this move would shift the debate from whether open theists reject a dogma cherished by the universal church to an academic matter on which Christian theologians are free to disagree. But this shift is not as innocent as Boyd would have us believe. It expects us to believe that such questions as the nature of divine omniscience and the scope of reality can be settled in isolation from a comprehensive doctrine of God. But this cannot be done. The answers we give to

---

[32]According to Boyd, God cannot determine a free creaturely act. Hence God's present determinations concern states of affairs that can be brought about without the cooperation of creaturely freedom.

these questions will be determined by our view of divine eternity, omnipotence, simplicity and all the rest of the traditional divine attributes. Boyd's theories of divine knowledge and the scope of the real make sense only in view of his revision of the other traditional attributes.

Boyd's rejection of divine eternity erases the qualitative distinction between God's time (eternity) and our time, so that the scope of the real is the same for God as it is for us. However, if God is eternal, God's present (eternity) includes as *real* what for us is past and future. The tradition has affirmed divine eternity because thinking of God as temporal, even if everlastingly temporal, attributes imperfection to God. But as we discovered above, Boyd and other open theists consider God to be temporally everlasting instead of eternal and this change is decisive for his view of divine knowledge. Additionally, his view of divine knowledge must also be set in relation to his revision of the notion of divine power. According to Boyd, God does not know everything he will do, because what God will do depends on what free agents do. But this means that free agents act apart from divine power and action; otherwise God would know what free agents will do by knowing what he will do. But for the tradition, asserting that an event could happen independently of divine power and action denies God's omnipotence. If an event can occur apart from God's cooperation, God is not omnipotent in the long-understood sense of that term. Hence Boyd's strategy of defining omniscience abstractly and in isolation from the entire doctrine of God merely hides the imperfection in divine knowledge by shifting it to other aspects of the doctrine of God not currently under discussion.

Traditional theology does not isolate divine omniscience from the other divine attributes. Each and every one asserts that God suffers no defects and possesses every positive quality to the infinite maximum. And all of them together witness that God is "that than which nothing greater can be conceived."[33] In analogy to divine omnipotence, which asserts that God can do whatever is logically possible to do, the tradition asserts that there are no defects in God's knowledge and God knows everything that it is logically possible to know. Benedict Pictet defines God's knowledge in this formal way: "The object of this knowledge is everything that can possibly be known or understood, whether it be God himself, or all other things which can be

---

[33]Saint Anselm, *Basic Writings*, p. 7.

conceived in or outside of God (*extra Deum*)."[34] The tradition restricts those things that God cannot do or know to those things that cannot be at all because the terms in which they are expressed are contradictory and the properties they mention are incompatible. Such incoherent terms as *married bachelor* or *square circle* refer to nothing at all and are simply meaningless.[35] But the tradition saw no contradiction in God knowing everything that will happen in the future, including free actions. Free actions and the things that come to be in dependence on them do not contain incompatible properties; hence they fall within the scope of divine omniscience.[36]

The question to be answered is not whether a being with the attributes and scope of knowledge Boyd attributes to God would be omniscient according to his definition. *The decisive question is whether that being would be perfect.* To say that God is omniscient because God knows everything about everything real leaves out the decisive issue of whether there can be anything real that does not depend on God for its being. The tradition asserts that divine perfection demands that God be completely free of all dependence and that everything other than God be dependent on him for its being and action. Hence a being who cannot know the future actions of free beings because those acts occur independently of divine power is not perfect, is not "that than which nothing greater can be conceived."[37]

Confidence in God's care? Can Boyd's view of providence found a level of confidence in God's care of us consistent with the Scriptures? Let's remind ourselves of what open theism promises. At some point in time God decided to create a world containing free creatures with whom to share his life and love. God never willed creatures to experience sin and suffering, although

---

[34]*PRRD* 3:396.

[35]A square circle is not an especially-difficult-to-make circle; it is nonsense. One cannot frame a meaningful sentence either by saying that God can or that God cannot make a square circle. The expression "square circle" does not refer to something God cannot make; it is meaningless, has no referent and no referential possibility. The words do not form a concept at all. Likewise, it makes no sense to say that God knows where and when all the square circles exist, those that have been, those that are and those that will be.

[36]Boyd argues that such knowledge is a logical contradiction because God could attain it only by knowing himself as the causal determiner of those actions. And a free action determined by an external force is a logical contradiction, or so argues Boyd. Boyd here mistakes determination for coercion. Divine determination means that God directs an action to one effect even if its created cause is not determined to one effect. But there is no logical contradiction in this direction if God can direct without coercing.

[37]Saint Anselm, *Basic Writings*, p. 7.

God knew it was possible, given their freedom to choose. But it was also possible for every creature to choose to love God in every decision. At that point God could not know which possibility would become actual. Given the great good God intended for creatures, God judged the risk worth taking. Although God could not know what every free creature would do under every possible circumstance, God did know what every creature could do and might plausibly do. Being infinitely wise, God also knew what he would do in response to any decision a creature might make. God's action in response to creatures will always be the wisest action in keeping with God's goal of helping as many people as possible—and the salvation of all was possible—come to know him and live with him eternally. So we can take the great comfort of knowing that (1) God loves us and wants to give us eternal life, (2) God has known from the moment he decided to create the world that we might come to exist, (3) God knows everything we might do or that might happen to us, (4) God knows the wisest possible response to whatever happens and (5) God knows how and possesses the power and will to minimize the evil we suffer and maximize our chances of achieving eternal life. In response to this vision we need not belittle the level of love, power and wisdom Boyd attributes to God. A being who exercises such detailed foresight and meticulous care over the world certainly deserves our praise and gratitude. But we must ask whether Boyd's view can inspire the confidence we need.

Consider the divine plan and God's decision to create our world. As I pointed out, God cannot know what will happen but only what might happen. As God considers whether creating a world with free beings is worth the risk, God must take into account worst-case as well as best-case scenarios and every possibility in between. What is the worst case? The worse case is the complete failure of God's plan: everyone sins and no one accepts the remedy of forgiveness and sanctification in Jesus Christ. As we saw in our exposition of his view of providence, Boyd asserts that God could rule out the worst-case possibility and other unacceptable outcomes. But Boyd does not explain how God can do this without interfering with free choice or acquiring middle knowledge of the truth value of counterfactuals of creaturely freedom. Until Boyd explains how this is possible on open theist premises it's difficult to take his assurances seriously. Clearly on open theist principles God does not know how much evil will plague his world or how many free creatures will be saved. How, then, does God decide? On what basis does God know that he can keep evil to a minimum and persuade enough people to love him in return to make it worth the risk? I do not see how this is possible on open theist premises. If God knows he can persuade even one person, why doesn't God know how to persuade all? But if God cannot know that he can persuade even one person, how can we find adequate comfort and confidence in his goodwill toward all?

Second, let's deal with creation, evil and salvation with reference to individuals. Boyd frames God's goals in general terms, speaking only of "a people" whom God aims to create. But what comfort is there in this promise unless we have assurance that we will be among those people? Return to the moment of God's decision to create the world. At that time God knows that Ron Highfield could and might exist, but not that he *will* exist. God cannot know this because innumerable free decisions must be made before my mother and father produce me. Hence in my self-understanding in relation to God I cannot think of myself as chosen and planned by God from all eternity. I cannot hold on to the thought that I am irreplaceable and have an indispensable part to play in God's eternal plan. I am just one of those lucky *might* possibilities that became actual, but for all God knew I might have remained only a possibility left behind by actuality.

According to Boyd, God knows what evil might come upon us, but God does not know or control what evil will actually befall us. Hence God cannot know how long we will live, how much we will suffer or how we will die. Indeed, God does not know the course of our lives for the next few minutes. It follows that God cannot have a single plan for our lives that makes them a unique part of the world God is creating. Instead God has innumerable plans for our lives, each different in view of the differing free decisions that might affect our lives. Now some of these plans must be better than others, and surely God wants the best for us. But we cannot take comfort in believing that God will achieve the best for us; our only comfort is that God will achieve the best outcome for us within his power. But given the negative possibilities of free evil decisions the best "within his power" might not be something to rejoice in.

Concerning salvation, God knows that if I come to exist God will do everything consistent with my freedom to persuade me to love God. And God will do everything consistent with the freedom of other creatures to protect me from evil and temptation. But who will protect me from myself? I will deal with freedom at length below, but let us be clear here that open theism excludes creaturely freedom from the sphere of divine sovereignty. No matter what God does or what persuasive powers God brings to bear, I still have the power to choose against God and my eternal happiness. But in our self-consciousness in choosing or in our reflection on our past choices do we ever know that we choose with full and clear knowledge of what we are choosing? Do we ever get exactly what we thought we were choosing? Have we not experienced a thoughtless kind of certainty in our choice only to find that we later regret our decision and find it mystifying? Does our ultimate salvation depend on a series of stupid little choices? Unless God has sovereignty over our freedom, it would seem to be so. How can we find comfort in God's care unless we believe God can protect us from our own free choices?

15

CREATION, PROVIDENCE AND FREEDOM

THE QUESTION OF HUMAN FREEDOM arises in every dimension of Christian theology. I deferred discussion of freedom of creation until I could discuss it along with freedom in providence. In my exposition of the foreknowledge and open theist models of providence we saw that the Molinist and open theist views of providence find their genesis and are driven in part by the conviction that genuine human freedom cannot exist within the sphere of divine sovereignty. God cannot control the outcome of a free decision. Molinists argue that through his middle knowledge God can foreknow free decisions and work with them to achieve God's goals. Open theists do not believe that foreknowledge of free decisions is possible because foreknowledge implies foreordination and foreordination would destroy freedom. Thinkers working within foreknowledge and open theist models accuse theologians adhering to omnipotence models of providence of sacrificing human freedom to preserve divine sovereignty.

This is a very old debate, dating at least as far back as the time of Plato and Aristotle. And much of the discussion has been carried on by philosophers and philosophically inclined theologians. Far too often the conversation is carried on among Christian theologians in exclusively philosophical terms to the neglect of biblical categories. It seems to me that the question turns on two basic issues: the nature of freedom and the nature of divine action. I believe that a biblically based theology can shed much light on these two questions. My contribution to this discussion will be an attempt to clarify the questions and bring biblical categories to bear in a way that is often missing.

THE BIBLICAL AND THEOLOGICAL IDEA OF FREEDOM

The general concept of freedom. The word *freedom* has a broad scope of meaning, and in many discussions speakers use the term imprecisely, or unknowingly use it in different senses. Hence some analysis must be done before we begin our theological discussion. Its primary meaning is negative, and whenever we use it we are declaring something to be absent. In almost any sentence that uses the word *absent* or *missing* we can substitute the word *free* for those words without changing the sentence's meaning. But it also has a positive side. We don't use the word *free* simply to declare something missing; rather we use it to say that one thing is absent *to another thing.* It's not the absent thing that is "free" but the thing from which it is missing. It is now experiencing freedom, that is, a state of being free from the absent thing. Moreover in most contexts freedom is a very positive word. The removal of the now absent thing improves the self's condition. It allows the one experiencing the absence to possess a good that had been compromised by the now absent thing.[1] The free person can now experience or pursue goods that had been blocked. It is important to note here that freedom is always relative to the absent thing. One can be free from one thing while enslaved by another. Absolute freedom is reserved for God alone. Hence to describe human freedom fully we have to specify from what we are free and to what extent.

In his analysis of the concept, Mortimer Adler discovers four factors operative in every view of freedom: self, other, power and exemption. In every situation where the concept of freedom applies, we will find a *self* that has been made *exempt* from the *other* by a *power.*[2] Views of freedom are differentiated from each other by their differing understandings of these four aspects. Adler finds five different views of freedom in the history of Western philosophy and theology, two of which are variants the others: (1) circumstantial freedom of self-realization; (2) natural freedom of self-determination; (3) acquired freedom of self-perfection; (4) political liberty, which is a variant of self-realization; and (5) collective freedom, which is a variant of self-perfection. In the circumstantial view, you are free when circumstances permit you to do what you wish. In self-determination you exercise

---

[1]How could one ever think of God as one of those things one would be better off without?
[2]Mortimer Adler, *The Idea of Freedom: A Dialectical Examination of the Conceptions of Freedom* (Garden City, NY: Doubleday, 1958), 1:611.

power over the whole self (not just external action) to decide what you do and become. The freedom of self-perfection is achieved when you arrive at a condition in which you will only the good. Political liberty is the individual's circumstantial freedom over against the political order. Collective freedom is the political community's power to determine itself over against other political entities or unruly individuals. According to Adler, for all their differences these understandings possess a common idea of freedom that unites them: "*A man is free who has in himself the ability or power whereby he can make what he does his own action and what he achieves his own property.*"[3] As I want to show below, the Bible understands these four factors in a distinct way that differentiates its view of freedom from others.

I will develop my understanding of freedom by exploring three biblical themes: (1) true freedom as an eschatological blessing, (2) freedom and moral responsibility and (3) freedom and the possibility of repentance and forgiveness.

Freedom as an eschatological blessing. The nature of human freedom, like all issues concerning human beings, finds a place in all three aspects of the God-human relationship: creation, reconciliation and eschatological consummation. Consistent with my method I will keep the contributions of each aspect distinct but hold them in a relation of mutual enlightenment. When discussing freedom under the heading of creation we focus on God's act of bringing about and maintaining the external conditions that make life and human action possible and the internal conditions that exempt the human will from the forces at work on *lower* levels of being, that is, physical, chemical, biological or psychological mechanisms.[4] The topics of reconciliation and eschatological consummation must be considered together, for the former is the beginning of the latter. In this context freedom means the beginning, process and completion of human liberation from sin and death, and this use of the word *freedom* with certain adjustments corresponds to Adler's "freedom of self-perfection." In my view, to grasp the

---

[3] Adler, *Idea of Freedom*, 1:614. Gerald C. MacCallum Jr. understands the general concept of freedom similarly to Adler. In freedom, an agent does or does not have to do something, or become or not become something ("Negative and Positive Freedom," *Philosophical Review* 76 [July 1967]: 312-34).

[4] These two aspects of the freedom of creation correspond to Adler's circumstantial and self-determining freedom. As I will unfold it, freedom always means freedom from something lower. It never means from something higher. The higher liberates from the lower. Freedom frees.

nature of freedom given in creation, it is necessary to begin at the end, with eschatological freedom. Many recent theological discussions of freedom give the impression that the Bible speaks directly and often about a freedom of the will given with our creation and shared by every human being. But such is not the case. Now and then it alludes to political freedom, which is no concern of this study (Acts 24:23). Paul refers to the Gentiles' freedom from the legal requirements of the law of Moses (Gal 2:4; 5:1). But when the topic is freedom itself, the freedom under discussion is exemption from the power and guilt of sin and sin's consequences, whose trajectory is death.[5] In the Gospel of John, Jesus teaches that "everyone who sins is a slave to sin. . . . So if the Son sets you free, you will be free indeed" (Jn 8:34-36). In Romans 6, Paul describes sin as a power that enslaves all human beings and our union with Christ as the power that frees us from sin's power (Rom 6:7-18, 20, 22). In Romans 7 Paul mercilessly drives home the point of the natural human incapacity to free oneself from sin. However much we struggle to extract ourselves from our corruption we merely dig ourselves deeper. Only Christ can "deliver" and "rescue" us from "this body that is subject to death" (Rom 7:24).

In Romans 8 Paul proclaims that we have been freed from "sin and death" through the work of Christ and the Spirit (Rom 8:2). Consider Romans 8:1-4:

> Therefore, there is now no condemnation for those who are in Christ Jesus, because through Christ Jesus the law of the Spirit who gives life has set you free from the law of sin and death. For what the law was powerless to do be-cause it was weakened by the flesh, God did by sending his own Son in the likeness of sinful flesh to be a sin offering. And so he condemned sin in the flesh, in order that the righteous requirement of the law might be fully met in us, who do not live according to the flesh but according to the Spirit.

The apostle to the Gentiles has already established that sin and guilt are universal (Rom 1–5) and that human beings possess no power to free them-

---

[5]In his chapter surveying the freedom theme in the Bible, Richard Bauckham sees the biblical theme of freedom as "multidimensional," including liberation from personal, social and political oppression and from the powers of sin and death. Freedom is both negative and positive. It is freedom from those powers that keep us from loving service to God and the neighbor. But he does not mention eschatological freedom. This omission keeps Bauckham from articulating a core Christian meaning of freedom by which to measure all partial instantiations. See Richard Bauckham, *God and the Crisis of Freedom: Biblical and Contemporary Perspectives* (Louisville: Westminster John Knox, 2002), pp. 7-25.

selves from its power (Rom 6–7). Here he shows that in Christ's death and resurrection we have been liberated from the guilt of sin and that, through the Spirit, Christ's resurrection power is made available to us here and now.[6] But Paul weaves together present and future liberation, speaking not only of the here-and-now but also of a future liberation that will include not only human beings but also the whole creation. Creation itself groans while it waits to be freed from its "bondage to decay and brought into the freedom and glory of the children of God" (Rom 8:19-21). In the middle part of the chapter, Paul uses the terms "adoption," "salvation" and "redemption" to describe that eschatological consummation (Rom 8:22-25). The definitive salvation to be brought about at the end is framed as liberation, as the most radical realization of freedom possible. Importantly, this freedom is conceived as freedom *from* sin and the conditions that give sin its power, not as freedom to sin or not sin. In this state sin has no appeal or deceptive power. According to Hans Dieter Betz, Paul's "doctrine of salvation is very clearly and consciously formulated as a doctrine of freedom."[7]

In the New Testament, freedom is also described in terms of the liberated self. John and Peter speak of a new birth (Jn 1:13; 3:3-8; 1 Pet 1:3). Paul speaks of a new creation and a new man:

> So from now on we regard no one from a worldly point of view. Though we once regarded Christ in this way, we do so no longer. Therefore, if anyone is in Christ, the new creation [kainē ktisis] has come: The old has gone, the new is here! All this is from God, who reconciled us to himself through Christ and gave us the ministry of reconciliation. (2 Cor 5:16-18)

> For we know that our *old self* [ho palaios hēmōn anthrōpos] was crucified with him so that the body ruled by sin might be done away with, that we should no longer be slaves to sin—because anyone who has died has been set free from sin. (Rom 6:6-7)

---

[6]Wolfhart Pannenberg, in his essay "Human Nature and the Individual," in *Human Nature, Election, and History* (Philadelphia: Westminster, 1977), p. 19, affirms: "In Paul, as in John, the Spirit of Christ is conceived of as a liberating power. . . . But it is not primarily a liberation from any oppressive social system that is needed so that a natural freedom of man can be exercised fully without crippling impediments. The human heart itself is considered the impediment."

[7]Hans Dieter Betz, *Paul's Concept of Freedom in the Context of Hellenistic Discussions About Possibilities of Human Freedom*, Protocol Series of the Colloquies of the Center for Hermeneutical Studies in Hellenism and Modern Culture, 26 (Berkeley, CA: The Center 1977), p. 7. See also Lincoln E. Galloway, *Freedom in the Gospel: Paul's Exemplum in 1 Cor 9 in Conversation with the Discourses of Epictetus and Philo* (Leuven: Peeters, 2004).

> You were taught, with regard to your former way of life, to put off your *old self*, which is being corrupted by its deceitful desires; to be made new in the attitude of your minds. (Eph 4:22-23)

> Do not lie to each other, since you have taken off your *old self* with its practices and have put on the *new self*, which is being renewed in knowledge in the image of its Creator. (Col 3:9-10)

In these texts the internal and external conditions of the self have been so altered that the "freed" self, undeniably the same creature, emerges as a new being. This new human being is really the true or perfected form of the created human being, now conformed to "the image of its Creator," its archetype and end (Col 3:10). Now we have before us the way in which the New Testament views the four aspects of freedom laid out by Adler. The new self (the self) has been exempted (the exemption) from sin and death (the other) by the act of Christ and the Holy Spirit (the power). The new self has been forgiven, sanctified, healed and raised from the dead so that the self wills only to be God's image and to do his will. But the human willing of God's will is a genuine human act of will and desire and love and admiration. The new self "has in himself the ability or power whereby he can make what he does his own action and what he achieves his own property."[8] The New Testament, then, views freedom in the fullest sense of that word as a supernatural transformation inwardly and outwardly so that every limitation that blocks human beings from becoming and living as children of God is removed. It involves union with Christ, sanctification and resurrection to eternal life by the life-giving Spirit. For Christian theology at least, the word *freedom* should be reserved for conditions that in some way are analogous to the eschatological freedom as defined above.

Christian theology understands freedom from the perspective of the end. This eschatological vision provides theology with a clear and fundamental concept of freedom by which to measure theories of freedom developed within the topics of creation, providence and reconciliation. To quote my definition again, freedom is exemption from "every limitation that blocks human beings from becoming and living as children of God." Although this vision is of a qualitatively different and supernaturally enabled freedom, it can guide negatively by warning us away from views of created freedom

---

[8]Adler, *Idea of Freedom*, 1:614.

fundamentally at odds with it. It rules out views that see created freedom as indifference in relation to God and views that claim that created freedom fully realizes the idea of freedom.[9] It can guide us positively by providing the perfect end toward which created freedom must be directed. Or put in another way, eschatological freedom is created freedom brought to full actuality within the sphere of divine freedom.[10] However, our theological sources do not allow us to give a precise answer to the question of how much freedom human beings possess by virtue of being human creatures. And the temptation is almost irresistible to enter into philosophical speculation based on some abstract idea of freedom. Although I will take into account such philosophical speculation I will judge such theories by the Scripture-based idea of freedom developed above. Since there are no theoretical discussions of created freedom in the Bible we will need to argue indirectly.

Freedom and moral responsibility. Now that we have isolated the biblical view of eschatological freedom, which must serve as the norm for all other freedoms, we can ask about the nature and scope of the freedom given along with our creation. The first observation to make is that we are not created or born into a state of eschatological freedom. So at the outset of our investigation we know that the freedom we have by virtue of being God's human creatures is less than perfect. We are not born free from "every limitation that blocks human beings from becoming and living as children of God." At this point in the discussion thinkers often make a distinction between the state of human beings before and after the fall, between the state of integrity and the state of corruption. In its favor this distinction makes clear the difference between our created nature and its distortion by sin and it prevents us from inferring that God creates sin and so relieving human beings of responsibility for sin.[11] I will leave the full discussion of the fall, sin and evil

---

[9]Indifference toward God is one of the limits that needs to be removed for us to experience the fullest freedom. Anyone who is indifferent to God is ignorant of their own good and the identity of that good, which is God!

[10]There is another sense in which freedom is relative. It is relative to the nature of the agent. A creature can get no freer than the full actualization of its nature. A nature cannot actualize potential it does not possess; hence its joy cannot be diminished by not becoming what it cannot become or imagine or feel as missing.

[11]Augustine attributes to Adam a higher level of freedom before he sinned than afterward. Before his fall he possessed the "power not to sin." But after the fall Adam and his descendants "do not possess the power not to sin." In the resurrection, the saved will "not possess the power to sin." In the Augustinian reasoning, Adam could have refrained from sinning but did not. Because he sinned he lost his power not to sin. His will was turned toward sin, and he passed this orienta-

until a later section and observe here that every human being ever born, with the exception of the Savior, was born vulnerable to sin and, assuming they live to the age of responsibility, is destined to sin personally. Although many theologians teach that created human freedom was diminished by the fall, especially in relation to God, very few have argued that native vulnerability and acquired guilt completely wipe away every vestige of rationality and freedom. Hence when I ask about the God-given freedom of the created human being I am speaking of human beings as we know them.

In Scripture, God addresses humankind with words, commands and threats. The fact of such divine speech is beyond doubt. What we make of it for the issue of freedom, however, is a matter of controversy. A few references will suffice to sample the hundreds of occasions of divine instruction, commands and warnings. In Genesis 1 the Creator blesses human beings and instructs them to multiply and fill the earth. They are told to rule over all spheres of the earth (Gen 1:28). In Genesis 2 the Creator commands the human pair not to eat of the "tree of the knowledge of good and evil" and warns them that doing so will lead to death (Gen 2:17). I could also list the Ten Commandments (Ex 20:1-17) or some of the hundreds of prohibitions and instructions of Exodus and Leviticus. The teachings of Jesus are full of instructions and warnings, as are the canonical writings of the New Testament. In the Wisdom literature (Job, Proverbs and Ecclesiastes) human beings are addressed as capable of wisdom in ordinary living and of using their reason to anticipate the consequences of their actions. In this second freedom-related theme, human beings are addressed as capable of understanding rational speech and as having reason enough to learn from experience and evaluate possible and likely outcomes of different actions in a range of circumstances. Usually to address people with negative commands assumes that they are capable of refraining from the forbidden actions.[12] Likewise with a positive command one assumes that the addressee is capable of doing what is commanded. It would also seem clear that a threat of punishment for disobedience assumes a level of responsibility and knowledge high enough to make actual punishment for transgression just.

---

tion, this inability, to his descendants. They never were able not to sin, but share Adam's responsibility for the sin he enacted when he was able not to sin (*On Rebuke and Grace* 31-35).

[12]Commands can also serve to expose human sin and powerlessness. We are commanded to love God with all our heart, soul and strength; to be perfect as our Father in heaven is perfect; and not to covet. For Paul, our relationship to the law demonstrates not our freedom but our slavery.

Freedom as noncompulsion. But theologians differ on just what kind and level of freedom satisfies the above conditions. Thinkers in the Augustinian tradition from Augustine himself to Bernard of Clairvaux (1090–1153) and beyond argue that what is required is noncompulsion or exemption from necessity. To represent this tradition we will briefly examine Bernard's influential essay "Grace and Free Choice."[13] In this essay Bernard distinguishes between three different levels of freedom that are coordinated with the three states of humankind: before the fall, after the fall and in the resurrection. The state of grace after the fall and before the resurrection forms a bridge between the second and third. In chapters one to six, Bernard lays out his basic conception of free choice. First, he speaks of *freedom from necessity*, which he labels "free choice." Free choice is the natural power of will with which human beings are endowed. We cannot be compelled to will or consent to a particular thing. Our bodies may be forced to endure experiences we do not will, but our wills cannot be forced. Bernard concludes: "Hence, where you have consent, there also is the will. But where the will is, there is freedom. And this is what I understand by the term 'free choice.' . . . But as to the will, since it is impossible for it not to obey itself—no one does not will what he wills, or wills what he does not will—so it is impossible for it to be deprived of its freedom."[14]

The Creator gave us free choice, or freedom from necessity, along with our human existence, and the fall in no way changed this power. Nevertheless, since the fall everyone is born unable to will the good apart from divine grace. The sinner is doing what he wants when he sins. The thief willingly takes an item rather than someone else putting it in his pocket without his knowledge. The fact that one wants to do the sinful act reveals something amiss in the heart, something shameful that needs reformation. The will is bent toward sin. On the noncompulsion view of created and sinful freedom, human beings are born vulnerable to temptation and possess a tendency to sin. So we need a second type of freedom, *freedom from sin*. Bernard explains:

> I think it has been clearly shown that even freedom of choice is to some extent
> held captive as long as it is unaccompanied or imperfectly accompanied by

---

[13]Bernard of Clairvaux, "Grace and Free Choice," in *The Works of Bernard of Clairvaux*, vol. 7, *Treatises III*, trans. Daniel O'Donovan OSCO, intro. Bernard McGinn (Kalamazoo, MI: Cistercian Publications, 1977), pp. 53-111.
[14] Ibid., p. 56.

the two remaining freedoms; and that from no other cause arises this frailty of ours of which the Apostle speaks: "So that you do not the things you would." To will lies in our power indeed as a result of free choice, but not to carry out what we will. I am not saying to will the good or to will the bad, but simply to will. For to will the good indicates an achievement; and to will the bad, a defect; whereas simply to will denotes the subject itself which does either the achieving or the failing. To this subject, however, creating grace gives existence. Saving grace gives it the achievement. But when it fails, it is to blame for its own failure. Free choice, accordingly, constitutes us willers; grace, willers of the good. Because of our willing faculty, we are able to will; but because of grace, to will the good. Just as, simply to fear is one thing, and to fear God, another; to love, one, and to love God, another—since to fear and to love, on their own, signify affections, but, coupled with the additional word "God," virtues—so also to will is one thing, and to will the good, another.[15]

Bernard concludes that free choice needs freeing; that is, the will, by nature free from necessity, needs enlightening and strengthened so that it wills the good.[16] Grace enables the will to consent to goodness and accomplish it. It does not interfere with the natural operation of free will.

But a third freedom is available, *freedom from sorrow*. Freedom from sorrow is freedom from all the punishments of sin, from death and from bad habits that channel us in wrong directions. This will be achieved by God's power in the resurrection when we are endowed with "the freedom and glory of the children of God" (Rom 8:21). In this life we have only brief moments that approach this freedom. I discussed this third freedom under the topic of eschatological freedom and made it my paradigm of true freedom.[17]

---

[15] Ibid., p. 72.

[16] From what is the will exempt here, if not from necessity? The will freely chooses the bad but is deceived, blind and weighed down by habits. Might we not think of grace as divine persuasion that liberates by shining divine light in the darkness? Certainly the action of grace is completely unlike necessity or coercion.

[17] Kathryn Tanner uses Bernard of Clairvaux, along with Thomas Aquinas and Karl Barth, as an example of a theologian who adheres to the proper rule for the relationship of God and creatures: "The most general rule for talk about created efficacy is as follows. It is simply a specification of our rule for discourse about the creature. The theologian should talk of created efficacy as immediately and entirely grounded in the creative agency of God. If a theologian's discourse conforms to our rule for talk about God's agency, he or she should say that divine agency is required for any power, operation and efficacy of created beings; conversely, created beings should be said to have power, to operate and produce created effects only as God's agency extends to them in those respects" (*God and Creation in Christian Theology* [Minneapolis: Fortress, 2005], p. 91).

Incompatibilist libertarian freedom. Other theologians have concluded that exemption from necessity or compulsion is not enough to ground the responsibility presupposed by divine commands, warnings, promises and punishments. Human beings must possess a more radical exemption. Not only must human beings be exempt from external compulsion or alien necessity; they must also be exempt from themselves, from their nature, reason, character, habits—and God.[18] According to this theory, the human activity of willing cannot possess a natural bent or acquired character that compels the direction of the act. This would seem to rule out the notion that God created human beings with an end as their greatest good or Bernard's idea that in the fall human beings lost the ability to will the good in truth. This "radical" view would rule out Thomas Aquinas's understanding of willing also. On his view, the will is an "intellectual appetite" that *by its very nature* wills total happiness or the truly good. According to Aquinas, it cannot do otherwise, for this is its activity. But happiness is never presented to us as one object among others. If it were present in this way, we would immediately and always will it. Aquinas says: "Something apprehended to be good and appropriate in any and every circumstance that could be thought of would, to be sure, compel us to will it, and this is the reason human beings compulsively will total happiness."[19] On the other hand, "something not found to be good in any and every circumstance that can be thought of will not compel the will, even as

---

[18]This view is often called libertarian freedom. But this label is ambiguous because it is also used of such theologians as Thomas Aquinas and Anselm of Canterbury who believe that the power of alternative choice is compatible with divine control. The thinkers I have in mind are theological incompatibilists. Their form of libertarianism demands the absence of divine control. Tanner traces this way of thinking back to the early modern rejection of the traditional rules for speaking of God's relationship to creation: "First, common claims made for human beings and the natural order in modern times encourage improper inferences from theological statements associated with the positive side of our rules: talk of the creature's power and freedom suggests a power and freedom vis-à-vis God's agency. Second, modern methods of analysis promote the distortion of both sides of the rules. Talk of the creature's capacities moves in a Pelagian direction while talk of God's sovereignty approaches advocacy of a divine tyranny" (*God and Creation in Christian Theology*, pp. 121-22). Michael Welker (*Creation and Reality*, trans. John E. Hoffmeyer [Minneapolis: Fortress Press, 1999], pp. 1-20), falls into the trap against which Tanner warns. He criticizes the tradition that emphasizes God's transcendence and the creature's dependence as involving "false abstractions" (p. 6). Welker opposes to the tradition a reading of Genesis 1-2 that emphasizes "God's self-binding to external events, actions, and presuppositions; God's interested observation; even God's learning" (pp. 9-10).

[19]Thomas Aquinas, *Quaestiones Disputatae de Malo* q. 6 (Thomas Aquinas, *Selected Philosophical Writings*, ed. and trans. Timothy McDermott, Oxford World's Classics [New York: Oxford University Press, 1993], p. 179).

regards determination of what to will; because then, even when thinking about it, someone can will its opposite, because of some other particular circumstance in which the opposite perhaps is good or appropriate."[20] In other words, though our constitutional will for total happiness underlies all particular acts of willing, our inability to see infallibly the relation of any particular object to our total happiness makes it impossible for any object to compel our consent. We can always plausibly imagine that another alternative is better even if to all appearances it is less good. Viewed from the side of the human subject, in this life we are unable simply to enjoy the things that make for total happiness but must make fallible judgments and take risk-laden actions in our quest to attain this state. On the one hand, the inability of objects to compel our consent is a good thing and a significant power when compared to other animals. It exempts us from the power of immediately present objects to appear falsely as total happiness and thus to compel us in our ignorance toward unhappiness. We are not like mice who, overwhelmed with desire for cheese and underequipped with reason, fail to see danger of the trap. On the other hand, the condition that makes it necessary to deliberate and choose among objects is less than ideal. We are alienated from the truly good, unaware of important facts, ignorant of the extent of our ignorance and fallible in our judgments. The conditions that make choices possible also make them necessary, and the conditions that make them necessary make them urgent.[21] And because of this ignorance, fallibility and urgency, every decision demands risk and every choice is a chance.[22]

Such theological advocates of self-determining freedom (also called the freedom of indifference, autonomy and libertarian freedom)[23] as William Lane Craig and Gregory Boyd may seem to be saying no more than what

---

[20]Aquinas, *Quaestiones Disputatae de Malo* q. 6 (Aquinas, *Selected Philosophical Writings*, p. 179).

[21]This condition is higher than the state of not having the ability to judge indifferently objects presented to us. But it is not as high as the state of possessing total happiness, which renders judgment and choices unnecessary, since the thing they intend is already possessed.

[22]See Thomas F. Tracy, "God and Creatures Acting: The Idea of Double Agency," in *Creation and the God of Abraham*, ed. David Burrell et al. (Cambridge: Cambridge University Press, 2010), pp. 221-37, for an assessment of how a creaturely cause or a free human act could be considered to have both a creaturely cause and a divine cause on incompatibilist and compatibilist interpretations.

[23]The term "libertarian freedom" is ambiguous. In one sense Thomas Aquinas advocated libertarian freedom, since he rejected psychological determinism and accepted the possibility of alternative choice. Nevertheless, he still believed that we always will our final happiness and that God determines the free will to one end in a way that does not rule out alternative choice as psychological possibility.

Aquinas said when they argue that one is free only if at the point of decision one possesses the power to choose any of the alternatives being contemplated or none. Or, looking backward, one is free only if, though one chose *a*, one could have chosen *b* instead under the exact same circumstances.[24] But this is not true.[25] For Aquinas (1) the capacity for alternative choice is a power of a nature determined to seek complete happiness in whatever alternative it chooses, and (2) not even free will can escape God's control.[26] For the advocates of self-determining freedom, one is free not only in one's rational decisions but also in one's irrational ones. As Susan Wolf points out, "To want autonomy . . . is to want not only the ability to act rationally but also the ability to act *ir*rationally—but the latter is a strange ability to want, if it is an ability at all."[27] James Spiegel points out that such freedom boils does down to pure arbitrariness. How can I be responsible for an act that does not derive from my character and beliefs?[28] For radical libertarians, the final indifference must remain even if your reason assures you that *a* leads to infinite happiness and *b* leads to eternal misery.[29] The message of reason can be a very persuasive factor, but it cannot be determinative. For then the will would in principle be locked in one mode; that is, it would have to will the good in truth or total happiness. The reason I labeled this view of freedom "radical" is that it attempts to exempt the will from any natural or acquired direction such as Aquinas or Bernard discern. In this view the will's nature or end is to affirm or deny without any predetermined direction or set limitations. It is pure will.[30]

---

[24]The type of possibility under discussion here is not only logical possibility but also physical possibility. One must have power in one's nature to will either *a* or *b*.

[25]Even where they overlap the mood is different. As we saw above, for Aquinas, the indifference with which the will confronts objects is born from a combination of a fundamental determination for total happiness and ignorance of the relationship of finite objects of choice to total happiness. Libertarians seem to think of the indifference as a sort of complete self-mastery on the part of the subject that renders it immune from all compulsion, even by a vision of total happiness or a vision of God.

[26]Aquinas, *Quaestiones Disputatae de Malo* 6 (see Aquinas, *Selected Philosophical Writings*, p. 180).

[27]Susan Wolf, *Freedom Within Reason* (Oxford: Oxford University Press, 1990), p. 56.

[28]James S. Spiegel, *The Benefits of Providence: A New Look at Divine Sovereignty* (Wheaton, IL: Crossway, 2005), p. 69.

[29]We must not confuse logical possibility with psychological indifference. Even where there is logical possibility psychological possibility may be absent.

[30]What would persuasion mean to a pure will? Persuasion appeals to a natural tendency, but pure will would seem to possess no natural tendency or end. For a study of the background of the concept of "pure will" in theology and anthropology, see Michael Allen Gillespie, *The Theological Origins of Modernity* (Chicago: University of Chicago Press, 2008); and Gillespie, *Nihilism*

Autonomy theorists seem to think that such radical indifference of the will is a necessary ground for the level of human responsibility presupposed by divine commands and warnings and judgments of praise or blame recorded in Scripture. For several reasons I do not believe this is true. First, self-determining freedom attributes way more autonomy to human beings than such responsibility demands. Even if, as Aquinas teaches, we are compelled to will our own ultimate happiness, reason makes us responsible for using our knowledge of the moral law and the consequences of our actions wisely. We are not slaves to the object with the greatest immediate appeal to our passions. If we allow it to seduce us, we can be justly accused of laziness and thoughtlessness. We have no excuse for stealing bread from our neighbor's children. Divine commands and God-given reason enable us to know what we should do. Second, if the human will possesses no natural direction but is pure will, how can we be held responsible in blame or praise for anything we choose? Of course, we would be responsible in the sense that our act of willing was purely the product of self-determination exempt from compulsion. But since as pure will we can acknowledge no inner law or given direction within ourselves, we could not be held responsible for doing *what we knew we should not do.* Pure will possesses no inner guide to which it is answerable by nature and no given structure that some acts could contradict. Pure will knows no law but what it gives itself in the moment. Additionally, even if I acknowledge that my choice made yesterday was under my control then, since it is not under my control now there is no basis for holding me responsible for such past actions. For the continuity of my being is my own arbitrary construction. Hence, any opposition to our willing must come from outside, from other pure wills or from the omnipotent will of God. Pure wills can oppose each other with pure, external power only. But not even omnipotent power, understood as pure will, can impose obligations on lesser powers. We are justified, then, to ask whether such a view of freedom is consistent with the Christian doctrines of God and creation. The God who is Father, Son and Spirit cannot be rightly conceived as pure, arbitrary will. Omnipotence should not be made into a master attribute but understood only in relation to

---

Before Nietzsche (Chicago: University of Chicago Press, 1995). See also Vernon J. Bourke, *Will in Western Thought: An Historico-Critical Survey* (New York: Sheed & Ward, 1964).

divine love, justice and all the rest.[31] Nor is humanity the image of a divine, pure will. Its dignity cannot be made greater by expanding its arbitrary rule. God created humanity in such a way that only by loving God and willing his will can we perfect our God-given nature. If this is true, then "to love God is to love oneself in truth."[32] Within the framework of Scripture, to choose in opposition to God is also to contradict one's own being, to work in opposition to one's total happiness and true good.

There is a third reason I do not believe God's commands, warnings and promises to human beings demand self-determining freedom. Libertarians seem to imply that unlike other theories libertarianism attributes to the free subject complete responsibility for its action. But this is not true. Perhaps the subject is *solely* but it cannot be *completely* responsibility for its actions; for such breadth of responsibility requires more than freedom from all inner or outer necessity. It requires the full self-awareness and complete knowledge of the object of choice and all that will flow from that choice. Perhaps if I knew myself well enough to know my true end and the basis of my ultimate happiness, I would invariably choose that end. The range of options available also affects our choices, for if more options were on the table or I had deeper understanding of the options perhaps I would make a different choice. But even with the most robust view of self-determining freedom, we never have to hand all possible options. Everyone knows that one must know *something* about what one chooses before one can be held responsible for choosing that option. But in our actual choosing we never know exhaustively what we are choosing. Nor can we know the extent of our ignorance. In Georg Wilhelm Friedrich Hegel's words, "If you stop at the consideration that, having an arbitrary will, a man can will this or that then, of course, his freedom consists in that ability. But if you keep firmly in view that the content of his

---

[31]See my *Great Is the Lord* (Grand Rapids: Eerdmans, 2008), for my argument against extreme voluntarism. Does God possess libertarian free will or are God's actions *ad extra* determined by God's nature? In my view, the first alternative risks conceiving God's nature as pure, arbitrary will, while the second risks making God's actions *ad extra* necessary in a vicious sense. My instincts and sympathies lie with those thinkers who refuse to choose either horn of the dilemma. If we conceive of God's act of creation in a trinitarian way as an act of divine love that incorporates created others within the Father's love for the Son in the Spirit, we can get a hint of an act that transcends the dichotomy between necessity and freedom or between arbitrary creation and automatic emanation. Divine love is both totally free and a perfect expression of the divine nature. What other interpretation could we give to John's declaration that "God is love" (1 Jn 4:9)?

[32]Søren Kierkegaard, *Works of Love*, trans. Howard and Edna Hong (New York: HarperCollins, 2009), pp. 112-13.

willing is a *given* one, then he is determined thereby and in that respect at all events is free no longer."[33]

It seems to me that Bernard's "freedom from necessity" or Aquinas's mild libertarian freedom is fully adequate to explain moral responsibility. When a thief, murderer or liar is confronted with his sin he would have to admit that he did his deed willingly. No one made him do it, and the deed expressed what he wanted to do. A defense attorney would not be able to persuade a jury with an argument like the following: "Indeed my client wanted to kill victim *x*. But he did not freely choose to want to kill victim *x*. We don't know how he came to want to kill her. Perhaps he acquired this character by being mistreated as a child or being bullied in school. Or perhaps he mistakenly thought his action would bring him happiness. Hence my client did not freely kill victim *x*. And no one should be punished for something they did not freely do."[34]

Freedom and the possibility of repentance and forgiveness. A third biblical theme relevant to the issue of created human freedom is the complex of guilt, repentance and divine forgiveness. Paul voices the view of all Scripture when he declares, "All have sinned and fall short of the glory of God" (Rom 3:23). John makes the same claim: "If we claim to be without sin, we deceive ourselves and the truth is not in us" (1 Jn 1:8). The New Testament attests with certainty that sin is universal among human beings, but it is not present in the same way that rationality and embodiment are universal. Sin is a human act, not an aspect of human nature. Human beings sin willingly, *all* of them. This is a very strange situation that will require extensive discussion in another context. Here we merely need to be clear that Scripture understands that only willing acts can be sinful. This is made clear by its attribution of responsibility and guilt to the sinner. But responsibility and guilt are not the last word, for in Jesus Christ God offers forgiveness and renewal to those who repent. This offer bears witness that decisions to disobey and to give in to temptation are not definitive and irreversible. The New Testament speaks of only one sin as unforgiveable, blasphemy against

---

[33]*Hegel's Philosophy of Right*, trans. T. M. Knox (New York: Oxford University Press, 1967), pp. 230-31.

[34]Perhaps the most famous statement of this line of thought is the fifth-century Sophist Gorgias's *Encomium of Helen*. Gorgias defends Helen's act from blame by arguing that her actions were determined by circumstances and forces beyond her control, such as physical compulsion, being overwhelmed by love, the manipulation of a god or being deceived by clever arguments.

the Holy Spirit (Mt 12:32; Mk 3:9; Lk 12:10). In the New Testament, forgiveness and repentance are bound together. To repent is to condemn one's wrong decisions and sinful actions with regret and shame. In repentance one looks back on those sinful deeds with a new perspective that reveals their corruption. One, as it were, relives those sinful choices, judging them to be wrong and self-contradictory even at the time. The penitent says to himself: "I ought to have known better. I did know better!" But he also says, "How could I have been such a fool? Why didn't I think?" No matter how sincere, however, such revocation and sorrow cannot actually change the past, undo the harm or guarantee that we will not do it again. Hence forgiveness and renewal are also necessary. Forgiveness on the part of the one wronged (God) is the act of foregoing the just punishment for the wrong done in view of the sincere sorrow of the penitent person and in hope of a reformed life in the future. To forgive is an act of grace and hope. The grace of forgiveness balances accounts not by exacting revenge or pursuing legal justice but by absorbing the harm. But forgiveness is not an end in itself. Forgiveness can become the basis for reconciliation and a new life. But forgiveness deals only with punishment merited by sinful acts. In addition to forgiveness the sinner needs to be exempted from the power of sin, which drives us to sin again. In the New Testament the answer to this need is the sanctifying power of the Holy Spirit, who enlightens and strengthens us in this life and will raise us into eternal life.

I believe this theme sheds significant light on our quest for understanding created human freedom. Human beings are clearly declared guilty and in need of forgiveness, which is a way of saying that they are responsible and answerable for their acts and that their acts are defective in a way that they alone cannot repair. This in turn assumes that we sin willingly and hence are guilty. We cannot evade due punishment by denying responsibility. As I argued above, the traditional doctrine of free will, as explained in Bernard and Aquinas, seems an adequate presupposition for the responsibility and guilt attributed to human beings in Scripture. To some, however, it may still seem that a more radical freedom is required. But this hypothesis is disconfirmed by the New Testament teaching about repentance, forgiveness and renewal. The idea of repentance seems to harbor a paradox. On the one hand, unless one acted freely in the sinful act repentance makes no sense. You can't change an attitude that you never had. On the other hand, how can you repent for an act done freely? For doesn't freedom—in its libertarian form at least—imply self-mastery and self-knowledge? And doesn't self-knowledge imply clarity and stability in one's decisions? How then can one repent of one's free acts? The penitent looks back at his or her decisions with horror and shame and sees them as mistaken and presumptuous. In repentance something happens—new information comes to light or a new conviction about what is good or real is attained—to change our understanding and evaluation of our past deeds and the spirit and expectation with which we chose them and did them. We see not only our willingness, which now seems alien, but also our ignorance, arrogance, irrationality and mistakenness. To account for the experience of regret and repentance, our concept of free will must be able to include willingness *and* blindness, exemption from necessity *and* ignorance of ourselves and of the full consequences of our deeds.

The possibility of forgiveness follows the same logic as repentance. To forgive someone of an offense assumes that the offender has repented and, hence, that the offending deed does not express the indelible character of the guilty party. Forgiveness, then, assumes that the sinful act, though done willingly, also involves such mitigating factors as ignorance and deception.[35] Jesus begs his Father to forgive his executioners and uses their lack of understanding as a factor: "Father, forgive them, for they do not know what they are doing" (Lk 23:34). Paul also recognizes that forgiveness assumes both willingness and ignorance in acts of sin: "Even though I was once a blasphemer and a persecutor and a violent man, I was shown mercy because I acted in ignorance and unbelief" (1 Tim 1:13). Clearly the "free choice" by which those who killed Jesus acted and in which Paul rejected Christ and persecuted his followers was not yet exempt from every power and condition that limited its ability to judge what leads to the truly good and total happiness. In this case at least, Bernard seems to be right: natural free choice needs freeing from sin and sorrow.

Forgiveness not only presupposes repentance but also anticipates the liberation of the sinner from the conditions that make sin inevitable. We now enter a very perplexing maze of issues usually dealt with in treatments of the

---

[35]Scripture also describes the condition from which Christ saves as deception (Rom 7:11; Titus 3:3), ignorance (Eph 4:18), slavery (Rom 6:6; 7:14; John 8:34), blindness (2 Cor 4:4; 1 Jn 2:11), misery (Rom 3:16), corruption (2 Pet 2:10) and death (Eph 2:1).

doctrine of the fall and original sin. I will deal with this directly in another
section. Here it will serve our discussion of freedom better if I analyze the
issue in existential terms. Before we consider guilt, let us consider error.
Human beings are by nature finite and fallible. We are mostly ignorant of
our world, we lack perfect self-knowledge, and we are unable to predict the
consequences of our acts very far into the future. Mistakes do not happen
by necessity, but they are unavoidable. But we are not only fallible; we know
we are fallible. Not only do we make mistakes, we know we have and will
again make mistakes. The reflexivity of human consciousness adds a new
possibility beyond simply making correct or mistaken judgments about the
world. We make judgments about our judgments and ourselves as those who
make them. We make judgments about other people's judgments of our
judgments and ourselves. Understandably we do not want to be limited,
dependent, fallible and mortal. This is not yet an evil desire, for it is natural
to affirm our being and unnatural to negate it. If we were perfectly attuned
to our Creator, however, we would know that our finiteness rests secure in
his infinity. What we lack God has, and because God loves us we have what
God has. But under the present conditions we do not have this attunement
or security. We set about to secure ourselves and protect our sense of dignity.
We make ourselves our own infinity, an abyss of desire and imagination, of
boasting and lying. We deny our fallibility by making lying judgments about
our judgments and rejecting angrily other people's judgments of our decisions. We vehemently defend our dignity when others point out our mistakes or wrongs. And so the power of reflexivity coupled with finitude leads
us to cross the boundary between error and guilt. At the root of guilt is refusal of truth and substitution of a lie more in keeping with the false infinity
of desire and imagination.[36] From this root grows murder, lying, stealing,
adultery and all other sins.

The renewal spoken about in Scripture deals with the morbid condition
just described. This transformation is spoken of as a new birth (Jn 3:3-8),

---

[36]My terminology and some of these thoughts are inspired by Paul Ricoeur, *Fallible Man: Philosophy of the Will*, trans. Charles Kelbley (Chicago: Regnery, 1965); and Ricoeur, *The Symbolism
of Evil* (Boston: Beacon, 1986). I am not proposing an explanation of the transition from fallibility to fault. The fact of the transition is clear, but the moment of transition is obscure. The
Scriptures address it as always already having occurred. The story of the fall does not count as
an exception because the evil serpent is already present in the garden. Ricoeur concludes that
since human beings are fallible by nature "the *possibility* of moral evil is inherent in man's constitution" (*Fallible Man*, p. 203).

new creation (2 Cor 5:17; Gal 6:15), enlightenment (Heb 6:4), renewal (Tit 3:5), salvation (Rom 1:16; 2 Tim 2:10), adoption as God's children (Rom 8:15; Eph 1:5), reconciliation (2 Cor 5:18-20), liberation (Rom 8:21), justification (Rom 4:25; 5:18), sanctification (1 Cor 6:11; Heb 10:29), resurrection (Rom 6:5; Phil 3:11), new life (Rom 6:4) and eternal life (Jn 3:16; Rom 2:7). In all these transformations, actual and anticipated, our created humanity is preserved but freed from the power of sin and death, and brought to its definitive fulfillment and perfection. From the perspective of this liberating transformation, the freedom we have as created human beings appears less than perfect. As creatures blessed with reason, we are exempt from the necessity of following the passions aroused by the objects we meet. Even apart from our subjection to the power of sin, as creatures we are not exempt from all limits on our ability to realize perfectly our God-given nature. We are largely ignorant of ourselves and our world. We cannot foresee the full consequences of our decisions, and we do not have complete knowledge of how every action is related to the supreme good. As sinners, in addition to fallibility we suffer from an unthinking and habitual tendency toward pride, envy, selfishness and idolatry, and other spiritual sicknesses. To sum up, as creatures and as sinners we are free from necessity or compulsion in our choices for this or that. Yet such free choice cannot by its own power escape the finite human condition that makes error unavoidable or the lack of awareness of God that makes sin inevitable. I have to agree with Bernard when he says, "I think it has been clearly shown that even freedom of choice is to some extent held captive as long as it is unaccompanied or imperfectly accompanied by the two remaining freedoms."[37]

Summary. I have argued that the highest form of human freedom is still future for us. Eschatological freedom is *a supernatural transformation inwardly and outwardly so that every limitation that blocks human beings from becoming and living as children of God is removed.* Freedom under the conditions of creation and fall, which we now experience, is only analogously related to eschatological freedom. We have not yet been exempted "inwardly and outwardly" from all limits on our knowledge and love of God. True, our wills are free from necessity, reason informs our choices and imagination enables us to consider multiple possibilities. But we are not yet free from sin

---

[37]Bernard, "Grace and Free Choice," p. 72.

and fallibility and mortality. We do not yet know ourselves completely. We are not yet completely holy, and our vision of God and love for him are not yet perfect. We do not yet will the good wholly and irrevocably. We are not yet wholly free. In reconciling human freedom with the omnipotence model of divine providence we must keep clearly in mind the distinction between ultimate freedom of the "children of God" (Rom 8) and our present fragmentary experience of freedom.

THE COMPATIBILITY OF PROVIDENCE AND FREEDOM

Grasping the relationship between divine providence and human freedom depends on keeping clearly in mind which view of providence is being reconciled with which view of freedom. As we have seen, thinkers propose very different models of providence and widely varying understandings of human freedom. Misunderstandings and misplaced criticisms invariably arise when discussants do not make clear which models they are defending. It makes good sense, then, to begin this discussion by stating clearly my understanding of providence and the concept of human freedom I find compelling.

I understand providence as *that aspect of the God-creation relationship in which God so orders and directs every event in the history of creation that God's eternal purpose for creation is realized perfectly.* And I understand human beings' highest freedom as *a supernatural transformation inwardly and outwardly so that every limitation that blocks human beings from becoming and living as children of God is removed.* For Christian faith, human freedom, then, is not the natural power of self-determination that God has to respect even if it is blind, misguided and irrational. It is not the absence of circumstances that prevent you from doing whatever you wish. Freedom is not a power or a state we experience fully in the present. It is the goal of God's acts of creation, reconciliation and consummation. Given this understanding, it would be absurd to imagine that God's ordering and directing activity in providence could conflict with human freedom. The very purpose of providence is to free us from "every limitation that blocks human beings from becoming and living as children of God." Hence, even if it is difficult or impossible to perceive or grasp how God's ordering and directing our lives moves us and the world toward the goal of ultimate freedom, we may assume that it does.

Aquinas on primary and secondary causality. Can the kind of providential control I am advocating be reconciled with human freedom? The history of the doctrine of providence has produced many theories of reconciliation. I will briefly explain two such attempts. Many follow Thomas Aquinas in distinguishing the way God causes things to come into being and happen from the way these activities occur among creatures. The God-creature relationship is of a different order than the creature-creature relationship. God, who has no cause outside of himself, causes things to come into being from nothing. Creatures are second causes in that they have a cause outside and prior to them. They cannot create from nothing but can only, as it were, pass on the cause that caused them. Creatures operate in the same space, and their existence and activities mutually exclude each other. When one creature acts to create a certain effect in another, the effect cannot also be simultaneously the act of the second creature; the second creature is passive. But divine action, being of a completely different order, is not subject to these limitations. Just as God can be omnipresent without displacing creatures, God can be omnicausal (not meaning only-causal but everywhere-causal) without disrupting the causality of creatures. According to Aquinas, "The causality of God, Who is the first agent, extends to all things not only as to the constituent principles of the species, but also as to the individualizing principles. . . . Hence all things that exist in whatsoever manner are necessarily directed by God toward the end."[38] God's causality gives creatures their total being, which includes their causal powers and their acts. In our acts we create new states of affairs; but we must destroy the previously existing states of affairs to bring new ones into being. But God does not need to disrupt any existing thing to bring about something new. Hence God can "cause" an effect to come about through the undisrupted action of a free creature. As Aquinas puts it: "And just as by moving natural causes He does not prevent their actions from being natural, so by moving voluntary causes He does not deprive their actions of being voluntary; but rather is He the cause of this very thing in them, for He operates in each thing according to its own nature."[39]

Jonathan Edwards and Paul Helm on psychological determinism. A

---

[38]*SumTh* 1.22.2 (Anton C. Pegis, ed., *Basic Writings of Saint Thomas Aquinas* [New York: Random House, 1945], 1:232).

[39]*SumTh* 1.83.1 (Pegis, *Basic Writings*, 1:786). For fuller documentation of the thoughts in this paragraph, see my earlier treatment of Aquinas's view of providence in chap. 13.

second approach, famously advocated by Jonathan Edwards, defines freedom as willingness, and claims that freedom is "the power, opportunity, or advantage, that any one has, to do as he pleases. Or in other words, his being free from hindrance or impediment in the way of doing, or conducting in any respect, as he wills."[40] Paul Helm, a modern disciple of Edwards, explains that one can be free to do as one wills even when the will is "causally determined by desires and intentions."[41] For Edwards and his followers, freedom is completely compatible with psychological determinism. An all-encompassing divine providence is easily reconcilable with this view of freedom. God orders and directs human free action by ordaining "those factors which determine human agency."[42] If God wills agent x to turn left at time t, God makes sure that agent x finds himself in the situation at time t that will cause him to will to turn left. In this way when human beings do what they will, they also do what God wills. God can determine the entire course of human history without compelling and thus without destroying human freedom.

The advantages of Aquinas's view of freedom. It seems to me that Aquinas's reconciliation is fundamentally sound because it is based on the premise of creation from nothing, which is rooted in divine revelation. It holds steadily in mind the proper distinction between God and the world and divine and human action. It does not attempt to render the mystery of the God-creature relation transparent to reason; rather it unfolds the implications of the biblical understanding of God and creation for providence. I do not find the psychological determinism advocated by Edwards and Helm as helpful as Aquinas's approach. Unless psychological determinism is making the trivial point that once you decide on a course of action your decision can be accounted for by discovering the reasons that led to that decision, it must be accompanied by the claim that the qualities of the finite objects of our experience determine our wills to one effect. Aquinas refutes this idea to my satisfaction. We cannot help willing our ultimate happiness, but our minds always transcend finite objects and we can imagine defects in those objects and alternative ways to happiness. In this way we are free from the power of our immediate situations. For Aquinas, only the vision

---

[40]Jonathan Edwards, *Freedom of the Will* 1.5, in *The Works of Jonathan Edwards* (1834; repr., Peabody, MA: Hendrickson, 1998), 1:10. In this definition Edwards echoes John Locke, *Essay Concerning Human Understanding* and corresponds to Adler's freedom of self-realization.
[41]Paul Helm, *The Providence of God* (Downers Grove, IL: InterVarsity Press, 1993), p. 174.
[42]Ibid.

of God, who is the highest good and the necessary and sufficient basis for ultimate happiness, could determine the will absolutely. Aquinas makes a comparison between truths that compel the mind to assent and goods that compel the will to desire:

> Now in the same way there is also a certain good which is desirable for its own sake, namely, happiness, which has the nature of an ultimate goal, and the will is compelled to adhere to that good, for there is a sort of natural compulsion on every one to want to be happy. . . . If then there were goods which were the *sine qua nons* of happiness, these too would compel desire.[43]

But according to Aquinas, no finite good can compel desire. Only the infinite good is the "*sine qua nons* of happiness."

If I thought psychological determinism were true as an anthropological fact, which I do not, I would have no theological objection to the notion that God brings about the conditions under which we are attracted to the course of action God wills. Indeed, this divine act would be liberating; for our decisions and our actual course of action would be guided by the infinitely wise God!

Freedom as a gift only God can give. Whatever possibilities the above ways of reconciliation possess, I want to pursue another route, keeping clearly in mind that only eschatological freedom as described above is freedom in the fullest sense. Taking eschatological freedom as the norm, how can we evaluate our acts in the here-and-now? Given this norm, an act is free insofar as it manifests the absence of all those limitations that hinder us from loving God with our whole hearts, from willing God's will absolutely. An act of will is unfree insofar as it is unlike that act. It makes no difference whether we are acting in the power of alternative choice or are determined by the appealing or unappealing qualities of our circumstances. What matters in terms of freedom is whether we are unlimited in our love for God. In this life we strive for many things and we use our intelligence and power to achieve them. But Bernard is correct to remind us that we labor under the burdens of sin, ignorance and bad habits. In our willing acts, in our wisest choices, in those most free by ordinary standards, even free by the standard of self-determination, we strive for false goods, want destructive

---

[43]Thomas Aquinas, *Commentary on Aristotle's* De Interpretatione (Aquinas, *Selected Philosophical Writings*, p. 284).

things, travel futile ways and seek empty goals. If God were to leave achieving our highest happiness, the true human end, and the glorious freedom of the children of God to the cumulative effects of our "free" choices, we would inexorably work our destruction. Where would we be if God let stand our arrogant "self-determination" and our misguided "willing"? True human freedom is not a means of attaining the goal of eternal life; it is the goal.

The biblical doctrine of eschatological freedom assumes that we were created for God, that God is our end. God created us so that our deepest desire is for ultimate happiness, and such happiness can be attained only by union with God. And God's goal for us is eschatological freedom and eternal life in union with him. Notice what we have here. We will our ultimate happiness and God wills our ultimate happiness. But God knows perfectly what will make for our ultimate happiness and glorious freedom, and God knows the way to attain it. We do not. The conclusion follows: God wills for us in every act and event what we would freely will for ourselves if we could see it as it really is. Knowing this, does it make sense for us to insist that God refrain from so strongly influencing what we will and do that we actually attain ultimate happiness? Had we rather be autonomous and fail than be under God's loving control and succeed? Does it make sense for us to pray to God a prayer like the following? "Dear Father in heaven. I cannot guide my steps so that they will lead me home to you. Please take charge of my life. Channel my willing action in the direction you know is right, work in my mind and my heart to protect them from error and sin, give me right choices and overcome my bad ones. Set before me your will in all its goodness so that I may know and will it." Yes, such a prayer would make perfect sense, because you would be asking for more freedom not less and a greater happiness not a lesser.

How might the answer to this prayer work out? What if God placed us in situations in which we would make certain choices or placed thoughts in our minds or affections in our hearts that gave a certain direction to our wills; or what if through a mysterious causality God moved our wills in a certain direction or through any other noncoercive means channeled our lives toward our ultimate joy? Would God's loving action destroy human freedom? Absolutely not; for it is precisely in these actions that God is answering our prayers for divine help in attaining that glorious liberty! And even for those who have not prayed for divine help, God knows what we would pray if we only know what to pray.

Part Three

THE
CHALLENGE
OF EVIL

CREATION, PROVIDENCE AND EVIL

T HE ISSUE OF EVIL ARISES in the doctrine of creation as the question of its nature and consistency with God's act of creation. If in creation God establishes the being, life, time and space of creatures, what room is there for error, sin and corruption? Is evil an eternal reality or a necessary aspect of creatures? Does the presence of evil undermine the doctrine of creation? The issue of evil arises in the doctrine of providence in a different way: How does God use and overcome evil in bringing about the ultimate perfection of his creation? Does God's working through evil acts taint God with the evil in those acts, justify evil or exonerate sinful human beings? The first half of the chapter will deal with the first set of questions and the second half with the second set.

The Christian doctrine of creation concerns itself with the origination, being and telos of the created world. The Scriptures make clear that God is the creator of *all things*. Nothing exists, comes into being or remains in being except by the will and power of God. And Scripture unambiguously declares that the divine work of creation is good and all things God creates are good. But Scripture also acknowledges what everyone knows: something is wrong with God's world. Sin and suffering and violence and death plague the good creation. In large measure biblical religion and theology focus on salvation from sin and death. The Mosaic moral and religious laws were given to prevent sin from working out its anarchic logic, to make the sinner conscious of the seriousness of sin and to secure atonement for sin. Jesus Christ

came into the world "to save sinners" (1 Tim 1:15) and bring "life and immortality to light" (2 Tim 1:10). In the final consummation Christ will destroy sin and death "so that God may be all in all" (1 Cor 15:28). Christianity
takes sin, suffering and death as a problem only God can solve. God reconciled the world to himself in Christ, unleashed the Spirit to begin a new
creation and raised Christ from the dead. These divine acts in the middle of
history anticipate the end when all evil will be overcome, creation will be
brought to perfection and eternal life will reign.

Reconciliation and eschatological consummation focus on the ground
and means of the final victory over sin and death. Little attention is given
to the origin and metaphysical nature of evil. Sin and death are variously
conceived as corruption, rebellion, the malice of diabolical powers,
lostness, slavery, deception, blindness and death. Human beings face
spiritual enemies that tempt, deceive and blind them: the "god of this
world," the devil, demons, and principalities and powers. These forces
possess power human beings cannot withstand unaided. But these beings
have no power in relation to God. The New Testament is rife with dualistic language: good and evil, light and darkness, above and below, angels
and demons, God and the devil. But the dualism found in the New Testament between the good powers and the bad ones is moral not metaphysical in nature. It makes a distinction between beings that desire and
do good and beings that desire and do evil, not between good substances
and bad substances.

DUALISM—MORAL OR METAPHYSICAL?

It is important to get clear on the nature of the opposition between good and
evil in Christian teaching. It may be that there is a tendency in human
thought to understand ourselves and the world in terms of the opposition
of two kinds of things. After all, thought requires subject and object. If we
try to reduce the infinite variety of things to one undifferentiated substance
we destroy thought itself, because thought requires difference and relation.
On the other hand, if we attempt to understand the world as an infinite
number of unrelated substances, thought becomes overwhelmed with chaos
and confusion, for there exists no subject to unify experience. Thought
tends to reduce the multiplicity of the world to as much unity as possible
without losing itself in absolute simplicity, and this tendency gives the

number two its special significance.[1] Hence we find dualistic patterns in every area of thought: between being and becoming, mind and body, being and time, God and the world, and good and evil.[2]

As I indicated above, the Bible also speaks the language of dualism. We can discern two dominant types of oppositions in the Scriptures, that is, between God and creation and between good and evil.[3] The former we can call creational dualism and the latter moral dualism. God is the eternal, self-existent, unchanging Creator. Creation is temporal, changing and dependent on God for its existence. Even though God and the world cannot be reduced to one another or both to a higher unity, the God-creation dualism is not eternal. Creation exists because of the power, will and love of God. And as the result of God's decision and act, creation is good and does not stand in a relation of opposition or competition to God. But in some extrabiblical philosophics and religions the dualism between the divine and the nondivine becomes what I shall call metaphysical dualism. We find a mild, rational form in Platonism and extreme, mythological forms in Gnosticism and Manichaeism. For Plato, the creator of our world shaped unformed matter into the best possible material world; however, matter is incapable of being completely brought under the control of mind. The reason I call Plato's dualism "mild" is that matter, as the principle of evil, is not virulent and personal or as real as the highest good.[4] The more radical dualisms reduce the diversity of things we experience to "two equally primordial and mutually opposed principles," one of which is identified as good and the other as evil.[5] Radical dualism combines dualism of being with moral dualism to produce not only two kinds of principles but also two kinds of wills, interjecting a fundamental antagonism into reality. It sees these good and evil principles as such unyielding antagonists that they cannot be derived one from the other or reduced to a more fundamental principle. Hence it postulates that both are eternal and capable of existing on their own; that is, each is a *substance*. In

---

[1]We can also see the significance of the number three in the need for a third thing to mediate between one and two.

[2]Robert Audi, ed., *Cambridge Dictionary of Philosophy*, 2nd ed. (Cambridge: Cambridge University Press, 1999), s.v. "Dualism."

[3]We also find historical dualism, the opposition between the present age and the coming age.

[4]Plato, *Timaeus*, trans. Donald J. Zeyl, in *The Complete Works of Plato*, ed. John M. Cooper (Indianapolis: Hackett, 1997).

[5]Eberhard Simons, "Dualism," in *Encyclopedia of Theology: The Concise Sacramentum Mundi*, ed. Karl Rahner (New York: Crossroad, 1975), pp. 370-74.

Gnosticism and Manichaeism, the two struggle against each other for dominance in the world. One is light and the other dark. One is spiritual and the other material, heavy and light, or high and low. In metaphysical dualism evil is a real thing like a tree, a fluid or a mind. It possesses its own power, resistance, causality and substance. Redemption from evil is accomplished by escaping from the mixed world into the world of pure light or good. Dualistic religions propose various methods of escape or purification.[6]

The Bible does not characterize the God-creature relationship as good versus evil. The moral dualism of the Bible distinguishes between the community of beings who give themselves to doing good and the community of those who do evil. The community given to evil sets itself against God and those who serve him. The community of darkness includes Satan, fallen angels, demons and some human beings. Moral dualism differs from metaphysical dualism in its conviction that created substances (humans and angels and all others) are metaphysically good (that is, they can be directed toward a good end), and that even when they become morally evil creatures remain metaphysically good.[7] Morally evil beings desire and do evil habitually. They may give themselves to evil to such an extent that the natural tendencies of their created natures have been subverted from their natural telos toward an unnatural one. Evil has become second nature to them.

If evil is neither an eternal substance nor a created substance, what is it? Is it an illusion? From the patristic period onward, when the church faced challenges from dualistic religions and philosophies, Christian theologians argued that evil is disorder or privation of a good thing, not a substance. That is to say, evil cannot exist and have effects except in the defective activity of a good substance. If you remove evil from a corrupted creature, the good creature remains; but if you remove the good from a creature, nothing remains. Augustine of Hippo spent many years of his adult life as an adherent of the Manichaean religion, believing that evil is a dark, fluid-like substance that corrupts good things by invading them. Enlightened by reading Neo-Platonic books, he came to believe that evil is defect and disorder in a substance but not a substance itself. Speaking to God, Augustine

---

[6]See James Kevin Coyle, *Manichaeism and Its Legacy* (Leiden: Brill, 2009), and the literature on Manichaeism cited there.

[7]It follows from this that all created beings, no matter how morally corrupt they become, remain redeemable.

says, "For you evil does not exist at all, and not only for you but for your created universe, because there is nothing outside it which could break in and destroy the order which you have imposed upon it."[8] And further, "I inquired what wickedness is; and I did not find a substance but a perversity of will twisted away from the highest substance, you O God, toward inferior things, rejecting its own inner life."[9] And in another place Augustine says, "So every being, even if it is corrupt, insofar as it is a being is good, and insofar as it is corrupt, is evil."[10] This view of evil is not limited to Augustine or to Western Christianity. Basil the Great states it with great clarity:

> Do not consider God the cause for the existence of evil, nor imagine evil as having its own existence. For evil is not subsistent the way an animal might be; nor can we see its essence substantiated. For evil is the absence of good. . . . Evil does not have its own subsistence, but it follows upon the wounds of the soul. For it is neither uncreated, as the impious say who hold evil in the same honor as the good nature, considering both of these to be without beginning and superior to generation; nor is it created, for if all things are from God, how can evil be from good. For nothing that is vile comes from the beautiful, nor does evil come from virtue.[11]

Are we, then, to conclude that, since evil is not a substance and possesses no power of causality, evil is an illusion and has no effects? No, that would be a hasty conclusion. According to the biblical story of salvation God takes sin and evil very seriously. But how does evil disturb God's world and call forth such a dramatic divine response if, as Augustine maintained, it does not exist the way other things exist? How can we make some sense of this seeming paradox?

---

[8]Augustine, *Confessions* 7.13 (Saint Augustine, *Confessions,* trans. Henry Chadwick [New York: Oxford University Press, 1991], p. 125).

[9]*Confessions* 7.16 (Augustine, *Confessions*, p. 126).

[10]*The Augustine Catechism: The Enchiridion on Faith, Hope, and Love* 4.13, ed. John E. Rotelle, OSA (New York: New City Press, 1999), p. 43.

[11]Basil the Great, *God Is Not the Author of Evil* 8, quoted in Dumitru Staniloae, *The Experience of God: Orthodox Dogmatic Theology,* vol. 2, *The World: Creation and Deification* (Brookline, MA: Holy Cross Orthodox Press, 2000), p. 148. Anselm of Canterbury spends three sections in his essay *On the Fall of the Devil* demonstrating the nonsubstantial nature of evil. He compares the word *evil* to the word *nothing.* He says, "It is in this way that 'evil' and 'nothing' signify things, that is, what is signified is not something in reality but only in grammatical form. 'Nothing' signifies simply non-being or the lack of all that is real. And evil is only non-good or the absence of good where good ought to be found." Finally, he gives evil and nothing the status of "quasi-realities" (*On the Fall of the Devil* 11; *Anselm of Canterbury: The Major Works,* trans. Brian Davies and G. R. Evans [New York: Oxford University Press, 1998], p. 210).

SIN, ERROR AND CORRUPTION

To gain clarity about what sin and evil are and how they disturb God's world, it will be helpful to distinguish among fallibility, peccability and corruptibility, and correspondingly among error, sin and death. It is important to note that these three "abilities" are really not abilities at all but disabilities. They are our inability to preserve ourselves from error, sin and corruption. Fallibility is the natural human condition. We do not possess perfect knowledge of everything within the space in which we must act; nor do we have the power to gain it. Although we possess the idea of truth, goodness and happiness, in our actual existence we are not identical to them. Hence we must act without knowing all the consequences of our action but in the knowledge that we do not know. We make a mistake when we act on an incorrect understanding of the way things are. We can distinguish mental errors such as one might make on a math or logic exam from mistakes involving actions that fail to achieve what we intend in the extra-mental world. We turn into the lane to our right, thinking we are a safe distance from the nearest car in that lane, when in fact there is a car in our blind spot. Every finite being is by definition fallible.[12] The only way a finite being can be certain never to err is to be guided by an omniscient being in all its actions.[13] Pure mistakes are morally neutral. By pure mistakes I mean erroneous conclusions or actions where there is no neglect or presumption. Some of these mistakes arise out of pure ignorance in which no defect of perception or thought is involved. The facts were impossible to perceive or think so that even their absence created no suspicion; yet they were in a position to affect the outcome of the action. No culpability accrues to such mistakes. Other mistakes occur through inaccurate perception or thought but without neglect or presumption. Though no culpability can be charged to our account for this type of mistake, nevertheless great suffering and loss can be caused even by such innocent errors. But some errors seem to slide over into the category of sin. Knowing our fallibility, we should take it into account in our calculations, considering the consequences of being wrong, not arrogantly

---

[12]The preceding and much of the following is inspired by—though not guided by—Paul Ricoeur's phenomenological and existential analysis of the will in *Fallible Man: Philosophy of the Will*, trans. Charles Kelbley (Chicago: Regnery, 1965); and Ricoeur, *The Symbolism of Evil: Philosophy of the Will* (Boston: Beacon, 1967).

[13]In this case there would be no need for the finite being to be omniscient itself, for it cannot in any case act simultaneously in all possible situations.

overestimating our powers or showing off to others. To act presumptuously even if no evil actually happens may offend against our proper place in relation to God. But in the end we cannot require people to possess absolute knowledge and therefore absolute certainty before they act.

Peccability or vulnerability to sin is not the same condition as fallibility, just as sin is not the same kind of act as error. Peccability presupposes the human capacity for moral action, action in awareness of a moral order to which we should adhere but can imagine disobeying. The moral imperative is clear or clear enough, but we are aware of other possibilities, some of which have the appearance of desirability. Human beings are naturally peccable. The impetus of the moral law must compete with other motives for action. Apart from divine grace illuminating the mind and strengthening the will, human beings, being naturally fallible, can err in judgment about the moral law and fail to perceive its wisdom. We are liable to allow ourselves to be guided by our most urgent passion rather than listening to the whisper of reason.

Sin and error are closely related, though not identical. We may err without sinning, but we cannot sin without erring. Hence if we could not err we could not sin.[14] Both sin and error presuppose voluntary action, for we cannot be compelled either to sin or err. Sin is willing action that transgresses the moral law. A mistake is a voluntary act that fails to accomplish what was intended. One cannot make a mistake intentionally.[15] How then can sin also be error, since in sinning one deliberately does what is forbidden? The answer lies in the insight that sin has a double aspect that error does not. Sin not only intends a morally forbidden thing but also questions the law that forbids it. In sinning we judge that the law forbids something good or withholds something we are due. This judgment is mistaken. The sinner intentionally acts in a certain way and so does not make a mistake in that respect; one accomplishes what was intended. But it doubly errs in presumptuously pitting its judgment against that of the moral lawgiver and

---

[14]Ricoeur concludes that since human beings are fallible by nature "the *possibility* of moral evil is inherent in man's constitution" (*Fallible Man*, p. 203). We come to know our fallibility through actual fault: "Fallibility is the *condition* of evil, although evil is the revealer of fallibility" (p. 221). "Man is the Joy of Yes in the sadness of the finite" (p. 215).

[15]Mistakes must be distinguished from accidents. An accident is something that happens to us whereas a mistake is something we do. When something goes wrong in our acting we call it a mistake. An accident happens apart from a judgment that failed to guide our actions to success. In an accident another line of causality unexpectedly crosses our path and brings us harm.

in thinking an end is good when it is not. Blaise Pascal puts his finger on the two-layered nature of the sinful act: "It is certainly an evil to be full of faults; but it is a still greater evil to be full of them and to refuse to recognize them, since that is to add the further fault of voluntary illusion."[16]

Like other creatures human beings are corruptible by nature. All creatures are composed of other things that had and can have an independent existence and come together in ways to form other composite things. And whatever is composed of parts can decompose. Death is loss of integrity and unity, failure of the natural systems that sustain life. Since God is the creator, God could have protected creatures from corruption by continuing to create and give unity and life to them. Instead, God allots each creature its time and preserves it throughout its allotted life span. I shall have more to say about death and its relation to sin later; but at this point we can say that death is more than merely the inherent corruptibility of the creature. For it does not follow that corruptibility must of necessity lead to corruption. Why, then, doesn't God preserve creatures continually instead of allowing them to deteriorate and die?

To sum up, human beings are naturally fallible, peccable and corruptible. It is not evil that this is so; rather it is part of the goodness of creation. Only if creatures actually err, sin and fall prey to corruption do fallibility, peccability and corruptibility appear in a negative light.

THE FALL AND ORIGINAL SIN

But human beings do err. They sin and fall prey to corruption. In an inexplicable act of will, the crown of God's good creation transitions from innocence to guilt, affirms nothingness over being and chooses falsehood instead of truth. Suffering, death and destruction result from this fall. In traditional theology, the original sin of Adam and Eve serves to explain the universality and inevitability of sin and provides a theodicy exonerating the Creator from responsibility for sin and evil. God created human beings good in body and soul. And they remained innocent until through a misuse of freedom they sinned and brought death and corruption into the world. Even though there has never been a consensus about how Adam's children shared solidarity with him in sin—biologically, through imitation or as sharing in

---

[16]Blaise Pascal, *Pensées* no. 100, quoted in Ricoeur, *Fallible Man*, p. 24.

a covenant—the fact of that solidarity was taken for granted.[17] Despite any mitigating factors, the tradition has been clear that human beings as a whole and individually are responsible for their sin and the evil that flows from it.

The fall. Providence makes no sense apart from history, and history begins with the fall. Genesis 1–2 deals with the creation and commissioning of humankind but does not record a history of human acts before the fall. There is no sense of lapsed time between the creation of woman (Gen 2:21-25) and the fall (Gen 3:1-7). The paradisiacal garden produces no history, for simple human needs are abundantly met, no danger threatens, no glory beckons, no greed lures and no lust drives. No prospect of death gives urgency to action or opportunity for courage. There are no conflicts, no wars and no competitions. There is no need for cities, fortresses or kings; no need for irrigation or trade or writing. There is no thought of art, architecture or music; and there is no religion or ethics. In paradise human beings add nothing to creation, create nothing new and become nothing different.

This is not to say that the Bible looks at the fall as a positive development. Perhaps it could have gone the other way. Perhaps if the human beings had resisted temptation and obeyed the Creator another history could have unfolded in which humans created beauty, achieved glory and attained virtue in other ways. Either way, paradise was never meant to last. In the "fall" narrative the possibility of evil appears in the form of a snake. There is no indication that the snake is the embodiment of a fallen angel. It is simply part of nature, though it is said to be more "crafty" than the other animals God created. The snake articulates what could as well have arisen in Eve's heart. It questions the reasonableness and even the sincerity of God's command. Perhaps God wants to keep you down, reserving wisdom for himself alone. When Eve looks at the fruit with these thoughts in her heart she sees its beauty and its apparent goodness. She eats and gives to her husband and he eats. Innocence is gone and knowledge attained. But this knowledge brings guilt, shame, lies and conflict. Everything in the world now becomes an instrument for good or evil; judgments must be made about the correct order of things. Natural fallibility has become enacted error, peccability has become manifested in sin and corruptibility has become corruption in fact.

---

[17]See Peter C. Bouteneff, *Beginnings: Ancient Christian Readings of the Biblical Creation Narratives* (Grand Rapids: Baker Academic, 2008), for the patristic interpretation of the creation and fall narratives up to the late fourth century.

Humanity is exiled from paradise and from the divine presence.

It is often underappreciated that, after this first disobedience, Genesis rehearses a series of ever worse sins: fratricide, gratuitous murder and finally the descent of heavenly beings (the "sons of God") to mate unnaturally with human women.[18] The great flood and a new beginning for humanity become necessary because "The LORD saw how great the wickedness of the human race had become on the earth, and that every inclination of the thoughts of the human heart was only evil all the time" (Gen 6:5). The sin of Adam and Eve, though trivial in comparison with those that follow, opens the door through which other evils enter, and culminates in catastrophic divine judgment. This history pictures godforsaken humanity as subject to a downward pull toward moral degradation destined to end with the un-doing of creation.[19] Or God may initiate new beginnings as in the postflood generation, the election of Abraham, the exodus and the giving of the law. The definitive answer, however, is the new creation initiated and promised in Jesus Christ. In the broadest sense, then, providence is God's hidden presence and activity in the apparently godforsaken world, working to bring creation to its eschatological consummation according to the divine counsel.

It seems to me not quite on target to view the biblical story of the fall as an *explanation* of why the world is plagued with evil. Or if it is explanation, it explains the facts by a reference to a mystery. No doubt Genesis and Paul (Rom 5:12-21) view sin and death as having entered the world through an event within history rather than simply being a perennial structure of nature. Yet the story of the fall does not explain why this event occurred. Sin entered the human world through a free act of Adam, but the act was facilitated by the serpent, which was already there before the sin of Adam. Whence the serpent? Why did Adam sin? Why do all his descendants sin? If the story does not suffice as an explanation of why the first human sinned, neither can it suffice as an explanation for why all subsequent human beings sin.

---

[18]In much of the pseudepigraphical Jewish literature, the evil described in Gen 6 is of much greater importance in explaining the origin of the evils in the world than Gen 3 is. Heavenly beings brought immorality, false teaching and the technology of war to human beings.

[19]Richard J. Clifford, "The Hebrew Scriptures and the Theology of Creation," *Theological Studies* 46 (1985): 507-23. See the section on Genesis: pp. 520-23. Stanley K. Stowers also emphasizes importance of the thematic unity of Gen 1-11 in interpreting Paul's thought in Romans. See *A Rereading of Romans: Justice, Jews, and Gentiles* (New Haven, CT: Yale University Press, 1995), pp. 88-89.

The transition from fallibility to fault cannot be explained by a line of necessary causality or trajectory toward fulfillment of created human nature.[20] To say Adam fell through a free act is to say he sinned for no good reason, against reason, truth and goodness. The story of the fall is more articulation than explanation. One thing is clear, however: The story affirms clearly the difference between God's good creation and the evil that plagues it. Sin originates in creaturely freedom, and the evils that follow are the just, and possibly remedial, punishments for sin.

Original sin. For the topic of evil and divine providence, the doctrine of original sin is relevant only insofar as it touches the issue of the origin and continued presence of evil in the world. So much of one's theology (election, predestination, justification and baptism) is determined by the position one takes on this subject. The traditional Western doctrine, definitively formulated by Augustine of Hippo, teaches that each of Adam's descendants, generation after generation, inherits the guilt of Adam's "original" sin and the irresistible tendency to sin called concupiscence.[21] It is said that somehow all human beings were included in Adam or represented by him, so that his sinful act counts as (or actually is) the sin of each and all. All are born guilty of the sin of Adam and subject to the punishment due for this terrible sin.

The doctrine of original sin as the double inheritance of guilt of sin and concupiscence was not a widely held view before the time of Augustine. The classic proof text is Romans 5:12: "Therefore, just as sin entered the world through one man, and death through sin, and in this way death came to all people, because all sinned."

Augustine, however, used Jerome's Latin translation in which the last crucial phrase, *eph' hō pantes hēmarton* ("because all sinned," NIV, NRSV) is translated *in quo omnes peccaverunt*, which means "in whom all sinned." Augustine took this phrase to mean that all Adam's descendants were "in" him and participated in his sin and guilt.[22] Paul's emphasis is different. He

---

[20]Ricoeur addresses the problem of the transition from fallibility to fault in *The Symbolism of Evil*. He sees the doctrine of original sin as a tertiary formulation. The first level is agonized confession and the second is myth.

[21]For an exhaustive history of the dogma of original sin, see Julius Gross, *Entstehungsgeschichte des Erbsündendogmas: Ein Beitrag zur Geschichte des Problems vom Ursprung des Übels*, 4 vols. (Munich: Reinhardt, 1960).

[22]I do not want to imply that Augustine's doctrine was based merely on a Latin mistranslation of this text. He incorporates many other texts, theological arguments and ecclesiastical practices—especially infant baptism—into his argument. See his treatises *On the Merits and the Remission*

is interested in the parallel and contrast between Adam and Christ. He aims not to illuminate our relationship with Adam and develop a doctrine of inherited sin; rather he wishes to exalt Christ's work to universal status and to show the superiority of divine grace over the most powerful and ancient spiritual enemies. Everyone understands the negative consequences of Adam's transgression, reasons Paul. Adam was so placed that his transgression brought all human beings under the power of sin and death. Likewise, Christ is so placed that his act of obedience brings life to all. If the translation "because all sinned" is correct, Paul may be saying that Adam brought sin into the world and sin brings death to everyone because everyone sins.[23] In other words, because of Adam, sin came into the world and reigns so powerfully that it inevitably seduces all into sin. But Paul more or less takes this idea for granted and uses it to highlight the universality and superior efficacy of the obedience of Christ.

A contemporary of Augustine, the Eastern bishop John Chrysostom (ca. 347–407), expresses his incomprehension of the notion of inherited guilt:

> For the fact that when he [Adam] had sinned and become mortal, those who were of him should be so also, is nothing unlikely. But how would it follow that from his disobedience another would become a sinner? For at this rate a man of this sort will not even deserve punishment, if, that is, it was not from his own self that he became a sinner.[24]

Karl Barth rejected the Augustinian/Calvinist doctrine of inherited guilt as a contradiction in terms "in the face of which there is no help for it but to juggle away either one part or the other."[25] In Barth's view the truth in the doctrine of original sin is that "we are dealing with the original and radical and therefore the comprehensive and total act of man, with the im-

---

of Sins and the Baptism of Infants, On the Grace of Christ and On Original Sin, On Marriage and Concupiscence, Against Two Letters of the Pelagians and others.

[23]Douglas J. Moo, after surveying many interpretations of Rom 5:12, concludes that, though it is not clear how Paul conceives the connection between Adam's sin and his death or between his sin and the subsequent reign of death over all, this much is clear: "No one escapes the reign of death because no one escapes the power of sin" (The Epistle to the Romans, New International Commentary on the New Testament [Grand Rapids: Eerdmans, 1996], p. 323). Perhaps Paul uses "death" in a dual sense, of spiritual death (alienation, futility, guilt, etc.) and physical death. The connection between sin and spiritual death is readily perceived, whereas the connection between sin and physical death is obscure.

[24]John Chrysostom, Tenth Homily on Romans (NPNF¹ 11:403). His comments refer to Rom 5:19.

[25]CD IV/1, p. 501.

prisonment of his existence in that circle of evil being and evil activity."[26] This "imprisonment" does not come upon us as a fate, the inherited legacy of our first parents; rather we are sinners even in this original sense willingly. We know that we are sinners in this radical sense from the beginning only because we are declared to be so by the word of God. According to Barth we are declared to be those "whose free will and commission and omission, whose actualizations of their good human nature, always follow the rule and perverted order which, according to the prophetic witness, is manifested at once at the very beginning of world-history."[27] Adam "has not poisoned us or passed on a disease. What we do after him is not done according to an example which irresistibly overthrows us, or in an imitation of his act which is ordained for all his successors. No one has to be Adam. We are so freely and on our own responsibility."[28]

What about the other aspect of the traditional doctrine, inherited concupiscence? Are we born with a fallen nature, a vulnerability to sin so deep that avoiding personal sin is impossible? Paul speaks of sin producing "in me all [kinds of] intense desires" (Rom 7:8, author's translation: *en emoi pasan epithymian*). The Latin Vulgate translates this Greek phrase as *in me omnem concupiscentiam*. Both traditional Roman Catholic and Protestant theology understand indwelling concupiscence as a result of the fall rather than being a created attribute, though they differ in certain other ways. According to the decree of the fifth session of the Council of Trent (1546), the sacrament of baptism removes the guilt of original sin but concupiscence remains. The remaining concupiscence is not "truly and properly" sin in the baptized but merely "inclines to sin."[29] In other words, concupiscence, though not sinful as such, is not a wholly natural or innocent state either. The Protestant Reformers and their heirs rejected the Roman Catholic distinction between original sin and concupiscence and hence tended to identify them. Because of this identification the Reformers understood the "intense desires" Paul speaks of in Romans 7:8 not only as "inclining" us to sin but also as themselves sin.[30] According to the Protestant Heidelberg

---

[26]Ibid., p. 500.
[27]Ibid., p. 510.
[28]Ibid., p. 509.
[29]Jaroslav Pelikan and Valerie Hotchkiss, eds., *Creeds and Confessions of Faith in the Christian Tradition* (New Haven: Yale University Press, 2003), 2:826.
[30]See Heinrich Heppe, *Reformed Dogmatics*, rev. and ed. Ernst Bizer, trans. G. T. Thompson (1950;

Catechism, as a result of the disobedience of "our first parents" our natures became so "poisoned that we are all conceived and born in a state of sin." We are "by nature prone to hate God and my neighbor." Moreover, "the wrath of God is revealed from heaven, both against our inborn and our actual sins."[31] For all their differences, both Roman Catholic and Protestant dogma teach that our native and habitual inclination toward sin cannot be identified with our divinely created nature. The passions of the flesh should be under the rule of reason, which should be directed toward God and guided by his law.

It seems to me that the tradition is correct to differentiate concupiscence from what we can call a "vulnerability" to sin or a weakness in the face of temptation. Concupiscence possesses an affirmative character, a positive drive toward or compulsion to sin. We must take care, however, not to confuse concupiscence with natural desires. No doubt that our natural desires for life, food, sex, children, pleasures and comforts of all kinds are disordered and unruly and often come to dominate our lives in the face of reason and divine law. But our bodies and souls and their drives and desires are aspects of God's good creation. They can be used for good when rightly ordered and made subservient to reason, moral law and God, who is the telos of human life. They are not evils to be rooted out totally.

I asked above why God did not protect his fallible, peccable and corruptible creatures from actually falling into error, sin and corruption. In my view any but the most general answer to this question would lead us into speculation. I think, however, something needs to be said on the subject because some may be tempted to doubt God's wisdom and goodness or his foreknowledge and power. If God is perfectly good, what reason could God have for allowing such evil to enter the world? Was God unable to prevent it? To answer these questions, it will be helpful first to remember my earlier argument that the human condition of fallibility, peccability and corruptibility in no way calls into question the goodness of creation or the wisdom of the Creator. Indeed, as I pointed out above, these qualities are aspects of the good of creation. Second, God's act of creation is completely gratuitous

---

repr., London: Wakeman Great Reprints, n.d.), pp. 331-41, for the post-Reformation Reformed view of original sin and concupiscence. The Reformed Orthodox considered original sin under two aspects, imputed and inherent. Imputed original sin is the guilt of Adam's sin and inherent sin is the actual impulse to sin that dominates human life.

[31]The quotes are taken from questions 5-10 (Pelikan and Hotchkiss, *Creeds and Confessions*, 2:430).

and in no way necessary or owed. But if God has no obligation to create creatures, it follows that God has no obligation *to sustain* them in being, although the act of creating and sustaining are so connected that creation would be meaningless apart from sustenance. Creating without sustaining would rob the creature of time, and apart from time creatures cannot exist. However, even if the act of divine creation includes sustenance and time, it does not bind God to give the creature unending time or any specific span of time. Creation is not rendered meaningless simply because sustenance is limited in time. Likewise, God is under no obligation to protect creatures from corruption, error or sin. Just as God's act of creation does not determine how long God must sustain creatures, his acts of sustenance and cooperation do not obligate him to protect them from all deterioration and disorder. But divine creation, sustenance and cooperation would become meaningless if God exercised no protection from corruption, error and sin. Apart from some level of divine protection creatures would fall into nothingness immediately. But God has no obligation to exercise any particular level of protection. Hence God's choice to withhold from creatures complete protection from error, sin and corruption does not violate an obligation implicit in his decision to create and sustain creatures, which, as I made clear above, is a free and gracious decision. Hence we can conclude that if God wished to bring the world to its appointed end by allowing the fall and its consequences, God was completely free of any obligation to refrain on moral grounds or grounds of self-consistency.

A third point must be made. In view of God's redemptive act in Christ, we know God will not leave his creatures forever subject to sin, error and corruption. And if God will redeem his creatures from sin and make them invulnerable forever to sin and corruption, we know that God did not leave us vulnerable originally because God was powerless to do otherwise. And it would be absurd to think of this omission as arbitrary. God must have had a good reason, even if we do not know what it is. Can we say on the basis of what God has done in Christ that it is better for creatures to have fallen into sin, error and corruption and be redeemed through Christ than never to have fallen? The dangers of such a view are obvious, but the consequences of denying it are also significant. First, let's deal briefly with the consequences of denying that God intentionally allowed humanity to fall, even though God could have prevented it. It seems that one would have to accept the idea that humanity fell against God's intentions. Was God not able to protect his creatures from error, sin and death? Or perhaps the Creator did not know that creation would fall. It seems to me that these conclusions are unacceptable in a Christian doctrine of God. What about the dangers of the thesis that God allowed human beings to fall as a part of his plan? The most urgent danger is that we make God complicit in the guilt of sin and make error and corruption necessary means to God's end. Additionally we run the risk of relieving human beings of moral responsibility for sin or of the obligation to use reason to avoid error and withstand corruption.

Does God sin by allowing human beings to sin so that God could redeem them in Christ for their greater good and his greater glory? Does God err by allowing human beings to err? Does God become corrupt for allowing human beings to become corrupt? No. We know this not only because the Scriptures assure us that God cannot sin, err or become corrupt but also from the line of reasoning I unfolded above. God has no obligations to creatures that God does not take on himself. Apart from God's active protection, human beings *will* err, sin and become corrupt. When God withholds that protection from creatures, God does not "cause" them to sin, err and become corrupt; they simply fall toward disorder for lack of power to maintain the proper order. Sin, error and corruption are nothing in themselves; hence they need no cause and can cause nothing else. Nor, as I argued above, does God owe creatures such grace, even granted their existence. There is, then, no question of divine guilt in human sin even though sin is inevitable apart from grace.

How Do We Know Sin and Evil?

Many discussions of sin and evil suffer from lack of clear definitions and specification of the source of knowledge from which they are derived. And many debates end in futility because the parties use different definitions of the central terms in which the conversation takes place. Is our knowledge of good and evil derived from our subjective reaction to events and states of affairs we encounter? Can a theologically adequate definition of evil be derived this way? Aristotle defines good as "what all things seek."[32] If we defined evil in analogy to the way Aristotle defined good, we would define it

---

[32] Aristotle, *Nicomachean Ethics* 1.1: *ou pant ephietai.*

as "that which no one seeks" or "that which all seek to avoid." We could of course learn much about human psychology and sociology by studying those specific things that human beings wish to avoid, but this definition does not bring us into the theological realm. For evil to become a theological issue God must be implicated in some way in its definition. But the above definition of evil speaks only of the subjectively negative aspects of our relation to the rest of creation. It would be a serious error to incorporate this subjective definition into theology by, for example, identifying human wishes with God's will. This move would set up the theological problem in this way:

1. The world is such that people are forced to suffer great evil (defined subjectively).

2. Since God is the creator of the world, either

3. God wills evil, or

4. Evil happens despite God's will.

This argument is posed in such a way that its target audience would most naturally opt for the second horn of the dilemma (4). God has set the world up in such a way that God allows events to happen that God wills not to happen. The hearer would recoil in horror at the thought of affirming the first horn, that God wills evil (3). God does not will evil but wills only that there be a world where such evil is possible for the sake of the good this kind of world makes possible.

But this argument contains a fallacy. There is an equivocation in the use of the word *evil* that makes the argument work. As used in the first proposition the term *evil* names a subjective reaction to negative experiences. It's just a matter of observation that "evil" (subjectively defined) happens. But the term *evil* in propositions three and four changes definitions to something like, "that which ought never to be done." The first premise (1) uses evil as synonymous with the undesirable, whereas three and four use it as if it were not only undesirable but also wrong. Apart from this change the first horn of the dilemma would lose its force. If we were to substitute the expression "that which all human beings desire to avoid" for the word "evil," the first horn of the dilemma (3) would read "God sometimes wills that which all human beings desire to avoid." Now this assertion does not sound as horrifying as the previous one; and the reason is obvious. We can think of all sorts of situations where the perfectly good God might will something unpleasant for human beings. However, because the argument uses the same word for two different meanings, the switch is hidden. The dilemma is created by substituting a metaphysical/moral definition of evil into propositions three and four in place of the physical/psychological one used in the first premise. Only in this way can the fact of suffering, which we fear and dread, force us to choose between God's goodness and his power.

Assumed in the above fallacious argument is the idea that the true nature of evil can be known from human experience. The experience of pain and suffering, violence and death, especially in their most dramatic forms, on the most massive scales, and to the most innocent victims, causes the feeling of despair or revulsion or offense or rebellion. The thought is irrepressible that these horrible things ought not to happen and that there is no possible justification for them. This feeling then sets up the modern problem of theodicy: how to absolve God of the charge of injustice for creating a world where these things can happen and then allowing or causing them to happen. I will deal with the theodicy problem below. Here we are addressing the ground of our knowledge of evil and seeking to understand how evil relates to God as creator.

Sound Christian theology cannot admit that subjective human experience is an adequate basis for understanding the true nature of evil, that is, evil as defined by its relationship to God. All Christian knowledge of God and of God's relationship to creation—creation, reconciliation and consummation—is derived from God's revelation in his acts toward creation with their center in Jesus Christ. Theologically speaking, we know evil only as that from which God redeems us in Jesus Christ, as that which God rejects, condemns and overcomes in Christ. Sin is that for which Christ atones and from which he liberates. Pursuing this idea with rigor, Karl Barth argues that we see sin in its "absolutely pure and developed and unequivocal form" revealed in the event of the crucifixion of Jesus Christ.[33] Indeed, the obedience of Jesus in accepting the cross is the "mirror in which we see the man of sin as such."[34] In that event human beings are unmasked as those who hate God and their neighbors.[35] It would contradict what we learn from God's act in

---

[33]CD IV/1, p. 397.
[34]Ibid.
[35]Ibid., p. 398.

Christ to seek to learn what sin and evil are from experience and then look for a savior suited to deal with those problems. Nor can the law teach us all we need to know about sin. Though the divine law may have been a teacher to "lead us to Christ" (Gal 3:24 NASB) and did great service by giving us the knowledge of sin (Rom 3:20; 7:7), it could not anticipate God's answer to sin given in Christ. Since the law meets us as a demand, it might lead us to think that sin is something we can avoid or escape by sheer will power. What we learn about sin from reflecting on the work of Christ can be stated in a series of paradoxes. Sin is both an act and a condition, something we have no power to escape but for which we are responsible, something universal but not essential to human beings. Sinners are both rebels and slaves, dead and alive. Sin can be forgiven and its power broken, but only by God. In the resurrection, sin will no longer be possible because the conditions that facilitate a peccable creature's entry into actual sin will no longer exist. God in Christ brings about those new conditions.

In the Bible, sin is conceived in moral and theological terms as a condition of the will and a quality of acts. In sin the will fails to adhere to the true order of good but substitutes another order, which is really disorder. Hence when sinners act according to their disordered wills they destroy order within creation. Disorder is a kind of corruption, and corruption works death. Sin is, therefore, associated with death. Since we know that in Christ death will be abolished, we can conclude that death is evil in some respects. How so? As we noted earlier, corruptibility is an intrinsic aspect of the created order and so is good. Corruptibility possesses a positive side in that to be corruptible is to be changeable and to be changeable is to be perfectible. Death as corruption, as returning to nothing and losing the good of being, is evil because it contradicts the creative will of God and the resurrection of Christ. But death as transformation to perfection is not evil; for it leads not to the dissolution of creation but to a new being of a higher order. Jesus Christ overcomes death not to return creation to its innocent state but to bring it to its originally intended telos. Apart from sin, "death" might be translation or rapture into this higher order.

From the perspective of eschatological perfection we can see that God rejects sin and evil absolutely. Creation is destined to be completely purified of them. Disorder of every kind will be removed. In view of their eternal destiny as absolutely nothing we can see why Karl Barth referred to sin and evil as "nothingness." Disorder cannot exist except in and among creatures. When God reorders and perfects creatures, disorder will not "exist"; it will be nothing. In relation to God, sin and evil are absolutely nothing and possess no power. However, in relation to us sin and evil are threatening, deadly and destructive. Threat of disintegration into nothing is nauseating, anxiety-provoking and terrifying. But nonexistence is threatening only if you exist. Only because God created a good world can evil exist and possess this menacing character. Only in creatures with sentience and consciousness can evil find a foothold, and only in self-conscious and free beings can evil become an act.

How are sin and evil as they are understood in eschatological perspective related to creation? As I have argued many times already in this study, creation is the beginning of what ends in eschatological redemption. Creation is not an end in itself but from the beginning anticipates its perfection in Christ. Human beings are fallible, peccable and corruptible by nature. Even this imperfect creation is good, good in the sense that it can be ordered according to God's good will. Hence we must not confuse the imperfections, needs and vulnerabilities of creation with sin and evil. Sin and evil are not good and cannot be ordered according to God's will. For they are defined as rejected, opposed to God, totally defeated in Christ and destined to be absolutely nothing. There is no necessary link between being a fallible, peccable and corruptible creature and being a sinner. Yet sin and evil possess a factual existence. Human beings actually erred, sinned and became corrupt unto death. So we have to tread a fine line between the imperfection of creation and evil.

Perhaps we can think of the relation this way. The coming into existence of creatures depends absolutely on the creative act of God. Apart from this relation creatures would be nothing at all. Apart from God's continual sustaining and concurring, creatures would cease to be or act. Creatures have no power in themselves to resist corruption, error or sin. In the promise of eschatological redemption we are assured, however, that God will at that time protect and perfect creatures so that they will not err, sin or fall into any form of corruption. And the doctrine of providence teaches that God will guide history to its ultimate goal of the perfection of creation and the redemption of sinners. Hence the genuine question we face is this: Why does God call into being and sustain creatures but not protect or perfect them at the same time? Why, instead, does God leave them vulnerable for a time to error, sin and corruption?

There is no answer to this question available to us in this life that can satisfy human reason and moral intuition. The answer for now is that God knows. And we must be satisfied to believe that there is a good answer and trust God to work all things for good (Rom 8:28).

THE PHILOSOPHICAL
PROBLEMS OF EVIL

WHEN WRITERS SPEAK OF THE "problem" of evil they usually mean an intellectual problem occasioned by the apparent incompatibility between one or more strongly held beliefs and the factual existence of evil. In ancient polytheist cultures, evil was not the intellectual problem that it became later. Natural disasters, accidents, insanity, illness and early death were the results of fate or the caprice and malice of the gods. In their myths these ancient religions accounted for the origin of evil and the miseries of human life. One hoped to escape them but gave no thought to understanding them systematically or teleologically.[1]

ANCIENT FORMS OF THE PROBLEM

As Greek philosophers began to seek a rational understanding of the world, they looked for a unifying principle to make sense of the complex and changing world of appearances. Stoic philosophers developed a worldview in which an all-encompassing divine providence or fate determined the course of events. This conviction frees the mind from all worry about what might happen, since it is already determined. The Stoic sought freedom and serenity by uprooting the passions from the soul. Marcus Aurelius, the Stoic Roman Emperor, found in this philosophy "power to live free from all compulsion in the greatest tranquility of mind, even if all the world cry out against you as much as they choose."[2]

---

[1]Paul Ricoeur, "Evil: A Challenge to Philosophy and Theology," *Journal of the American Academy of Religion* 52 (1985): 637.
[2]Marcus Aurelius, *Meditations* 7.68 (*Marcus Aurelius and His Times*, intro. Irwin Edman [Roslyn, NY: Black, 1945], p. 78).

Other philosophers discerned a problem with such dogmatic confidence in all-determining divine providence. The Epicurean philosophers responded that one should reject the idea that god cares or even knows about human beings. It is best simply to avoid what evil we can and put the rest out of mind. If we are not dead we are alive, so we need not worry about death, and after we die we will experience nothing, so we need not worry about death.[3] The classic trilemma, which clearly formulates an intellectual problem, may have been first formulated by the skeptic Sextus Empiricus in his *Outlines of Pyrrhonism* (A.D. 200). It is stated in this way:

> Those who firmly maintain that god exists will be forced into impiety; for if they say that he takes care of everything, they will be saying that god is the cause of evils, if they say that he takes care of some things only or even of nothing they will be forced to say that he is either malevolent or weak, and manifestly these are impious conclusions.[4]

As a skeptic, the author's goal was not to argue for dogmatic atheism but for complete indifference to all dogmatic claims. In this way the skeptic can achieve an untroubled mind free from all vacillation and doubt. As Mark Larrimore points out, all three of these schools of thought (Stoicism, Epicureanism and Pyrrhonism) agree in seeking in the face of evil "happiness in this life."[5]

The advent of Christianity changed the terms of the debate. It taught that there is only one God, the omnipotent, benevolent creator of everything that is not God. As we discussed in the previous chapter, for early Christianity evil was not an eternal counterpart to God. Sin and corruption entered the world through freedom, and hence the Christian experience of stubborn evil is understood in terms of moral rather than metaphysical dualism. If it is accurate at all to speak of a Christian understanding of evil's "origin" we must think of that origin as an irrational and inexplicable act of creaturely freedom. And Christianity's answer to the presence of evil is the work of Jesus Christ and the creator Spirit whose work will be completed in the future resurrection of the dead and the redemption of creation. In the early

---

[3]Mark Larrimore, "Evil," in *New Dictionary of the History of Ideas*, ed. Maryanne Cline Horowitz (Detroit: Charles Scribner's Sons, 2005), 2:748.

[4]Benson Mates, trans., *The Skeptic Way: Sextus Empiricus's Outlines of Pyrrhonism* (New York: Oxford University Press, 1996), p. 175; quoted in Mark Larrimore, *The Problem of Evil: A Reader* (Malden, MA: Blackwell, 2011), pp. xx-xxi.

[5]Larrimore, "Evil," 2:748.

centuries, Christian theologians did not experience their greatest challenge from atheistic arguments from evil or the responses of Stoicism, Epicureanism or Pyrrhonian skepticism. Until the modern era the greatest challenge to the Christian view of evil and providence arose from radical dualism.[6] Tertullian, in refuting the dualistic teaching of Marcion, summarizes one of his arguments against the worship-worthy nature of the God of the Old Testament:

> Now then, ye dogs, whom the apostle puts outside, and who yelp at the God of truth, let us come to your various questions. These are the bones of contention, which you are perpetually gnawing! If God is good, and prescient of the future, and able to avert evil, why did He permit man, the very image and likeness of Himself, and, by the origin of his soul, His own substance too, to be deceived by the devil, and fall from obedience of the law into death? For if He had been good, and so unwilling that such a catastrophe should happen, and prescient, so as not to be ignorant of what was to come to pass, and powerful enough to hinder its occurrence, that issue would never have come about, which should be impossible under these three conditions of the divine greatness. Since, however, it has occurred, the contrary proposition is most certainly true, that God must be deemed neither good, nor prescient, nor powerful. For *as* no such issue could have happened had God been such as He is reputed—good, and prescient, and mighty—*so* has this issue actually happened, because He is not such a God.[7]

Tertullian responds to Marcion by defending God's goodness in creating the world. God graciously gave human beings freedom so that they could add to their created goodness their voluntarily chosen virtue sharpened against the temptation to evil.[8] In this way the Latin theologian defends God's decision to give human beings freedom, even though it was through freedom that humankind fell.

As we saw above, Augustine, too, developed his understanding of evil in response to the radical dualism of Manichaeism. His point was to defend the creator of this world from the charge of imperfection and wrongdoing. Augustine did not conceive the problem of evil in humanistic terms; that is, it did not arise from the feeling of injustice and outrage at underserved or

---

[6]Larrimore, *Problem of Evil*, p. xxi.
[7]Tertullian, *Against Marcion* 2.5 (*ANF* 3:300-301).
[8]Tertullian, *Against Marcion* 2.6.

excessive suffering in the world. Augustine and his successors for hundreds of years believed that human beings stood under the curse of the fall and fully deserved all the suffering they had to endure. Only by the mercy of God was that suffering not greater than it was.

THE MODERN PROBLEM OF EVIL

Only after Renaissance thinkers had advocated the goodness and natural nobility of humanity and the doctrine of original sin had been rejected by many could the idea of innocent human suffering become a widespread belief.[9] As advances in medical science prolonged life and medicines were developed to relieve pain, people came to expect to live long and healthy lives. Larrimore observes that "the problem of evil became acute only once suffering no longer seemed a necessary part of life, but exceptional."[10]

Almost every modern theologian and philosopher of any significance has found it necessary to address the problem of evil: Leibniz, Voltaire, Hume, Rousseau, Kant, Schelling, Hegel, Schopenhauer, Nietzsche—these are just the most well known, but the list could go on and on. I will briefly examine two contemporary forms of the problem discussed primarily among analytic philosophers before I consider at greater length a third form of the problem that almost anyone can feel.

THE LOGICAL PROBLEM OF EVIL

In the mid-twentieth century several atheist philosophers articulated the logical problem of evil. J. L. Mackie, for example, argues that the problem of evil is more than a challenging problem. Mackie claims that it can be demonstrated "not that religious beliefs lack rational support, but that they are positively irrational, that several parts of the essential theological doctrine are inconsistent with one another."[11] Specifically, one cannot consistently affirm the three following propositions, each essential to the religious view that the perfect God exists, and at the same time admit the existence of evil:

---

[9]Ernst Cassirer, *The Philosophy of the Enlightenment*, trans. Fritz C. A. Koelln and James P. Pettegrove (Princeton, NJ: Princeton University Press, 1951), argues for the connection between the declining belief in original sin and increasing sensitivity to the problem of evil (pp. 137-60).

[10]Larrimore, *Problem of Evil*, p. xxi. See also Odo Marquart, "Unburdenings: Theodicy Motives in Modern Philosophy," in *In Defense of the Accidental: Philosophical Studies*, trans. Robert M. Wallace (New York: Oxford University Press, 1991), pp. 8-28. Marquart argues that theodicy is specifically a modern phenomenon.

[11]J. L. Mackie, "Evil and Omnipotence," *Mind* 64 (1955): 200.

1. God is omnipotent.

2. God is omniscient.

3. God is perfectly good. And

4. Evil exists.

For if God is omnipotent, God would have the power to prevent or remove all evil, and if God is omniscient, God would know every instance of evil and how to prevent or remove it. And if God is perfectly good, God would will the nonexistence of evil. But evil exists, the argument continues, so either God is omnipotent and omniscient but not good or omniscient and good but not omnipotent or good and omnipotent but not omniscient. Hence the religious belief that God exists as a being perfect in every respect is logically incoherent and therefore untrue or meaningless.

This argument evoked discussion and invited refutation from theists. One could admit the argument's soundness but attempt to retrieve belief in God by modifying one or more of the propositions. Process theism has no problem denying God's omnipotence.[12] Open theism, as we've seen, modifies divine omniscience so that God can be said to be omniscient but not know the future or how to prevent every instance of evil. Alvin Plantinga affirms all three of the propositions in question but challenges Mackie's interpretation of God's goodness. Is it really true that God's goodness requires that God never allow evil? Could God have a morally sufficient reason for allowing evil? Plantinga offers the following as morally sufficient reason:

> A world containing creatures who are sometimes significantly free (and freely perform more good than evil actions) is more valuable, all else being equal, than a world containing no free creatures at all. Now God can create free creatures, but cannot cause or determine them to do only what is right. For if he does so, they are not significantly free after all. . . . To create creatures capable of moral good, therefore, he must create creatures capable of moral evil.[13]

We can summarize briefly Plantinga's extensive and tightly reasoned defense of this thesis. To be morally significant, an action cannot be causally determined by God or anything outside the person doing the action. Hence

---

[12]Charles Hartshorne, *Omnipotence and Other Theological Mistakes* (Albany: State University of New York Press, 1984).
[13]Alvin Plantinga, *The Nature of Necessary* (Oxford: Oxford University Press, 1974), pp. 166-67.

it is logically impossible for God to create a world in which there are persons with morally significant freedom who are also causally determined always to perform good actions. Freedom to perform good actions entails freedom to do evil. God cannot be blamed for failing to do the logically impossible. It is certainly possible, then, that the good of a world that contains persons with morally significant freedom outweighs the risk of the evil made possible by giving persons such freedom. But does this answer deny God's omnipotence to save divine goodness? Plantinga reminds his readers of the generally accepted idea that God's "inability" to do logical contradictions does not diminish his omnipotence. So if the two properties of morally significant freedom and being causally determined always to choose good cannot be possessed simultaneously by the same subject, affirming these two properties of the same subject would be a logical contradiction. It would be like saying, "$S_{(a)}$ and $S_{(\sim a)}$." Hence affirming that God cannot create a world in which free creatures are causally determined always to perform good actions does not contradict divine omnipotence. Therefore, the logical problem of evil is answered: there is no logical contradiction between the three classic affirmations of divine attributes and the factual existence of evil.[14]

Clearly those who believe that God is omnipotent, omniscient and perfect in goodness do not accept the logical argument from evil to atheism. But this does not mean we need to accept every aspect of Plantinga's freewill defense. Is Plantinga correct to say that it would be a logical contradiction to affirm that God can create a world where persons with morally significant freedom always do right? If the only way to state this divine action is in a logical contradiction (*a* and ~*a*), then, of course we should not say that God can do this. But why must God's act of creating such a world be stated as a contradiction? Where is the logical contradiction in beings with morally significant freedom always choosing to perform good actions?[15] Even granted the logical possibility of alternative choice, of choosing evil as well as good, there is no con-

---

[14]For a convenient summary of Plantinga's argument, see James R. Beebe, "Logical Problem of Evil," in *Internet Encyclopedia of Philosophy*, ed. James Fieser and Bradley Dowden, www.iep.utm.edu/evil-log (accessed July 2, 2014).

[15]James S. Spiegel argues against Plantinga that morally significant freedom need not include the real power to do evil. One may have only the logical possibility of choosing evil but be given only choices of ways to do something good. See *The Benefits of Providence: A New Look at Divine Sovereignty* (Wheaton, IL: Crossway, 2005), pp. 190-92.

tradiction in agents with the power of alternative choice always choosing good actions. Plantinga opposes free actions (i.e., not causally determined) to causally determined actions. But there is no need to speak this way. One can ask the simple question: can God create a world where persons with morally significant freedom always perform good actions? And if you do not prejudice the issue by assuming that *the only way* for God do this is through "causal determination," there is no logical contradiction in affirming on the basis of the belief in divine omnipotence that God could have done this.

Even if we assume that Plantinga has made an effective response to the logical problem of evil, where does that leave us in relation to other formulations of the problem? Many nonexistent things are logically possible! So it is logically possible to affirm the omnipotence, omniscience and goodness of God in face of the evil in the world. This does not prove that a perfect God exists. And it does not tell us why we should actually believe the good that will be achieved is worth the price of my suffering and all the suffering in the history of the world.

THE EVIDENTIAL PROBLEM OF EVIL

This last issue has been called the evidential problem of evil or the evidential argument from evil. In the 1970s William Rowe formulated an evidential argument from evil to atheism that James Sennett considers "the clearest, most easily understood, and most intuitively appealing of those available."[16] Rowe argues as follows:

1. There exist instances of intense suffering which an omnipotent, omniscient being could have prevented without thereby losing some greater good or permitting some evil equally bad or worse.

2. An omniscient, wholly good being would prevent the occurrence of any intense suffering it could, unless it could not do so without thereby losing some greater good or permitting some evil equally bad or worse.

3. (Therefore) There does not exist an omnipotent, omniscient, wholly good being.[17]

---

[16]James Sennett, "The Inscrutable Evil Defense Against the Inductive Argument from Evil," *Faith and Philosophy* 10 (1993): 220.

[17]William Rowe, "The Problem of Evil and Some Varieties of Atheism," in *Contemporary Perspectives on Religious Epistemology*, ed. R. Douglas Geivett and Brendan Sweetman (New York: Oxford University Press, 1992), p. 34.

As we see here, the evidential argument from evil takes up where the logical argument from evil ends. It argues that even if it is logically possible for an omnipotent, omniscient and good God to have morally sufficient reasons for permitting suffering, there is so much and such horrendous suffering in the world that it seems unlikely that any good achieved would be worth such cost. The argument attempts to place the burden of proof on the believer. Since the argument is valid so that accepting 1 and 2 requires one to accept 3, the critic must challenge the truth or meaningfulness or knowability of 1 or 2 or both. Take 1 first. We could deny 1 by arguing that there are no instances of suffering—even horrendous suffering—that could be prevented without losing a greater good. But how could one demonstrate this? If you argue from the premise that God is perfectly good to the thesis that God would not allow such gratuitous suffering, the author would accuse you of assuming a positive answer to the very issue that is being debated. But if you attempt to offer evidence that horrendous suffering can be justified by the good it produces you will look heartless; present suffering seems so much more real than far-future good. A third response to 1 seems much more promising. Premise 1 assumes that we, whether believer or unbeliever, have access to a perspective that would allow us to make the judgment it demands. Only a mind with a vision of all time and space and the sum total of all experience could presume to make this judgment. Hence we need not take 1 as a serious rational claim; rather we should view it as an emotional outburst masquerading as a rational claim and treat it accordingly.

Premise 2, often called the "theological premise," seems on its face to be acceptable to most theists. A loving, all-powerful and omniscient God would not allow gratuitous suffering, suffering with no point. Denying premise 2 would seem to be admitting the possibility that God may allow world history to be punctuated by irrational and absurd events, some of which tower on a massive scale. William Hasker, however, challenges this consensus and denies premise 2. Hasker asks us to consider the implications of making individual cases of horrendous suffering necessary and sufficient conditions for an "offsetting good."[18] This connection would imply that no one can come to any lasting harm no matter what befalls them, and possibly imply that our efforts to reduce the suffering an indi-

---

[18]William Hasker, *The Triumph of God over Evil: Theodicy for a World of Suffering* (Downers Grove, IL: IVP Academic, 2008), p. 191.

vidual must endure might actually cause them harm. Or even worse, inflicting suffering on others might actually bring more good than evil into the world. Hasker finds this view of suffering "absolutely appalling."[19] Premise 2 conflicts with a principle Hasker labels "the principle of divine intention," which is "implicit in the biblical picture of God and God's relationship to the world."[20] It reads, "It is an extremely important part of God's intention for human persons that they should place a high priority on fulfilling moral obligations and should assume major responsibility for the welfare of their fellow human beings."[21]

For this principle to have its intended effect, God must allow some gratuitous evils to occur. If human beings are obligated not to inflict suffering on others and to relieve suffering when possible, the evil avoided and the suffering relieved must have been gratuitous; otherwise acting according to this principle rather than accomplishing good would actually make matters worse.[22] Hasker, then, instead of defending against the evidential argument from evil by denying premise 1 and allowing premise 2, denies 2 and allows 1.

A full response to Hasker's interesting move would take us down a side trail. But I appreciate his questioning of premise 2, and his critique raises issues that need to be highlighted and will become important in my own treatment of the problem of evil sketched below. Hasker is correct to point out that the New Testament does not frame the problem of evil as the problem of gratuitous evil. Rather than assuring believers that they will never suffer in a way that will not also produce a greater good in this life or the next, the New Testament promises that God will overrule or achieve victory over evil. Hasker exploits this difference to develop his theodicy. God will achieve victory over evil, not in the sense that God will work every instance of evil into the final and supreme good; rather, "victory" means the end of evil's power and activity. Hasker and I differ on how this overcoming and victory is to be conceived, but his point is nevertheless well taken. Second, we need to ask an obvious question, which Rowe asked in rebuttal to Hasker's critique: how can we consider an instance of evil gratuitous (i.e., leading to no greater good) when according to Hasker's principle of divine

---

[19]Ibid.
[20]Ibid.
[21]Ibid.
[22]One could ask how an evil could be gratuitous when its presence becomes the occasion for virtuous acts in response.

intention gratuitous evils are necessary for moral action and compassion and, hence, necessary for the goods such action brings into the world? This question surfaces an ambiguity in the evidential argument from evil. Within what framework is a particular evil declared gratuitous or not: within the life span of the victim, within the lifetimes of everyone affected, within the entire history of the world, or absolutely? Of course, the wider the framework given for an instance of intense suffering to work out its results, the less access we have to its final end.

In my view, neither the evidential argument from evil nor the defense against it gets us much further than the intuitions with which they began. We simply do not have the God's-eye viewpoint needed to make the judgment the argument asks of us. If you believe in God you will also hold on to the hope that God will make all things right and overcome all evil in the end. If you do not believe in God you are just as likely to see horrendous evil as the ultimate defeater for belief in God. I am not saying that experiencing horrible evil might not sway one toward unbelief, but other factors may be more decisive than rational inference from the presence of evil. And this raises the question of what I shall call the rhetorical problem of evil.

The Rhetorical
Argument from Evil

THE RHETORICAL ARGUMENT FROM EVIL is a special form of the evidential argument.[1] The first premise of the evidential argument must be supported by data from human experience. The rational form of the argument gives the misleading impression that we have here an intellectual problem that could be discussed dispassionately and with a kind of mathematical precision. However, when the intellectual form and dispassionate mood of the evidential argument gives way to passion, it becomes the rhetorical argument from evil.[2] The rhetorical argument from evil begins by rehearsing in exquisite detail excruciating and nauseating accounts of horrendous cruelty and suffering. It does not move to a conclusion by way of inference but ends with a challenge, explicit or implicit: how dare you diminish the horror of human suffering by making it a means to greater good or a higher harmony! Such exclamations are more agonized protest than rational argument. It forces believers to choose between shame-faced silence and playing the role of the cold-hearted theodicist. Another form of the rhetorical argument from evil adopts the mood of irony and sarcasm. It pictures the theodicist as a ridiculous optimist who surveys the most horrible suffering whistling a tune about this being the best of all possible worlds. The contrast between the image of unspeakable suffering and the

---

[1]This argument/problem is often called the "emotional problem of evil." I prefer to call it the rhetorical argument because I am focusing on the inference it asks us to make.

[2]Even in the most philosophical discussions highly charged language often comes to the surface. This is inevitable since the evidence for gratuitous evil is partly the emotional impact the evidence has on us.

muddle-headed theodicist blinded by adherence to an optimistic meta-physical theory makes rational inference superfluous. In my view the rhetorical argument from evil is the most popular and most potent form of the argument from evil. And it is the most difficult form to respond to. We will examine a classic example from each form.

Leibniz's Theodicy and Voltaire's Satire

In 1710 Wilhelm Gottfried Leibniz (1646–1716) published his *Theodicy*.[3] This work is an elaborate defense of God as the creator of this world. He argues not from experience and empirical knowledge of the world but from the idea of God as powerful, wise and good that our world must be the best of all possible worlds God could have created. As we shall see, the phrase "the best of all possible worlds" is easily caricatured by those unsympathetic to or unwilling to think through Leibniz's argument. That same possibility for caricature holds for Alexander Pope's (1688–1744) famous poem *Essay on Man* (1732–1738). In the first epistle of the poem Pope chides those who criticize the created order of nature and the course of divine providence.

> Go, wiser thou! and, in thy scale of sense,
> Weigh thy opinion against providence;
> Call imperfection what thou fanciest such,
> Say, here He gives too little, there too much;
> Destroy all creatures for thy sport or gust,
> Yet cry, if man's unhappy, God's unjust;
> If man alone engross not Heaven's high care,
> Alone made perfect here, immortal there:
> Snatch from His hand the balance and the rod,
> Re-judge His justice, be the God of God.

Pope continues to correct the pride of those blind to providence in a mood reminiscent of the God from the whirlwind in Job 38–41. He concludes the first epistle with the refrain that will appear throughout the poem, "whatever is, is right":

> All nature is but art, unknown to thee;
> All chance, direction, which thou canst not see;
> All discord, harmony not understood;

---

[3]The full title is *Theodicy: Essays on the Goodness of God, the Freedom of Man, and the Origin of Evil.*

> All partial evil, universal good:
> And, spite of pride in erring reason's spite,
> *One truth is clear, whatever is, is right.*[4]

Just like Leibniz's expression "the best of all possible worlds," Pope's refrain "whatever is, is right" is easily made the object of derision by those ignorant of the poem or unwilling to think through Pope's argument. It is important to note that Leibniz and Pope argue from confidence in God's power, wisdom and goodness to their optimistic conclusions; in their view the force of our deeply held confidence in God should overcome the momentary and fragmentary appearances to the contrary. Those who make the rhetorical argument from evil argue in the opposite direction, from negative experiences to pessimistic or atheist conclusions. And they find views like those of Pope and Leibniz easy targets for satire.

In November 1755, a catastrophic earthquake and tsunami destroyed Lisbon, Portugal, killing up to sixty thousand people. One month after that horrible event the French intellectual Voltaire wrote a poem titled "The Lisbon Earthquake: An Inquiry into the Maxim 'Whatever Is, Is Right.'"[5] In catharsis and protest, Voltaire describes the horrors of the scene and punctuates his lament with questions designed to silence Pope and his ilk. Would you take all comfort from the dying wracked with pain and despair and deprive them of their right to complain "to find relief"? To those who say that God "rules, not by partial, but by general laws," Voltaire replies,

> Say what advantage can result to all,
> From wretched Lisbon's lamentable fall?

To those who speak of the good that will result from this ill, the poet retorts,

> No comfort could such shocking words impart,
> But deeper wound the sad, afflicted heart.
> When I lament my present wretched state;
> Allege not the unchanging laws of fate.

When the optimists exclaim,

---

[4]Alexander Pope, *An Essay on Man: Moral Essays and Satires* (London: Cassell, 1891), www
    .gutenberg.org/files/2428/2428-h/2428-h.htm, www.gutenberg.org/files/2428/2428-h/2428-h
    .htm (emphasis added).
[5]All quotes from "The Lisbon Earthquake" are taken from Mark Larrimore, *The Problem of Evil:
    A Reader* (Malden, MA: Blackwell, 2011), pp. 204-9.

> Others' enjoyments from your woes arise,
> To numerous insects shall my corpse give birth,
> When once it mixes with its mother earth.

Voltaire replies,

> Yet in this direful chaos you'd compose
> A general bliss from individuals' woes?
> Oh worthless bliss in injured reason's sight,
> With faltering voice you cry, "Whatever is, is right"?

The French philosopher lists a series of comfortless explanations for suffering: original sin, divine indifference, matter's defect and others. Only God could explain, and he does not.

> Leibniz can't tell me from what secret cause
> In a world governed by the wisest laws,
> Lasting disorders, woes that never end
> With our vain pleasures real sufferings blend.

Nor would believing that death is the end bring comfort now; for . . .

> Sad is the present if no future state,
> No blissful retribution mortals wait.

Hope of another state is all the comfort we have in this life. But in this life none but the self-deceived can seriously maintain that "whatever is, is right."

> All may be well; that hope can man sustain,
> All now is well; 'tis an illusion vain.

In 1759 Voltaire wrote an extended story ridiculing the idea that this is the "best of all possible worlds."[6] *Candide, or Optimism* recounts the adventures of Candide, the supposed illegitimate son of the sister of his patron, the baron of Thunder-ten-Tronckh. Candide is the compliant student and admirer of Dr. Pangloss who holds that this is the best of all possible worlds and everything that happens is for the best. Dr. Pangloss is a ludicrous caricature of Leibniz: "Pangloss was professor of metaphysico-theologico-cosmolo-nigology. He proved admirably that there is no effect without a cause, and that, in this best of all possible worlds, the Baron's castle was the

---

[6]Voltaire, *Candide* (New York: Boni and Liveright, 1918).

most magnificent of castles, and his lady the best of all possible Baronesses."[7]

Candide's first misfortune was to be banished from the castle for kissing the baron's daughter, the beautiful Cunegonde. Pining for his beloved Cunegonde, the naive Candide was soon kidnapped and pressed into the military service of the "Bulgarians" (i.e., the Prussians) in their war against the king of Abares (i.e., the King of France). Voltaire describes the scene with biting sarcasm:

> There was never anything so gallant, so spruce, so brilliant, and so well disposed as the two armies. Trumpets, fifes, hautboys, drums, and canon made music such as Hell itself had never heard. The cannons first of all laid flat about six thousand men on each side; the muskets swept away from this best of worlds nine or ten thousand ruffians who infested its surface. The bayonet was also a "sufficient reason" for the death of several thousands. . . . At length, while the two kings were causing Te Deum to be sung each in his own camp, Candide resolved to go elsewhere and reason on effects and causes. [In the destroyed villages] old men covered with wounds, beheld their wives, hugging their children to their bloody breasts, massacred before their faces; there, their daughters, disembowelled and breathing their last after having satisfied the natural wants of Bulgarian heroes; while others, half burnt in the flames, begged to be despatched. The earth was strewed with brains, arms, and legs.[8]

Having escaped the Bulgarians, Candide meets the destitute and syphilis-blinded Dr. Pangloss, who greets him with the news that Candide's beloved Cunegonde is dead. The two set off on a voyage to Lisbon only to be overtaken by a terrible storm and shipwrecked in sight of Lisbon harbor. As soon as they have set foot on land an earthquake and tsunami overwhelm the city. Thirty-thousand inhabitants are crushed under the ruins. Dr. Pangloss comforts the survivors with assurances that things could not have been different: "All that is, is for the best. If there is a volcano in Lisbon, it cannot be elsewhere. It is impossible that things should be other than they are, for everything is right."[9] Two Jews are among the group burned at the stake, "for it had been decided by the University of Coimbra that burning a few people alive by a slow fire, and with great ceremony, is an infallible secret to hinder

---

[7]Ibid., p. 2.
[8]Ibid., pp. 9-10.
[9]Ibid., p. 21.

the earth from quaking."[10] Additionally Dr. Pangloss was hanged for heresy.

Candide's misadventure continues as he finds Cunegonde alive and travels with her and others to South America. During this trip Cunegonde and the old woman accompanying her tell of their tragic tails of rape, slavery and mutilation. In South America Candide again loses Cunegonde and has to flee for his life. On the way he is almost eaten by cannibals and eventually stumbles into El Dorado, a utopia where the pebbles are diamonds and the soil is gold. Candide leaves El Dorado with one hundred red sheep loaded with supplies and riches. With his newfound wealth he sets out to find and rescue his lady Cunegonde. His plan requires that he set sail to meet her in Venice. Before he leaves he becomes friends with an unfortunate philosopher, Martin, who remains with him for the rest of the story.

Given his sufferings and what he has come to know of the world, Candide has begun to waver in his conviction that all is for the best in this best possible world. Still inclining to optimism, Candide asks Martin his opinion of good and evil. Martin announces, "I am a Manichean."[11] Surprised that there are still Manicheans in the world, Candide accuses Martin of being possessed by the devil. At which the pessimist Martin says,

> He [Satan] is so deeply concerned in the affairs of this world that he may very well be in me, as well as in everybody else; but I own to you that when I cast an eye on this globe, or rather on this little ball, I cannot help thinking that God has abandoned it to some malignant being. . . . I scarcely ever knew a city that did not desire the destruction of a neighbouring city, nor a family that did not wish to exterminate some other family. Everywhere the weak execrate the powerful, before whom they cringe; and the powerful beat them like sheep whose wool and flesh they sell. A million regimented assassins, from one extremity of Europe to the other, get their bread by disciplined depredation and murder, for want of more honest employment. Even in those cities which seem to enjoy peace, and where the arts flourish, the inhabitants are devoured by more envy, care, and uneasiness than are experienced by a besieged town. Secret griefs are more cruel than public calamities. In a word I have seen so much, and experienced so much that I am a Manichean.[12]

Surely there is also much good and a few good people in the world,

---

[10]Ibid., p. 23.
[11]Ibid., p. 99.
[12]Ibid., pp. 99-100.

Candide replies timidly. "That may be, but I know them not," responds Martin.[13] The philosopher challenges Candide to find even one happy person. In Paris, Candide is cheated out of much of the money he had not already lost. Arriving in Venice, Candide and Martin meet two of Candide's old friends who tell him their very sad tales, which of course confirms Martin's pessimism. They meet person after person, including a Venetian noble with a fabulous villa and wonderful library, whom Candide expects should be happy, only to find that they are miserable. Learning that his beloved Cunegonde has been sold as a slave, taken to Constantinople, and has lost her looks, he sets sail to find and ransom her. Along the way he discovers to his great surprise and delight that his mentor Dr. Pangloss has survived his hanging and is still alive and is serving as a galley slave on board the very ship on which he has secured passage.

Dr. Pangloss tells of how the stormy weather prevented the inquisitor general from burning him at the stake. The hangman, being used to burning people rather hanging them, bungled the job allowing Pangloss to survive. The hanged man, pretending to be dead, was sold to a physician for dissection. Upon being sliced in the belly Pangloss woke up, frightening the poor physician out of the room. As Dr. Pangloss finishes his tangled story detailing how he came to be a galley slave, Candide asks,

> "Well, my dear Pangloss when you had been hanged, dissected, whipped, and were tugging at the oar, did you always think that everything happens for the best?"
>
> "I am still of my first opinion," answered Pangloss, "for I am a philosopher and I cannot retract, especially as Leibnitz could never be wrong; and besides, the pre-established harmony is the finest thing in the world, and so is his *plenum* and *materia subtilis*."[14]

In the end Candide finds Cunegonde and marries her, reluctantly as it turns out. He buys a little farm with the last of his money he got in El Dorado. He, Cunegonde, Pangloss, Martin and other friends live together on their little plot. Toward the end of the story Candide and his friends meet a Turkish farmer who explains that he never bothered with the great events of the world but remained on his land content to cultivate his garden. On the way back to his house Candide reflects on the farmer's philosophy of life,

---

[13]Ibid., p. 100.
[14]Ibid., pp. 157-58.

concluding that that farmer's situation is preferable to anyone they met in their long journeys. Candide, Cunegonde and Pangloss settle into their roles in their new life on their little farm. Reflecting with satisfaction on their new lives, Pangloss, still a convinced optimist, says,

> There is a concatenation of events in this best of all possible worlds: for if you had not been kicked out of a magnificent castle for love of Miss Cunegonde: if you had not been put into the Inquisition: if you had not walked over America: if you had not stabbed the Baron: if you had not lost all your sheep from the fine country of El Dorado: you would not be here eating preserved citrons and pistachio-nuts.[15]

The story ends with Candide's reply: "All that is very well, but let us cultivate our garden."[16]

Ivan Karamazov and Protest Atheism

The Karamazov challenge. Russian novelist Fyodor Dostoevsky believed that his character Ivan Karamazov of *The Brothers Karamazov* had voiced the most powerful objection ever made to God, providence and the goodness of creation. Indeed Dostoevsky considered "irrefutable" Ivan's argument from "the senseless suffering of children" to "the absurdity of all historical reality." That is, he considered it irrefutable on a theoretical level. It can be refuted only in the practical experience of loving and being loved, being forgiven and forgiving, and allowing an inward love of life to teach us to love the Creator. Hence Dostoevsky's refutation of Ivan's protest comes not in the form of an argument but as the story of a life told in "the last words of Elder Zosima."[17] In what follows I will attempt to summarize Ivan's complaint and Dostoevsky's refutation before I venture my own response.

Ivan's protest against God's world comes in book five, chapter four, "Rebellion," and chapter five, "The Grand Inquisitor." In the previous chapter Ivan has been explaining to his younger brother Alyosha "my essential nature, that is what manner of man I am, what I believe in, and for what I hope."[18] Ivan explains that he is not concerned with the theoretical question

---

[15]Ibid., p. 168.
[16]Ibid., p. 169.
[17]Letter 782, May 10, 1879, quoted in Fyodor Dostoevsky, *The Brothers Karamazov: A Norton Critical Edition*, trans. Constance Garnett, ed. Susan McReynolds Oddo, rev. Susan McReynolds Oddo and Ralph E. Matlaw, 2nd ed. (New York: W. W. Norton, 2011), pp. 658-59.
[18]Ibid., p. 204.

of whether the existence of God can be proved; indeed human beings cannot solve such questions and we are wise not to bother with them. Ivan's concern lies elsewhere: "Yet would you believe it, in the final result I don't accept this world of God's, and although I know it exists, I don't accept it at all. It's not that I don't accept God, you must understand, it's the world created by Him I don't and cannot accept!"[19] Ivan's confession prompts Alyosha to ask, "Will you explain why you don't accept the world?"

Ivan begins by denying that one can really love one's neighbors. Perhaps one can love humanity in the abstract or "at a distance, but almost never close up."[20] One person can't really understand another's suffering and tends to resent suffering in others. Quite the opposite, human beings enjoy inflicting pain on others. Think of the rage and rape unleashed in war and religious persecution. And the more innocent the victim, the more savage the torturer. Observation of humanity clearly reveals that "in the heart of every man . . . a beast lies hidden, a beast of rage, a beast of sensual inflammation from the screams of the victim, a beast without restraint, let off the chain."[21] In illustration of this view of humanity Ivan tells the stories of two children made to suffer horribly by adults. An eight-year-old boy was torn to pieces by hunting dogs at the behest of their owner, a general in the army. And what was the child's crime? Throwing a rock at the general's favorite dog. We must hear Ivan's own words as he describes the suffering of another child, a little girl of five:

> These educated parents subjected this poor five-year-old girl to every possible torture. They beat, thrashed, kicked her, not knowing why themselves, turning her whole body into bruises; finally they reached the highest refinement: in the cold, in the frost, they shut her up all night in the outhouse, because she wouldn't ask to be taken out at night (as though a five-year-old child, sleeping its angelic sound sleep, could be taught to ask)—for that they smeared her whole face with her excrement and made her eat that excrement, and it was her mother, her mother who made her! And that mother could sleep at night, hearing the groans of that poor little child, locked up in that vile place! Can you understand that a little being, who still can't even comprehend what is being done to her, in that vile place, in the dark and cold, beats herself with

[19]Ibid., p. 203.
[20]Ibid., p. 205.
[21]Ibid., p. 209.

her tiny little fist on her strained little chest and cries her bloody, unresentful, meek little tears to "dear God" to protect her—can you understand that nonsense, my friend and my brother, my pious and humble novice, do you understand why this nonsense is necessary and created? Without it, they say, man could not have existed on earth, for he would not have known good and evil. Why should he know that diabolical good and evil, when it costs so much? The whole world of knowledge is not worth the little tears of that little child to "dear God."[22]

Perhaps, continues Ivan, there will be harmony in the end and the lamb will lie down with the lion and the mother will embrace the murderer that tore her son to pieces with his dogs. Perhaps there is hell and retribution. But nothing can redeem the tears of that little child crying to her "dear God" in that stinking outhouse or that little boy being ripped apart by those dogs. It happened, it's a fact, it cannot be changed! The mother of that little boy has no right to forgive his murderer, even if the little boy forgives him! No one has that right, no one! No ultimate harmony is worth one of their little tears. "I don't want harmony, for the love of humanity, I don't want it. I would rather remain with unavenged suffering. I'd rather remain with my unavenged suffering and unquenched indignation, *even if I am wrong.*"[23]

Alyosha is completely subdued and agrees that no human being has the right to sacrifice one child to purchase happiness even for the whole world. But Alyosha ventures to remind Ivan that "there is a Being and He can forgive everything, all *and for all,* because He gave His innocent blood for all and everything."[24] Ivan retorts that he has not forgotten him and had been expecting Alyosha to bring him into the discussion even earlier. Thus prompted, Ivan relates his latest composition, "The Grand Inquisitor."

The events take place in sixteenth-century Seville. On the day before, the cardinal and Grand Inquisitor has ordered one hundred heretics burned at the stake. Jesus comes among them, and though he says not a word everyone knows instinctively who he is. As he walks slowly through the streets he is met by a funeral procession bearing the body of a little girl who has just died. The mother of the child kneels and begs Jesus to bring her child back to life. Looking compassionately at the little white coffin, he says, "Maiden, arise!"

---

[22]Ibid., p. 209.
[23]Ibid., p. 212. See Kenneth Surin, *Theology and the Problem of Evil* (Oxford: Blackwell, 1986), chap. 3, for his treatment of the Karamazov approach to the problem of evil.
[24]Dostoevsky, *The Brothers Karamazov*, p. 213.

The little girl at once sits up and looks around "holding a bunch of white roses they had put in her hand."[25] At that moment the Grand Inquisitor makes his way through the crowd to Jesus. The inquisitor motions to his guards and they arrest the visitor and put him in the dungeon of the palace of the Holy Inquisition.

That evening the Grand Inquisitor comes to visit Jesus in his dark cell. The conversation is completely one-way with the divine visitor not saying a word. The Grand Inquisitor's interrogation may be paraphrased as follows:

> So you won't answer me? Well, your silence is fitting, since you said everything you had to say when you were here those sixteen hundred years ago. You turned your work over to us and you cannot interfere now. You promised people freedom, but the kind of freedom you offered is too demanding; it demands that human beings live freely before God in their own good conscience without compulsion from miracle, mystery and authority. You want them to live in direct relation to God. You want love and faith freely given. What suffering! People can't endure such freedom, for they are rebels from birth and do not understand such horrible freedom. You should have listened to the "dread spirit" as he tempted you in the desert with the three questions. He told you how to succeed with this rabble. You must give them bread before you require virtue of them. After considering your dreadful freedom they will come to us and say, "Make us your slaves, but feed us."[26] Human beings can never have both freedom and bread because they will never share with one another unless forced. They will conclude that they "can never be free, for they are weak, vicious, worthless and rebellious."[27] Yes, perhaps you will have your thousands or even tens of thousands who will rise to your plane; but what about the millions of the weak, sniveling rabble? Who will guide them? We love *them* too and in your name relieve them of the burden of freedom. We allow them to sin and pronounce your forgiveness. Yes, we deceive them and that is the suffering we endure on their behalf.
>
> Why didn't you throw yourself off the pinnacle of the temple? Why didn't you come down from the cross when the rabble challenged you? Don't speak, for I know the answer. It was from respect to the spiritual freedom you wished to give the people. You hoped they would follow your example and trust God without miracles, without signs, clinging to God even when he is absent! But they are not like you; they are not gods. They cannot endure such temptation.

---

[25]Ibid., p. 216.
[26]Ibid., p. 220.
[27]Ibid.

The millions too need a way to still their consciences and die in peace. We understand their weakness and have corrected your work. And now I have said it, though you knew it all along. We do not love you. We are working with *him*, not with you. The dread spirit offered you all the kingdoms of the earth if you would worship him. You scorned worldly power and majesty. But you sorely misunderstood human nature. They crave the "universal unity" of the anthill; they fear freedom more than the whip. And we have promised them this universal unity and happiness while we alone endure in silence the unhappy mystery of "the knowledge of good and evil."[28] They will die in peace, your name on their lips, in the assurance we give them of eternal life, though we know that beyond the grave "they will find nothing but death."[29]

Dostoevsky's answer. As I indicated above, Dostoevsky does not answer the outrage and despair of Ivan Karamazov with a philosophical argument but with the life and words of Zosima the Elder, a monk and mentor of Alyosha. As Zosima nears the end of his long life he gathers his closest associates to share with them his parting words. He recalls his childhood in which the most significant event was the illness and death of his brother Markel, eight years his senior. As Markel grew weaker he began attending church, something he had refused to do earlier. Soon thereafter his physical condition deteriorated so that he could no longer leave his house. But a complete spiritual transformation had occurred in him. He would cough all night long but in the morning he would say to his mother, "Don't cry mother . . . life is paradise, and we are all in paradise, but we won't see it, if we would, we should have heaven on earth the next day."[30] Markel spoke so kindly to everyone: "Dear ones . . . what have I done that you should love me so, how can you love anyone like me, and how was it I did not know, I did not appreciate it before?"[31] His mother objected mildly to such self-accusation. But Markel responded, "Every one of us has sinned against all men, and I more than any . . . little heart of mine, my joy, believe me, everyone is really responsible to all men for all men, and for everything. . . . And how is it we went on then living, getting angry and not knowing?"[32] The doctor replied to Markel's enquiry whether he would live another day that he would live

---

[28]Ibid., p. 225.
[29]Ibid., p. 226.
[30]Ibid., p. 249.
[31]Ibid.
[32]Ibid., p. 250.

many days, months and years even. The dying boy exclaimed, "Months and Years! . . . Why reckon the days? One day is enough for a man to know all happiness. . . . Let's go into the garden, walk and play there, love, appreciate, and kiss each other, and glorify life."[33] The spring had brought the trees to bud and the flowers to bloom, and birds welcomed new life with song. Markel, looking out his window at the scene, prayed, "Birds of heaven, happy birds, forgive me, for I have sinned against you too. . . . [I] did not notice the beauty and glory."[34] To his mother, who urged him not to take so many sins on himself, he replied, "Mother, my joy, it's for joy, not for grief I am crying. Though I can't explain it to you, I like to humble myself before them, for I don't know how to love them enough. If I have sinned against everyone, yet all forgive me, too, and that's heaven. Am I not in heaven now?"[35] Markel died, Zosima wept—and forgot.

But at the right time, many years later and after much "drunkenness, debauchery and devilry," he remembered.[36] Zosima, who had become a military officer, challenged the husband of a woman he fancied to a dual. In a foul mood the night before the dual he flew into a rage and floored his orderly Afanasy with two punches to his face. Afanasy took the blows without a word. The next morning Zosima arose with an uneasy feeling, not because he might kill or be killed, but because of the beating he gave to Afanasy. As he stood looking out the window the sun rose, the birds sang and the gray became green. And as he wept, face in hands, he remembered Markel's words, "Why do you love me, am I worth your waiting on me?" For the first time in his life Zosima was forced to ask: "After all what am I worth, that another man like me, made in the image and likeness of God, should serve me?" Can it be, Zosima asked himself, that "in truth we are each responsible to all for all"? The truth of this statement flooded his soul: "I am more than all others responsible for all, a greater sinner than all men in the world."[37] Zosima knew what he had to do.

He found Afanasy in his room and said to him, "I gave you two blows on the face yesterday, forgive me." Afanasy looked uncomprehendingly at his commanding officer. Realizing that his orderly did not understand, Zosima

---

[33]Ibid.
[34]Ibid.
[35]Ibid.
[36]Ibid., p. 255.
[37]Ibid., p. 258.

dropped to his knees, bowed his head to the ground and said again, "Forgive me." Afanasy, completely shaken said, "Your honor . . . sir, what are you doing? Am I worth it?"[38] Zosima dashed out of the room while Afanasy buried his face in his hands and sobbed. Zosima arrived at the site of the dual in good spirits. Since Zosima had made the challenge, the rules gave his opponent the first shot. As Zosima stood there the bullet grazed his cheek but did no further damage. When it came Zosima's turn he threw the pistol aside and looked at his opponent and said, "Forgive me, young fool that I am, sir . . . for my unprovoked insult to you and for forcing you to fire at me. I am ten times worse than you and more, maybe."[39] Zosima's comrades were outraged at the disgrace he had brought on them. But Zosima said, "Gentlemen . . . look around you at the gifts of God, the clear sky, the pure air, the tender grass, the birds; nature is beautiful and sinless, and we, only we, are godless and foolish, and we don't understand that life is a paradise, for we have only to understand that and it will at once be fulfilled in all its beauty, we shall embrace each other and weep."[40] Once they heard that Zosima had resigned his commission and decided to become a monk, they were reconciled.

Zosima stands as the Grand Inquisitor's complete opposite and refutation. The Grand Inquisitor sees nothing good in humankind and cannot comprehend Christ's faith and way of life. In contrast, Zosima recommends that we

> have no fear of men's sin, love a man even in his sin, for that is the semblance of divine love and is the highest love on earth. Love all God's creation, the whole and every grain of sand in it. Love every leaf, every ray of God's light. Love the animals, love the plants, love everything. If you love everything, you will perceive the divine mystery in things. Once you perceive it, you will begin to comprehend it better every day. And you will come at last to love the whole world with an all-embracing love.[41]

As one contemplates the sin of humanity one "wonders whether one should use force or humble love. Always decide to use humble love."[42]

---

[38]Ibid., p. 258.
[39]Ibid.
[40]Ibid., p. 259.
[41]Ibid., p. 275.
[42]Ibid.

Humble love is "marvelously strong, the strongest of all things and there is nothing else like it":[43]

> My brother asked the birds to forgive him; that sounds senseless, but it is right; for all is like an ocean, all is flowing and blending; a touch in one place sets up movement at the other end of the earth. . . . It's all like an ocean, I tell you. . . . My friends, pray to God for gladness. Be glad as children, as the birds of heaven. And let not the sin of men confound you in your doings. Fear not that it will wear away your work and hinder its being accomplished. . . . Fly from that dejection, children! There is only one means of salvation, then take yourself and make yourself responsible for all men's sin, that is the truth, you know, friends, for as soon as you sincerely make yourself responsible for everything and for all men, you will see at once that it is really so, and that you are to blame for everyone and for all things.[44]

Zosima warns against blaming others, for we will end up complaining against God and becoming proud like Satan. There are many things we cannot comprehend, and we would be led astray completely "if it were not for the precious image of Christ before us."[45] Much is hidden from the eyes of our understanding, but

> we have been given a precious mystic sense of our living bond with the other world, with the higher heavenly world, and the roots of our thoughts and feelings are not here but in other worlds. . . . If that feeling grows weak or is destroyed in you, the heavenly growth will die away in you. Then you will be indifferent to life and even grow to hate it. That's what I think.[46]

As we can already see, Zosima's life answers not only the Grand Inquisitor but also Ivan Karamazov. Ivan not only feels pain and outrage at the injustices in life but is also determined not to let time assuage them: "I don't want harmony. . . . I'd rather remain with my unavenged suffering and unquenched indignation, *even if I am wrong*."[47] Dostoevsky does not offer a grand theodicy in response to Ivan; rather he pictures in Zosima someone who allows the mysterious beauty and innocence of creation to enter his soul and teach him to love life, to love all things and all people, to hope always and

---

[43]Ibid.
[44]Ibid., pp. 275-76.
[45]Ibid., p. 276.
[46]Ibid.
[47]Ibid., p. 212.

never despair. In direct refutation of Ivan, Zosima speaks about the book of Job to his friends. Satan cannot believe that a human being can really love God, but he did, he did: "The Lord gave and the Lord has taken away. Blessed be the name of the Lord forever and ever."[48] Many people, laments Zosima, complain about God, asking, "How could God give up the most loved of His saints for the diversion of the devil . . . for no object except to boast to the devil?"[49] But that is just the greatness of the book: "In the fact that it is a mystery—that the passing earthly show and the eternal verity are brought together in it. In the face of earthly truth, the eternal truth is accomplished."[50] God gave Job new riches and new children. But how could Job be happy with his new family while remembering the loss of his first children? Indeed, "How dare he be happy!" we can hear Ivan object. Zosima replies:

> *But he could, he could.* It's the great mystery of human life that old grief passes gradually into quiet tender joy. The mild serenity of age takes the place of the riotous blood of youth. I bless the rising of the sun each day, and, as before, my heart sings to meet it, but now I love even more its setting, its long slanting rays and the soft tender gentle memories that come with them, the dear images from the whole of my long happy life—and over all the Divine Truth, softening, reconciling, forgiving.[51]

Zosima does not refute Ivan with systems of thought and eschatological visions of harmony. Instead Zosima's life and words make clear that Ivan hates life and has made his heart impervious to life's mystery. He refuses to allow grief to pass "gradually into quiet tender joy." The story of Markel, a boy of seventeen, seems opposed to Ivan's stories of the deaths of children. Only with Markel we do not look on helplessly from an external viewpoint but we hear the internal experience of one dying young. Markel's mysterious exclamation, "Months and Years! . . . Why reckon the days? One day is enough for a man to know all happiness,"[52] sums up his spiritual transformation. Life does not need to be long or always pleasant to be affirmed as worth living. When one experiences oneness with God and all things, nothing else matters.

---

[48]Ibid., p. 252.
[49]Ibid.
[50]Ibid.
[51]Ibid. (emphasis added).
[52]Ibid., p. 250.

THE KARAMAZOV ARGUMENT AFTER DOSTOEVSKY

Dostoevsky, knowing the power of Ivan's rhetoric in "The Rebellion" and "The Grand Inquisitor," was aware of the danger that his refutation in the life of Zosima would not succeed. As he prepared it for publication, he wrote, "The whole novel is written for its sake, if it will only come off, that's what worries me now!"[53] Konstantin Pobedonostsev, head of the Russian Church, wrote anxiously to Dostoevsky inquiring about how he intended to refute the Grand Inquisitor. Dostoevsky replied that he would refute the atheist arguments in book six, "The Russian Monk." But the author admits, "I tremble for it in this sense: will it be answer *enough?* The more so as it is not a direct point for point answer . . . but an oblique one. Something completely opposite to the world view expressed earlier appears in this part, but again it appears not point by point but so to speak in artistic form. And that is what worries me, that is, will I be understood and will I achieve anything of my aim."[54] Judging from the history of criticism and use of *The Brothers Karamazov*, Dostoevsky was correct to tremble that he would be misunderstood. Vasily Rozanov's *The Legend of the Grand Inquisitor* (1894), Albert Camus, *The Rebel* (1951) and other modern works treat the "Rebellion" and "The Grand Inquisitor" independently from the rest of the novel and with no reference to "The Russian Monk" and the life of Elder Zosima. "The Grand Inquisitor" is often printed as an independent book, which gives the impression that this story contains the heart of *The Brothers Karamazov*.[55] Many modern authors repeat Ivan's potent rhetoric against Christianity with no reference to Dostoevsky's refutation and his "defense of Christianity as the only true way of life in our time."[56]

French writer and Nobel laureate Albert Camus (1913–1960) inhabited a world similar to that envisioned by Ivan Karamazov, an absurdity devoid of meaning and values, a world so cold and indifferent to the human reality that, judged from a human perspective, its indifference appears as diabolical cruelty. Like Ivan, Camus championed protest and rebellion against such a

---

[53]Letter 785, June 11, 1879, quoted in ibid., p. 660.
[54]Letter 817, August 24, 1879, quoted in ibid., p. 662.
[55]For a critical assessment of Rozanov's approach to the Grand Inquisitor, see Nathan Rosen, "Style and Structure in The Brothers Karamazov: The Grand Inquisitor and the Russian Monk," *Russian Literature Triquarterly* 1 (1971): 352-63; reprinted in Dostoevsky, *The Brothers Karamazov*, pp. 724-32.
[56]Dostoevsky, *The Brothers Karamazov*, p. 725.

world. However, unlike Ivan, Camus searches for a way to defeat the logic of nihilism, which moves from the indifference of a world without God to the unwarranted conclusion of the moral equivalence between suicide or murder and affirmation of life.[57] In his quest, Camus develops the concepts of the absurd and rebellion and attempts to show that the inborn human rebellion against the absurd is not purely negative but also contains a positive element that renders nihilism inconsistent with its origin in rebellion.

By the absurd Camus does not mean the world alone or the human reality alone; rather he means the *relationship* between humanity and the world. Human beings need meaning, value and truth. They want life and love. But the universe is indifferent to the human spirit. There is no transcendent foundation for human values and no life beyond the grave to balance accounts. Hence there is a contradiction between the dynamic of the human spirit and the spiritless world it inhabits. Confronted with this contradiction (the absurd), the human spirit rebels; it says no: "No, I will not let you determine me, absorb me, annihilate me! Not while I live!" Rebellion is the human "No!" to anything that dehumanizes, to any power that encroaches on individual or shared human dignity. It is the slave's "No!" to the master. It is Prometheus's "No!" to Zeus. It is Ivan Karamazov's "I don't accept it" to God's world. It is every human "No!" to suffering, futility and death. Yet rebellion as Camus understands it is not pure negation, for within the slave's "No!" is an affirmation of common humanity between master and slave. In Prometheus's "No!" lies an affirmation of justice, which binds Zeus and him alike. And Ivan's "No!" presupposes a way the world ought to be. Hence rebels betray the original affirmative impetus of rebellion if they allow rebellion to morph into pure negation, moral nihilism. Suicide contradicts the rebellious affirmation of one's own dignity, and murder (of the master or the enemy) contradicts the solidarity and unity of humanity that first moved the rebel to speak his "No!"

Camus fits the profile of a protest atheist only in a qualified sense, and his rhetorical argument from evil stands subdued in the background. His book *The Rebel* (1951) focuses critically on nihilists, who betray the original nature of rebellion. Only indirectly does he target God, that is, the idea of God as

---

[57]Ivan expressed his intention to commit suicide when he reached 30 years of age. He had also said that since there is no God "everything is permitted." And indirectly, in the person of Smerdyakov, he killed his own father.

the justification for the evil in the world. *The Rebel* divides itself into two parts, an analysis of "metaphysical rebellion" and of "historical rebellion." Metaphysical rebellion focuses on God understood as the justification of creation and the nemesis of humanity. Historical rebellion centers on the "no" to all political, social and economic structures that fall short of perfect justice. Among Camus's examples of perverted metaphysical rebellion are the Marquis de Sade (1740–1814) and Ivan Karamazov. Sade's rebellion against God and creation led him to reject morality, praise crime and embrace "universal destruction," that is, universal murder.[58] Camus tacitly approves of Ivan's rebellion against every attempt to justify a world in which innocent children suffer. In the name of human dignity and justice, Camus too advocates rebellion against God or any transcendent principle that would serve as justification for evil. But Ivan goes too far when he argues that, because there is no God, no immortality and no transcendental moral law, "everything is permitted," that crime must be "recognized as the inevitable and the most rational outcome of his position for every atheist."[59] For Camus, Ivan abandons true rebellion when he concludes that his love of life—his love of those "sticky little leaves in the spring"[60]—is irrational and that suicide is the most rational action a rebel can take.

RESPONSES TO THE RHETORICAL PROBLEM OF EVIL

John K. Roth. In response to protest atheism, philosopher John K. Roth developed a "theodicy of protest."[61] Roth's protest theodicy takes up the atheist protest at the unforgivable outrage of evil, but rather than denying God he directs its rage against evil itself and the systems that facilitate and justify it. According to Roth, the standard theodicies affirm divine goodness too quickly and in their attempts to absolve God of evil either deny or qualify divine omnipotence. But in their eagerness to preserve divine benevolence "they legitimate evil" and lessen the motivation for fighting against it.[62] Evil, or at least the possibility of evil, is made a necessity rather than the outrage

---

[58]Albert Camus, *The Rebel: An Essay on Man in Rebellion*, trans. Anthony Bower (New York: Vintage, 1991), pp. 36-37.
[59]Dostoevsky, *The Brothers Karamazov*, p. 65.
[60]Ibid., p. 199.
[61]John K. Roth, "A Theodicy of Protest," in *Encountering Evil: Live Options in Theology*, ed. Stephen T. Davis (Louisville, KY: Westminster John Knox, 2001), pp. 1-20.
[62]Ibid., p. 15.

we experience it to be. Roth reverses the usual strategy by calling into question God's unequivocal goodness and retaining divine omnipotence. Given the terrible evils in the world, the mass killings and episodes of genocide of the twentieth century, we must rethink the goodness of God. According to Roth, the freewill defense of divine benevolence—in which God's exercise of power is self-limited to make creaturely freedom possible— cannot cope with the horror and extent of destruction in history: "A protesting theodicy is skeptical because it will not forget futile cries. No good that it can envision on earth or beyond, is worth the freedom—enfeebled and empowered—that wastes so much life."[63] Roth denies that God can make up for all evil in an eschatological event of salvation. Nothing God can do will change the past. "Everything hinges on the proposition that God possesses—but fails to use well enough—the power to intervene decisively at any moment to make history's course less wasteful. Thus, in spite and because of God's sovereignty, this God is everlastingly guilty, and the degrees run from gross negligence to murder."[64] Some may think a suffering God can help. Not according to Roth: "But the mass of agony does not have to be, and if God is only a suffering God, then we do indeed need a God to help us."[65]

It's hard to say how seriously Roth intends his indictment of God as lacking perfect goodness. On many occasions his charges of divine injustice and waste are accompanied by such qualifications as "as far as we can envision" or "that we can comprehend" or "from a human perspective." Hence I do not think Roth is making a strong metaphysical claim about God's ultimate nature. He leaves this shrouded in darkness. Roth hints at his real aims when he refers to his protest theodicy as "antitheodicy," a term that better describes what Roth is doing. His antitheodicy aims to achieve at least three objectives. First, he wants to avoid legitimating evil by making it an "instrument" God uses to bring about good. His arguments against this procedure are purely rhetorical in the Karamazov style. The second goal is intimately related to the first. By protesting evil as excessive and useless we may be motivated to fight against it in acts of justice and compassion. Our protest against excessive and wasteful evil—which God unjustly allows to

---

[63]Ibid., p. 11.
[64]Ibid., p. 14.
[65]Ibid., p. 15.

occur—can motivate us to "be for God by being against God, and the best way that we can do that is by giving life in care and compassion for others."[66] Third, Roth seems to think that some people are psychologically incapable of relating to God any other way, given their experience or sensitivity to evil. He explains, "The antitheodicy outlined here is one for sick souls who know that their sickness cannot—must not—be cured, and who likewise refuse to acquiesce because to do so would accomplish nothing."[67]

Jürgen Moltmann. No contemporary theologian has given more thought to the rhetorical problem of evil than Jürgen Moltmann. His thought is conditioned by his experience as a prisoner of war during and after World War II and his response to such protest atheists as Albert Camus, Max Horkheimer and Ernst Bloch, and post-Holocaust thinkers Richard Rubenstein and Elie Wiesel. Like Roth, Moltmann rejects any theodicy that justifies the horror of human suffering. He agrees with Rubenstein, who says, "To see any purpose in the death camps, the traditional believer is forced to regard the most demonic, anti-human explosion of all human history as a meaningful expression of God's purposes. The idea is simply too obscene for me to accept."[68] In his book *Night*, which recounts his experiences in Auschwitz, Eli Wiesel sees both the impossibility of continued belief in the traditional doctrine of God and the necessity of hope in God as the ground of human dignity. Wiesel experienced the death of his faith in God in Auschwitz as he watched a young man die on the gallows; nevertheless he felt compelled to pray "to the God in whom I no longer believed. My God, Lord of the Universe, give me the strength never to do what Rabbi Eliahou's son has done."[69] Moltmann agrees with Horkheimer's assessment of the God of traditional theism: "In view of the suffering in this world, in view of the injustice, it is impossible to believe the dogma of the existence of an omnipotent and all-gracious God."[70] Moltmann finds appealing Horkheimer's refusal to let the protest against suffering fall silent or the longing for justice be deadened by

---

[66]Ibid., p. 16.
[67]Ibid., p. 19.
[68]Richard Rubenstein, *After Auschwitz* (Indianapolis: Bobbs-Merrill, 1966), p. 153.
[69]Elie Wiesel, *Night* (New York: Hill & Wang, 1960), p. 95, quoted in Richard Bauckham, "Theodicy from Ivan Karamazov to Moltmann," *Modern Theology* 4, no. 1 (1987): 88. In this quote Wiesel refers to a man in the death camps who abandoned his father to save his own skin.
[70]Max Horkheimer, *Die Sehnsucht nach dem ganz Anderen (Longing for the Totally Other)* (Hamburg: Furche, 1970), p. 56, quoted in Jürgen Moltmann, *The Crucified God*, trans. R. A. Wilson and John Bowden (New York: Harper & Row, 1974), p. 225.

optimistic theism or nihilistic atheism. Horkheimer's critical theory and Moltmann's "critical theology" find common ground "in the framework of open questions, the question of suffering which cannot be answered and the question of righteousness which cannot be surrendered."[71]

Already we can see the outline of Moltmann's response to protest atheism and the rhetorical argument from evil. Negatively, he wishes to avoid, on the one hand, the problems engendered by "theism," by which he means the traditional doctrine of God and, on the other, those problems inherent in "atheism," by which he seems to mean nihilism. Theism justifies evil and enervates human attempts to fight against it, and atheism destroys all hope that evil will be overcome and in another way undermines outrage at evil, depleting the energy needed to resist it. On the positive side, Moltmann develops a cross-centered theology in which God and the world are bound together in the same history of becoming and suffering.[72] In doing this he rejects "the distinction made by the early church between theology as the doctrine of God and economy as the doctrine of salvation."[73] Moltmann refuses to speak of God or speculate about the Trinity apart from the history of creation and salvation whose meaning becomes clear only in the event of the cross: "Anyone who really talks of the Trinity talks of the cross of Jesus and does not speculate in heavenly riddles."[74] There is no Lord God who is above it all; rather, God becomes *Lord* only in interaction with creation.[75] God experiences in his own being and inner life the pain, suffering and evil the world experiences; creation is not something purely external to God. God's actions in the history of creation and salvation correspond in character to God's own life. Using the traditional distinction between God's internal and external acts, Moltmann asserts that God's "outward acts correspond to inward suffering, and outward suffering corresponds to inward

---

[71]Moltmann, *Crucified God*, p. 226.

[72]I explain Moltmann's view of the God-creation relationship in much greater detail in Ron Highfield, "Divine Self-Limitation in the Theology of Jürgen Moltmann: A Critical Appraisal," *Christian Scholar's Review* 32 (2002): 47-71.

[73]Moltmann, *Crucified God*, p. 67.

[74]Ibid., p. 207.

[75]"The economic Trinity completes itself and perfects itself to the immanent Trinity when history and experience of salvation are completed and perfected. When everything is 'in God' and 'God is all in all,' then the economic Trinity is raised into and transcended in the immanent Trinity" (Jürgen Moltmann, *The Trinity and the Kingdom*, trans. Margaret Kohl [Minneapolis: Fortress, 1993], p. 161).

acts."[76] God's outward act of creation presupposes God's inward self-restriction to make room for creation. In a sense God negates or withdraws his omnipresence and Lordship so that creatures can have space and freedom. According to Moltmann, "It is the affirmative force of God's self-negation which becomes the creative force in creation and salvation."[77] Hence to become the creator, God had to cease being the omnipresent and omnipotent Lord within the sphere of his being designated as the home of creation. To become the creator, God had to change, suffer and enter into a history aimed at the redemption not only of creation from its suffering but also of God from his self-imposed negation.

Moltmann also puts this divine self-negation in trinitarian terms. If the Father continued to love the Son in the Spirit eternally and unchanged, creation would not be possible. The existence of creation requires that God begin to love something in addition to the Son and very different from the Son. Hence creation requires "an alteration in his [the Father's] love for the Son (that is to say through a contraction of the Spirit) . . . and through an alteration in his [the Son's] response to the Father."[78] Of course any "altering" of the Father's love for the Son would mean some sort of negation or distancing. The Father treats the Son as his opposite, as not God. The space opened between the Father and the Son is the created world. The world is "other" than God, and otherness exists only by negation. If God as Father, Son and Holy Spirit, alike in essence, remained bound in complete unity without negation, no otherness could exist. For Moltmann this divine self-negation reaches its maximum in the cross. The Father abandons the Son completely, for the Father wants not only to make room for the otherness of creation but also for sinners who stand in total opposition and hostility to God. God makes room for sinners among the Trinity by negating the Son to the point of treating him as a sinner. But since the Father's love for the Son in the Spirit is stronger than the Father's negation of the Son, in the resurrection the Father reunites himself with the Son and with sinners with whom the Son was identified. God "is acting in himself in this manner of suffering and dying in order to open up in himself life and freedom for

---

[76]Ibid., p. 98.

[77]Jürgen Moltmann, *God in Creation: A New Theology of Creation and the Spirit of God*, trans. Margaret Kohl (Minneapolis: Fortress, 1993), p. 87.

[78]Moltmann, *Trinity and the Kingdom*, p. 112.

sinners."[79] Creation, which began as God's self-negation or as an alternation in the Father's love for the Son, is now drawn into union with God, and God is reunited with himself. History ends not only in the salvation of creation but in the salvation of God, salvation from God's suffering and exile from his own Lordship.

How does Moltmann's theology enable him to respond to the rhetoric of protest atheism? First, we must note that his understanding of God's relationship to history does not undermine the validity and intensity of the protest against evil. Evil remains an outrage against humanity to be protested and resisted with all our strength. Rather than willing the suffering of creation as a means to some higher good, God protests, resists and suffers from evil. In protesting and resisting evil, we join with God in hope of bringing about the end of evil and suffering. In Moltmann's thinking, evil is the negation that must be negated, totally overcome; this is true for God as well as for human beings. Second, Moltmann criticizes atheism for its tendency to fall into nihilism and hopelessness. Like Camus he sees in the Ivan Karamazov style of protest atheism a tendency toward the despair of suicide or murder. Unlike Camus, Moltmann does not advocate assertion of humanistic values in an absurd relation to an uncaring universe; rather, he asserts faith in the God of the future, the God of resurrection, as the ground of hope for overcoming evil and for meaningfulness protest against it.[80] In Moltmann's theology one does not need to deny God to protest the evil in the world and work to overcome it. The suffering God disclosed in the events of the cross and resurrection can never provide justification for evil or reason to despair in the face of evil. Suffering is not something God *uses* to achieve a purpose; it is what God *endures* to make room for creatures and sinners. God does not *cause* or *will* the suffering of creatures but *co-suffers* it, taking it into God's life, thereby overcoming it.

Responding to the "Rhetorical Argument from Evil"

As I emphasized when introducing the rhetorical argument from evil, we must keep clearly in mind its rhetorical form. It is not a *rational* argument

---

[79]Moltmann, *God in Creation*, p. 192.

[80]Meaningful protest can be lodged only against a state of affairs that ought not to be. Its force would be lost were it understood as merely an expression of anger at having one's preferences denied.

but an expression of anguish seeking to create corresponding anguish in the hearer. Yet in the mouth of the protest atheist the emotional appeal is not merely a cry of anguish but also an argument against all justifications and defenses of God that would remove the grounds for protest. Its effectiveness as an *argument* depends on the mismatch between cold, rational responses to evil, deaf to the cries of others, and the expected response of outrage at evil and sympathetic identification with the suffering. It defeats its opponents in the eyes of the audience by making them appear ludicrous or uncaring or immoral in offering a defense for God's allowing horrendous evils to occur. As effective as it is in silencing rational defenses, however, it does not provide an alternative answer or defeat the rational theodicies on an intellectual level. And, as Camus and Moltmann observe, protest atheism risks driving people into the twofold despair of suicide or murder. These two thinkers want to keep the protest alive but avoid despair.

Responses to protest atheism must keep in mind its dual nature as an expression of anguish and an argument against God. Christian apologists need not be rhetorically naive. Given the emotional appeal of protest atheism, our first spoken response should meet the rhetorical expectations implicit in the anguished form of the protest. Expressions of hope and faith or any justifications or defenses of God should be set in the context of clear condemnations of cruelty and expressions of sympathy and solidarity with victims of crime and other sufferers; and these should be accompanied with determination to fight evil and help its victims. Camus and Moltmann provide help at this point. Even though believers do not accept Camus's assumption that the human condition is *metaphysically* absurd, they do at times experience it as *epistemically* absurd; we see no reason and have no answer as to why particular horrible evils occur. When asked why a certain evil occurred we can honestly say, "I do not know, and I am so sorry." Yet silence, protest and sympathy cannot be our only responses. When the sufferer cries out "Why," our sympathetic "I do not know" need not mean "There is no reason." Epistemic absurdity need not imply metaphysical absurdity. Camus recommends rebelling against the absurd, choosing life and asserting human values over death and destruction even if our rebellion is ultimately futile. Moltmann asserts hope grounded in the resurrection of the crucified one. Both offer their positive encouragement as a muted "nevertheless." The rhetorical situation of protest and rebellion demands that we express outrage at evil and solidarity with its victims. But it also demands that we not give in to despair lest we abandon the victims in the opposite way and give their murders the final victory. When the speaker cries out in anguish and protest at horrible cruelty and suffering, wouldn't it be just as insensitive to assert the final absurdity of the human situation as to express untroubled confidence that all accounts will be balanced in the end? Is nihilism more comforting to the dying than optimism? If optimism implies a moral equivalence between the murderer and the victim by making the murderer an instrument of God, nihilism destroys all moral categories by which we could distinguish between guilt and innocence. Here, then, is the rhetorical space to insert our "nevertheless" of hope and faith without seeming heartless. To hold to the goodness and strength of God in the face of unspeakable evil does not justify evil but affirms the value and meaning of life, which, as Camus shows, is the very basis of protest and rebellion. To believe in the God of Jesus Christ, the crucified and resurrected one, does not belittle the sufferings of the innocent but declares that their torturers will not have the final say. To hope that God will "dry every tear" does not imply that the tears were cried for nothing; rather it asserts that life is stronger than death and joy more lasting than sadness.

So far we have found Moltmann's response to protest atheism helpful. He points us to an appropriate rhetorical space for voicing faith and hope in God in a way that does not minimize outrage and protest at suffering. But is he correct that we can speak of faith and hope to the victims of the gas chambers in good conscience only if we reject the traditional doctrine of God and replace it with the suffering God of panentheism? We need to distinguish this question from the question of whether panentheism does justice to the biblical theology of God and God's relation to the world. My view is that it does not. But we will not address this question in this context.[81] In Moltmann's view the traditional doctrine of God pictures God as an unmoved and all-determining power. This Power determines every event that occurs in history and nature so that every instance of crime and suffering expresses the divine will in an obvious way. From this perspective it would be axiomatically true to say of any instance of individual suffering, even the most horrendous, "God willed it." Of course, Moltmann recognizes that

---

[81]I have dealt with this issue already in this book and in other places.

advocates of the traditional doctrine of God also affirm the goodness of God, so they would add some qualifiers to that statement. They would say, "Though we may not understand God's ways, we know that God willed this event for a good reason, as a means to achieving his good will in the end." Moltmann rejects such thinking because it justifies evil by making it a means to a noble end, and this implication makes the traditional doctrine vulnerable to the critique of protest atheism. How can you maintain the stance of outrage and protest against something that brings about a greater good? Since protest atheists think they cannot give up outrage and protest without dishonoring the victims of murder, disease and natural disaster, and accepting God seems to demand cooling outrage and quieting protest, they reject God in the name of common decency.

There is some plausibility in the assumption that the suffering God of panentheism will find a warmer reception by outraged, grieving and protesting people than the unfeeling God of "theism." To say to the sufferer, "God does not want you to suffer and himself suffers in his fight against the evil that plagues you" clearly sounds more compassionate than saying, "God cannot himself suffer but wills your suffering in view of a greater purpose." The first statement offers a sympathetic presence, while the second abandons the sufferer and subordinates their suffering to a distant goal. The first keeps outrage and protest focused on the evil itself. The second provokes outrage and protest against God as the one responsible for the evil. In the first, God joins the sufferer in suffering, outrage and protest, whereas in the second God remains aloof and demands an end to outrage and protest. If we focus only on which view of God is most likely to escape the anger of the protest against evil, the suffering God of panentheism will win the prize. However, Moltmann also wishes to ground hope for overcoming evil in the resurrection of Christ. How will the outraged and protesting people receive this message? Might they not respond something like this: "Are you saying the resurrection of Auschwitz's dead will make up for their suffering and erase its memory? And if the resurrection doesn't blot out the memory of their suffering, why be raised? If God can raise the dead and redeem all creation in the end, couldn't God have prevented Auschwitz and Dachau? And if God could not have prevented Auschwitz and Dachau, why should we think God can raise the dead and bring all creation into perfection in the end? You offer us the worst of atheism and theism; you would rob us of the hopeful despair of atheism and give us the despairing hope of panentheism. Yours is a pathetic God, more to be pitied than protested. At least with the omnipotent, all-determining God we had an idea worthy of protest, an object sufficient for our outrage." If in responding to protest atheism we cannot say anything that gives grounds for cooling the outrage and toning down the protest, how can we do anything but join the protest? Hope and outrage can't exist in the same heart without each moderating the other.

However, if we ask not which view of God will incur the least scorn from protest atheism but which will provide the surest foundation for hope, the suffering God of panentheism cannot compete with the traditional doctrine of God. Perhaps we should keep in mind that, though the sufferer can be helped by sympathetic accompaniment, the most urgent need and desire of the sufferer is relief from the cause of the suffering. The most urgent need of a dying world is renewal. The sinner needs forgiveness and liberation from the power of sin. But can the suffering God deliver us from evil? Once we postulate God's vulnerability to suffering and emotional anguish, how can we regain confidence that God can save us? If evil possesses power in the present to hurt God why should we think God can overcome it in the future? We are vulnerable to sin, suffering and death by nature. Nothing we can do can remove that threat. Moltmann explicitly denies that God is vulnerable to sin, suffering and death *by nature*. Instead, God voluntarily suffers so that the world might exist and God voluntarily abandons the Son so that sinners might be accepted. But this theory of divine self-limitation is a fiction; it is an all-too-convenient, ad hoc hypothesis designed to secure the benefits of a finite and suffering God without being burdened with its liabilities. If God can suffer willingly, it follows that God's nature is such that God can suffer, which is to say that God is by nature vulnerable to suffering. Otherwise it would not even be possible for God voluntarily to suffer.[82] I can suffer and die. Hence I can suffer and die willingly or unwillingly. If God can willingly suffer and die, does it not follow that God can suffer and die unwillingly? Where is the ground for unshakable confidence that God can and will be victorious over sin, death and suffering?

---

[82]I argue this at length in Highfield, "Divine Self-Limitation," pp. 47-71.

"Do Not Be Afraid"

A PART FROM FIRM CONFIDENCE in the all-encompassing providence of God we fall prey to anxiety and confusion, and are tempted
by despair or presumption. How else can we deal with regret for past
failings and sadness for lost opportunities? And on what other basis may
we rest contentedly in the joy of the present moment without distraction
from an unchangeable past, an unpredictable future or the manifold allure-
ments of the present? Despite the best human wisdom available we cannot
always make things turn out right. Many forces are at work in determining
the future, some of them mysterious and some of them hostile, but their
sheer number places the future beyond our control. The human condition
is such that we can imagine many possible futures, and the human tendency—since we are finite and mortal—is to imagine negative futures more
vividly than positive ones. Without a solid foundation for hope that the
future will ultimately bring good to us, we are deprived of joy and meaning
in the present. But confidence in God's power and will to bring about our
salvation and to make all things work together for our good liberates us
from the oppressive threat of negative futures and provides energy to continue working for good ends.

There is no need to repeat here the doctrine of providence developed in
previous chapters. I want to focus instead on what Jesus and the New Testament writers affirm as the mood that arises naturally from faith in God's
providential care. The fifth thesis on biblical theology of providence stated
that for those who love God the preceding affirmations about God's providence impart comfort, hope, courage and joy.

THE FAITH OF JESUS

And this mood is exactly what we find in Jesus' teaching. Jesus taught his disciples to trust God to provide what we need in life. As an example we will consider Matthew 6:25-34:

> Therefore I tell you, do not worry about your life, what you will eat or drink; or about your body, what you will wear. Is not life more than food, and the body more than clothes? Look at the birds of the air; they do not sow or reap or store away in barns, and yet your heavenly Father feeds them. Are you not much more valuable than they? Can any one of you by worrying add a single hour to your life?
>
> And why do you worry about clothes? See how the flowers of the field grow. They do not labor or spin. Yet I tell you that not even Solomon in all his splendor was dressed like one of these. If that is how God clothes the grass of the field, which is here today and tomorrow is thrown into the fire, will he not much more clothe you—you of little faith? So do not worry, saying, "What shall we eat?" or "What shall we drink?" or "What shall we wear?" For the pagans run after all these things, and your heavenly Father knows that you need them. But seek first his kingdom and his righteousness, and all these things will be given to you as well. Therefore do not worry about tomorrow, for tomorrow will worry about itself. Each day has enough trouble of its own.

Matthew sets the sayings of these verses in the context of sayings that urge disciples to unmixed loyalty to God. Immediately preceding them is the saying in Matthew 6:24: "No one can serve two masters. Either you will hate the one and love the other, or you will be devoted to the one and despise the other. You cannot serve both God and money." Loyalty to money (Greek: *mamōna*) is set in opposition to service to God. Money is being pictured as a god from whom one anxiously seeks certain goods. Jesus calls for a decision between God and all other supposed sources of good. Jewish hearers would have caught a reference to the Shema of Deuteronomy 6:4-5: "Hear, O Israel: The LORD is our God, the Lord alone . . ." (NRSV). Or they would have remembered Exodus 20:2-3: "I am the LORD your God. . . . You shall have no other gods before me." God alone is the source of life and the goods that sustain life.

Matthew 6:25 begins with "therefore" to indicate that the theme of the preceding verse is being continued and expanded. "Do not worry about your life" (Mt 6:25). These words can be considered a command or a deduction.

Because there is only one God, who is the sole source of all good, we must not look to other sources. There is no God but God, no other creator, no other Lord of nature and history. Because God faithfully feeds the birds and clothes the grass, which are of much less value than you, God will surely take care of you. Hence "do not worry about your life." Given God's sole divinity and his love for you, anxiety about the future makes no sense.

In Matthew 6:31-32, Jesus pictures pagan life. Pagans don't know that God alone is God and that God cares for the birds of the air and the grass of the field. Their loyalties are divided among the gods who stingily and capriciously dole out the goods of life to those who please their whims. Pagans never know whether or when fate will strike or luck will arrive. They worry about life, and they have reason to do so. They are distracted and divided. As Jesus pictures them, pagans live in despair or presumption. They despair of life, and in their despair frantically attempt to conquer the future through their futile activity. In commenting on this text, Søren Kierkegaard captures the mood of the pagan:

> He does not keep silent like the care-free bird, he does not talk like the Christian, who talks about his riches; he actually has and knows nothing to talk of except poverty and its anxiety. He asks, What shall I eat? and What shall I drink?—today, tomorrow, the day after, the coming winter, the following spring, when I am old, as for me and mine, and the whole land, what shall we eat and drink?[1]

But Jesus pictures a life lived in a different mood and at a different pace. With undivided devotion and undistracted attention, we can live in the present focused on the higher goods, on righteousness and the reign of God. The future is in God's hands, and hence we can view it in a mood of hope and expectation. Despair and anxiety about what may come does not rob us of peace and confidence in the present. God calls us to serve him alone, to seek his presence. And God will give us the means to do what he commands. Jesus does not say that we will face no hardships or persecutions. He assures us that we will. The confidence Jesus inspires is based on God's knowledge and care over us throughout whatever we must experience. Jesus leaves no space for doubt that God can and will take care of us. For Jesus, that is

---

[1]Søren Kierkegaard, *Christian Discourses*, trans. Walter Lowrie (Princeton, NJ: Princeton University Press, 1971), p. 22.

certain. And by dwelling in the faith of Jesus, we can hear his words "Do not worry about your life" not as a command to do the impossible but as a power that gives what it commands.

FAITH IN JESUS

Paul, too, expresses complete confidence in God's care. The apostle to the Gentiles harbors no illusions that we can escape suffering in this life. Indeed, Jesus suffered and we must share in his sufferings. But these sufferings "are not worth comparing with the glory that will be revealed in us" (Rom 8:18). The classic text for divine providence in Paul and the whole New Testament is Romans 8:28-30:

> And we know that in all things God works for the good of those who love him, who have been called according to his purpose. For those God foreknew he also predestined to be conformed to the image of his Son, that he might be the firstborn among many brothers and sisters. And those he predestined, he also called; those he called, he also justified; those he justified, he also glorified.

These verses express the entire doctrine of divine providence in a compressed form. God works "all things" for the good that God purposed for us, that is, our salvation and glorification. Hence we are "more than conquerors through him who loved us" (Rom 8:37). Paul does not say that we have no enemies or that things will be easy as we go through them; he says, rather, that since God is for us and works for us in all things we will win over all enemies.

Considering this text in its setting in the letter, we note that in Romans 5 Paul begins to shift from his argument for justification by faith in Jesus Christ to the conduct of life in the Spirit. Confidence in our justification through Christ and in our state of peace with God leads to a life of rejoicing "in the hope of the glory of God" (Rom 5:2). Rejoicing in view of our ultimate glorification is understandable, but Paul adds that we "also glory in our sufferings." Even these afflictions will be ordered toward higher ends and will produce good fruit in us: perseverance, character and hope (Rom 5:3-4). And God confirms this hope by pouring divine love into our hearts by the Holy Spirit.

Paul takes up these themes again in Romans 8. He tells us there that the Spirit witnesses to our spirits that we are truly children of God and coheirs with Christ. We will share in his victory, but only if we also share in his suf-

ferings (Rom 8:17). With this warning Paul shifts to the themes of suffering
and divine providence. In Romans 8:18 he asserts a bottom-line judgment
that colors the rest of the chapter: "I consider that our present sufferings are
not worth comparing with the glory that will be revealed in us." When compared to the calculus of the evidential argument from evil or the incon-
solable outrage of protest atheism or the tentativeness of open theism, Paul's
assertion is quite startling. He does not argue philosophically that each and
every instance of suffering is necessary for a "higher harmony" or that suffering is somehow good because it is instrumental in bringing about a
greater good. He reasons, rather, from a Christocentric and eschatological
point of view. Christ suffered and has been raised victorious over sin and
death. He has been glorified. Those united with Christ must follow the same
path, first suffering and then glory. And the joy of the glory infinitely outweighs the sorrow of the suffering. The hope of such glory and the help of
the Spirit enable us to endure patiently the suffering we must face.

In Romans 8:31 Paul asks about our reaction to God's work on our behalf.
"What, then, shall we say in response to these things?" What shall we say and
feel, and how shall we live if "God is for us"? If God gave his Son for us, won't
God give lesser things? Paul then lyrically rehearses a series of ever more
powerful and dreadful enemies who have no power to interfere with God's
loving care for us. In the face of all physical and spiritual threats, "we are
more than conquerors through him who loved us" (Rom 8:37). To be the
victor, the winner, the conqueror, is to be filled with glorious joy. The enemy
is vanquished, trampled and forgotten. But Paul says that we are "*more than*
conquerors." How can one be "more than" a conqueror? Perhaps this
statement should be coupled with his earlier claim that the suffering we
must endure is "*not worth comparing* with the glory that will be revealed in
us" (Rom 8:18). Romans 8:37 looks at the incomparable future glory of
Romans 8:18 from the other angle and with a different metaphor. We are
more than conquerors, that is, the victory is so triumphant that it makes the
enemy look insignificant and battle effortless, a perspective that exposes
Ivan Karamazov's "unavenged suffering and unquenched indignation" as
despair and hatred of life.[2]

---

[2]Fyodor Dostoevsky, *The Brothers Karamazov: A Norton Critical Edition*, trans. Constance Garnett,
ed. Susan McReynolds Oddo, rev. Susan McReynolds Oddo and Ralph E. Matlaw, 2nd ed. (New
York: W. W. Norton, 2011), p. 212.

We can summarize Paul's perspective in two brief statements. (1) The glory of our eschatological destiny will infinitely outweigh any suffering we must endure in this life. (2) God will order everything that happens, no matter what it is, to our good.[3] The first statement considers world events as a whole, and the second considers them individually and as they happen. The first statement allows us to contemplate our lives as a whole—or even the history of the world as a whole—from an eschatological perspective. Even before the end, in the unsettled thick of life's battles, we know the final judgment about suffering. It is *"not worth comparing* with the glory that will be revealed in us" (Rom 8:18). The second statement assures us that even in the moment of trial God will not allow suffering's absurdity to stand. It will be forced to serve God and fit into an order that leads to our glory.

If we think about our suffering in temporal terms, we must deal in our hearts with the negative memories of the past, the pain and confusion of the present, and anxiety about the future. For Paul and the rest of the New Testament writers, the unchangeable past is dealt with through forgiveness of sins secured for us in Christ, the healing of the Spirit and reordering of God's working. In the present we receive patience, courage, hope and love to face and endure trials through faith in God's hidden working. Anxiety about what might happen in the future is tempered by well-grounded hope of our ultimate glorious victory through Christ. In the words of Dostoevsky's hero Zosima:

> It's the great mystery of human life that old grief passes gradually into quiet tender joy. The mild serenity of age takes the place of the riotous blood of youth. I bless the rising of the sun each day, and, as before, my heart sings to meet it, but now I love even more its setting, its long slanting rays and the soft tender gentle memories that come with them, the dear images from the whole of my long happy life—and over all the Divine Truth, softening, reconciling, forgiving.[4]

---

[3]In a perspective similar to the one I am advocating, James S. Spiegel defends what he calls the "Greater Good Theodicy." See *The Benefits of Providence: A New Look at Divine Sovereignty* (Wheaton, IL: Crossway, 2005), pp. 193-207. See also Marilyn McCord Adams, "Redemptive Suffering: A Christian Solution to the Problem of Evil," *Rationality, Religious Belief, and Moral Commitment*, ed. Robert Audi and William J. Wainwright (Ithaca, NY: Cornell University Press, 1986), pp. 249-67; John Edelman, "Suffering and the Will of God," *Faith and Philosophy* 10 (July 1993): 380-88; and Eleonore Stump, "Providence and Evil," in *Christian Philosophy*, ed. Thomas P. Flint (Notre Dame, IN: University of Notre Dame Press, 1990), pp. 51-91.

[4]Dostoevsky, *The Brothers Karamazov*, p. 252.

Have no fear of men's sin, love a man even in his sin, for that is the semblance of divine love and is the highest love on earth. Love all God's creation, the whole and every grain of sand in it. Love every leaf, every ray of God's light. Love the animals, love the plants, love everything. If you love everything, you will perceive the divine mystery in things. Once you perceive it, you will begin to comprehend it better every day. And you will come at last to love the whole world with an all-embracing love.[5]

---

[5]Ibid., p. 275.

SELECTED BIBLIOGRAPHY

Adams, Marilyn McCord. "Redemptive Suffering: A Christian Solution to the Problem of Evil." In *Rationality, Religious Belief, and Moral Commitment*, edited by Robert Audi and William J. Wainwright, pp. 249-67. Ithaca, NY: Cornell University Press, 1986.

Adler, Mortimer. *The Idea of Freedom: A Dialectical Examination of the Conceptions of Freedom*. 2 vols. Garden City, NY: Doubleday , 1958.

Alston, William P. "Divine and Human Action." In *Divine and Human Action: Essays in the Metaphysics of Theism*, edited by Thomas V. Morris, pp. 257-80. Ithaca, NY: Cornell University Press, 1988.

———. "How to Think About Divine Action: Twenty-five Years of Travail for Biblical Language." In *Divine Action: Studies Inspired by the Philosophical Theology of Austin Farrer*, edited by Brian Hebblethwaite and Edward Henderson, pp. 51-70. Edinburgh: T & T Clark, 1990.

Anderson, Bernhard W. "Creation." In *The Interpreters Dictionary of the Bible*, edited by George A. Buttrick, 1:725-32. Nashville: Abington, 1962.

———. *From Creation to New Creation: Old Testament Perspectives*. Minneapolis: Fortress, 1994.

Anselm of Canterbury. *Basic Writings*. 2nd ed. Edited and translated by S. N. Deane. LaSalle, IL: Open Court, 1969.

———. *The Major Works*. Translated by Brian Davies and G. R. Evans. New York: Oxford University Press, 1998.

Aristotle. *The Basic Works of Aristotle*. Edited by Richard McKeon. New York: Random House, 1941.

Arnold, Matthew. *The Poems of Matthew Arnold 1840–1867.* 1913. Reprint, London: Oxford University Press, 1940.

Artigas, Mariano. *The Mind of the Universe: Understanding Science and Religion*. Philadelphia: Templeton Foundation Press, 2000.

Asselt, Willem J. van, J. Martin Bac and Roelf T. te Velde, eds. *Reformed Thought on Freedom: The Concept of Free Choice in Early Modern Reformed Theology*. Grand Rapids: Baker Academic, 2010.

Audi, Robert, ed. *Cambridge Dictionary of Philosophy.* 2nd ed. Cambridge: Cambridge University Press, 1999.

Augustine of Hippo. *The Augustine Catechism: The Enchiridion on Faith, Hope, and Love.* Translated by Bruce Harbert. Edited by John E. Rotelle, OSA. Works of Saint Augustine: A Translation for the Twenty-First Century. New York: New City Press, 1999.

———. *Confessions.* Translated by Henry Chadwick. New York: Oxford University Press, 1991.

———. *Eighty-Three Different Questions.* Translated by D. L. Mosher. Fathers of the Church 70. Washington, DC: Catholic University of America Press, 1977.

Ayala, Francisco J. "The Evolution of Life: An Overview." In *Evolutionary and Molecular Biology: Scientific Perspectives on Divine Action*, edited by Robert John Russell, William Stoeger, SJ, and Francisco J. Ayala, pp. 21-57. Vatican City and Berkeley, CA: Vatican Observatory and Center for Theology and the Natural Sciences, 1998.

Bammel, C. P. "Adam in Origen." In *The Making of Orthodoxy: Essays in Honour of Henry Chadwick*, edited by Rowan Williams, pp. 62-93. Cambridge: Cambridge University Press, 1989.

Barbour, Ian. *Religion in an Age of Science: The Gifford Lectures.* 2 vols. San Francisco: Harper & Row, 1990.

Barth, Karl. *Church Dogmatics.* 4 vols. Translated by G. W. Bromiley and T. F. Torrance. Edinburgh: T. & T. Clark, 1936–1969.

Basinger, David. *The Case for Freewill Theism: A Philosophical Assessment.* Downers Grove, IL: InterVarsity Press, 1996.

Bauckham, Richard. *God and the Crisis of Freedom: Biblical and Contemporary Perspectives.* Louisville KY: Westminster John Knox Press, 2002.

———. *God Crucified: Monotheism and Christology in the New Testament.* Grand Rapids: Eerdmans, 1998.

———. "Theodicy from Ivan Karamazov to Moltmann." *Modern Theology* 4 (1987): 83-97.

Beebe, James R. "Logical Problem of Evil." In *Internet Encyclopedia of Philosophy*, edited by James Fieser and Bradley Dowden. www.iep.utm.edu/evil-log/. Accessed July 2, 2014.

Behr, John. *Formation of Christian Theology.* 2 vols. Crestwood, NY: St. Vladimir's Seminary Press, 2001.

Berg, Jacob Albert van den. *Biblical Argument in Manichean Missionary Practice.* Leiden: Brill, 2010.

Bernard of Clairvaux. "Grace and Free Choice." In *The Works of Bernard of Clairvaux*, 7:53-111. Translated by Daniel O'Donovan. Kalamazoo, MI: Cistercian Publications, 1977.

Betz, Hans Dieter. *Paul's Concept of Freedom in the Context of Hellenistic Discussions About Possibilities of Human Freedom.* Protocol Series of the Colloquies of the Center for Hermeneutical Studies in Hellenism and Modern Culture 26. Berkeley, CA: The Center, 1977.

Blackwell, Richard J. *Galileo, Bellarmine, and the Bible.* Notre Dame, IN: University of Notre Dame Press, 1991.

Blowers, Paul. *The Drama of the Divine Economy: Creator and Creation in Early Christian Theology and Piety.* New York: Oxford University Press, 2012.

Boethius. *The Consolation of Philosophy.* Translated by W. V. Cooper. London: J. M. Dent, 1902.

Botterweck, G. Johannes, and Helmer Ringgren, eds. *Theological Dictionary of the Old Testament.* Translated by John T. Willis et al. 15 vols. Grand Rapids: Eerdmans, 1974–2006.

Bourke, Vernon J. *Will in Western Thought: An Historico-Critical Survey.* New York: Sheed & Ward, 1964.

Bouteneff, Peter C. *Beginnings: Ancient Christian Readings of the Biblical Creation Narratives.* Grand Rapids: Baker Academic, 2008.

Boyd, Gregory. *God at War: The Bible and Spiritual Conflict.* Downers Grove, IL: InterVarsity Press, 1997.

———. *The God of the Possible: A Biblical Introduction to the Open View of God.* Grand Rapids: Baker Books, 2000.

———. "Response to William Lane Craig." In *Four Views on Divine Providence*, edited by Dennis W. Jowers, pp. 123-39. Grand Rapids: Zondervan, 2011.

———. *Satan and the Problem of Evil: Constructing a Trinitarian Warfare Theodicy.* Downers Grove, IL: InterVarsity Press, 2001.

Boyer, C. B. *A History of Mathematics.* 2nd ed. New York: Wiley, 1968.

Broadie, Sarah. *Nature and Divinity in Plato's Timaeus.* Cambridge: Cambridge University Press, 2011.

Brom, Luco J. van den. "As Thy New Horizons Beckon: God's Presence in the World." In *Understanding the Attributes of God*, edited by Gijsbert van den Brink and Marcel Sarot, pp. 75-97. New York: Peter Lang, 1999.

———. *Divine Presence in the World: A Critical Analysis of the Notion of Divine Omnipresence.* Kampen: Kok Pharos, 1993.

Brower, Jeffrey. "Medieval Theories of Relation." In *Stanford Encyclopedia of*

Philosophy (spring 2014 ed.), edited by Edward N. Zalta, http://plato.stanford
.edu/entries/relations-medieval/.

Brown, Raymond. *The Gospel According to John 1–12*. Anchor Bible 29. Garden
City, NY: Doubleday, 1966.

Brown, William P. *The Seven Pillars of Creation: The Bible, Science and the
Ecology of Wonder*. New York: Oxford University Press, 2009.

Brueggemann, Walter. "The Loss and Recovery of Creation in Old Testament
Theology." *Theology Today* 53 (1996): 177-90.

Burnyeat, Miles F. "Eikôs Mythos." *Rhizai: A Journal for Ancient Philosophy and
Science* 2 (2005): 143-65.

Burrell, David. "The Act of Creation: Theological Consequences." In *Creation
and the God of Abraham*, edited by David Burrell, Carlo Cogliati, Janet Soskice
and William Stoeger, pp. 40-52. New York: Cambridge University Press, 2010.

———. *Aquinas: God and Action*. Notre Dame, IN: University of Notre Dame
Press, 1979.

———. "Creatio Ex Nihilo Recovered." *Modern Theology* 29 (2013): 6-21.

———. "Divine Practical Knowing: How an Eternal God Acts in Time." In *Divine
Action: Studies Inspired by the Philosophical Theology of Austin Farrar*, edited
by Brian Hebblethwaite and Edward Henderson, pp. 93-102. Edinburgh:
T & T Clark, 1990.

———. *Knowing the Unknowable God*. Notre Dame, IN: University of Notre
Dame Press, 1986.

Bussanich, John. "Plotinus's Metaphysics of the One." In *The Cambridge Companion to Plotinus*, edited by Lloyd P. Gerson, pp. 38-65. Cambridge: Cambridge University Press, 1996.

Calvin, John. *Calvin's Commentaries*. Edited by David W. Torrance and Thomas
F. Torrance. 22 vols. Grand Rapids: Eerdmans, 1960.

Camus, Albert. *The Rebel: An Essay on Man in Rebellion*. Translated by Anthony
Bower. New York: Vintage, 1991.

Cassirer, Ernst. *The Philosophy of the Enlightenment*. Translated by Fritz C. A.
Koelln and James P. Pettegrove. Princeton, NJ: Princeton University Press,
1951.

Childs, Brevard. *Biblical Theology of the Old and New Testaments: Theological
Reflection on the Christian Bible*. Minneapolis: Fortress, 1993.

Clarke, Norris. "A New Look at the Immutability of God." In *God Knowable and
Unknowable*, edited by Robert J. Roth, SJ, pp. 44-72. New York: Fordham
University Press, 1973.

Clifford, Richard J. "Creation in the Psalms." In *Creation Accounts in the Ancient Near East and in the Bible,* edited by Richard J. Clifford and John J. Collins, pp. 57-69. Catholic Biblical Quarterly Monograph Series. Washington DC: Catholic Biblical Association of America, 1994.

———. "The Hebrew Scriptures and the Theology of Creation." *Theological Studies* 46 (1985): 507-23.

Collingwood, R. G. *An Essay on Metaphysics.* Chicago: Regnery, 1972.

Copan, Paul, and William Lane Craig. *Creation Out of Nothing: A Biblical, Philosophical, and Scientific Exploration.* Grand Rapids: Baker Academic, 2004.

Copleston, Frederick. *A History of Philosophy.* 9 vols. Garden City: Image, 1962–1967.

Cowan, Stephen B. "The Grounding Objection to Middle Knowledge Revisited." *Religious Studies* 39 (2003): 93-102.

Cox, Ronald. *By That Same Word: Creation and Salvation in Hellenistic Judaism and Early Christianity.* Berlin: de Gruyter, 2007.

Coyle, James Kevin. *Manichaeism and Its Legacy.* Leiden: Brill, 2009.

Craig, Edward, ed. *Routledge Encyclopedia of Philosophy.* London: Routledge, 1988.

Craig, William Lane. *Divine Foreknowledge and Human Freedom: The Coherence of Theism: Omniscience.* Leiden: Brill, 1990.

———. "God Directs All Things: On Behalf of a Molinist View of Providence." In *Four Views on Divine Providence,* edited by Dennis W. Jowers, pp. 79-100. Grand Rapids: Zondervan, 2011.

———. "Middle Knowledge: A Calvinist-Arminian Rapprochement." In *The Grace of God and the Will of Man,* edited by Clark Pinnock, pp. 141-64. Minneapolis: Bethany House, 1989.

———. "The Middle Knowledge View." In *Four Views on Divine Providence,* edited by Dennis W. Jowers, pp. 119-43. Grand Rapids: Zondervan, 2011.

———. *The Only Wise God: The Compatibility of Divine Foreknowledge and Human Freedom.* Grand Rapids: Baker Books, 1987.

———. *The Problem of Divine Foreknowledge and Human Freedom from Aristotle to Suárez.* Leiden: Brill, 1980.

———. "Response to Ron Highfield." In *Four Views on Divine Providence,* edited by Dennis W. Jowers, pp. 170-75. Grand Rapids: Zondervan, 2011.

———. *Time and Eternity: Exploring God's Relationship to Time.* Wheaton, IL: Crossway, 2001.

———. "Timelessness and Omnitemporality." In *God and Time: Four Views,* edited by Gregory E. Gannsle, pp. 129-60. Downers Grove, IL: InterVarsity Press, 2001.

Davies, Brian. "Classical Theism and the Doctrine of Divine Simplicity." In *Language, Meaning, and God: Essays in Honor of Herbert McCabe*, edited by Brian Davies, pp. 51-74. London: Cassell, 1987.

Dawkins, Richard. *The God Delusion*. Boston: Houghton Mifflin, 2006.

Dekker, Eef. "Was Arminius a Molinist?" *Sixteenth Century Journal* 27 (1996): 337-52.

Dembski, William. *Intelligent Design: The Bridge Between Science and Theology*. Downers Grove, IL: InterVarsity Press, 1999.

Denton, Michael J. *Nature's Destiny: How the Laws of Biology Reveal Purpose in the Universe*. New York: Free Press, 1998.

Dillon, John. *The Middle Platonists: 80 B.C. to 220 A.D.* Rev. ed. Ithaca, NY: Cornell University Press, 1996.

———. Introduction to *The Enneads*, by Plotinus. Translated by Stephen MacKenna. London: Penguin, 1991.

Dionysius the Areopagite on the Divine Names and the Mystical Theology. Translated by C. E. Rolt. Berwick, ME: IBIS Press, 2004.

Dodds, Michael J. "St. Thomas and the Motion of the Motionless God." *New Blackfriars* 68 (1987): 233-42.

———. *The Unchanging God of Love: Thomas Aquinas and Contemporary Theology on Divine Immutability*. 2nd ed. Washington, DC: Catholic University of America Press, 2008.

———. *Unlocking Divine Action: Contemporary Science and Thomas Aquinas*. Washington, DC: Catholic University of America Press, 2012.

Dombrowski, Daniel A. "Must a Perfect Being Be Immutable?" In *Hartshorne, Process Philosophy and Theology*, edited by Robert Kane and Stephen H. Phillip, pp. 91-111. Albany: State University of New York Press, 1989.

Doran, Christopher. "Intelligent Design: It's Just Too Good to Be True." *Theology and Science* 8 (2010): 223-37.

Dostoevsky, Fyodor. *The Brothers Karamazov: A Norton Critical Edition*. Translated by Constance Garnett. Edited and revised by Susan McReynolds Oddo and Ralph E. Matlaw. 2nd ed. New York: W. W. Norton, 2011.

Edwards, Jonathan. *The Works of Jonathan Edwards*. 2 vols. 1834. Reprint, Peabody, MA: Hendrickson, 1998.

Eichrodt, Walther. *Theology of the Old Testament*. 2 vols. Translated by J. A. Baker. Philadelphia: Westminster, 1967.

Emery, Gilles. "Trinity and Creation." In *The Theology of Saint Thomas*, edited by Rik van Nieuwenhove and Joseph Wawrykow, pp. 58-76. Notre Dame, IN: University of Notre Dame Press, 2010.

Eslick, Leonard J. "The Material Substrate in Plato." In *The Concept of Matter in Greek and Medieval Philosophy*, edited by Ernan McMullin, pp. 39-54. Notre Dame, IN: University of Notre Dame Press, 1963.

Fatoorchi, Pirooz. "Creation ex Nihilo and the Compatibility Question." In *Creation and the God of Abraham*, edited by David Burrell, Carlo Cogliati, Janet Soskice and William Stoeger, pp. 91-106. Cambridge: Cambridge University Press, 2010.

Fee, Gordon D. *Pauline Christology: An Exegetical-Theological Study.* Peabody, MA: Hendrickson, 2007.

Fergusson, David. *Creation.* Guides to Theology. Grand Rapids: Eerdmans, 2014.

FitzGerald, John J. "'Matter' in Nature and the Knowledge of Nature: Aristotle and the Aristotelian Tradition." In *The Concept of Matter in Greek and Medieval Philosophy*, edited by Ernan McMullin, pp. 59-78. Notre Dame, IN: University of Notre Dame Press, 1963.

Flint, Thomas P. *Divine Providence: The Molinist Account.* Ithaca, NY: Cornell University Press, 1998.

Fretheim, Terence. *God and World in the Old Testament: A Relational Theology of Creation.* Nashville: Abingdon, 2005.

Funkenstein, Amos. *Theology and the Scientific Imagination: From the Middle Ages to the Seventeenth Century.* Princeton, NJ: Princeton University Press, 1986.

Furley, David. *The Greek Cosmologists: The Formation of the Atomic Theory and Its Earliest Critics.* Cambridge: Cambridge University Press, 1987.

Galileo Galilei. *Discoveries and Opinions of Galileo.* Translated by Stillman Drake. Garden City, NY: Doubleday, 1957.

Gannsle, Gregory E., ed. *God and Time: Four Views.* Downers Grove, IL: InterVarsity Press, 2001.

Garrigou-Lagrange, Reginald. *Predestination: The Meaning of Predestination in Scripture and the Church.* Rockford, IL: Tan Books, 1998.

———. *Providence.* Translated by Dom Bede Rose, OSB. 1937. Reprint, Rockford, IL: Tan Books, 1998.

Gavrilyuk, Paul. "Creation in Early Christian Polemical Literature: Irenaeus Against the Gnostics and Athanasius Against the Arians." *Modern Theology* 29 (2013): 22-32.

————. *The Suffering of the Impassible God: The Dialectics of Patristic Thought.* Oxford: Oxford University Press, 2004.

Gillespie, Michael Allen. *The Theological Origins of Modernity.* Chicago: University of Chicago Press, 2008.

Gilson, Etienne. *The Christian Philosophy of Saint Augustine.* Translated by L. E. M. Lynch. New York: Octagon, 1988.

————. *The Philosophy of Thomas Aquinas.* Translated by Edward Bullough. Edited by G. A. Elrington. New York: Dorset, n.d.

Gould, Stephen Jay. "Nonoverlapping Magisteria." *Natural History* 106 (March 1997): 16-22.

Green, Joel B. *Practical Theological Interpretation: Engaging Biblical Texts for Faith and Formation.* Grand Rapids: Baker Academic, 2011.

Griffin, David Ray. *Evil Revisited: Responses and Reconsiderations.* Albany: State University of New York Press, 1991.

————. *God, Power and Evil.* Philadelphia: Westminster, 1976.

————. "Process Theology and the Christian Good News: A Response to Classical Free Will Theism." In *Searching for an Adequate God: A Dialogue Between Process and Free Will Theists,* edited by John B. Cobb Jr. and Clark H. Pinnock, pp. 1-38. Grand Rapids: Eerdmans, 2000.

Gunton, Colin. *The Triune Creator: A Historical and Systematic Study.* Grand Rapids: Eerdmans, 1998.

Hankinson, R. J. "Determinism and Indeterminism." In *The Cambridge History of Hellenistic Philosophy,* edited by Keimpe Algra, Jonathan Barnes, Jaap Mansfeld and Malcom Schofield, pp. 513-41. Cambridge: Cambridge University Press, 2005.

Hanratty, Gerald. "Divine Immutability and Impassibility Revisited." In *At the Heart of the Real,* edited by Fran O'Rourke, pp. 135-62. Dublin: Irish Academic Press, 1992.

Hartshorne, Charles. *The Divine Relativity.* New Haven, CT: Yale University Press, 1948.

Hasker, William. "An Adequate God." In *Searching for An Adequate God: A Dialogue Between Process and Free Will Theists,* edited by John B. Cobb Jr. and Clark H. Pinnock, pp. 215-45. Grand Rapids: Eerdmans, 2000.

————. *God, Time and Knowledge.* Ithaca, NY: Cornell University Press, 1998.

————. *The Triumph of God over Evil: Theodicy for a World of Suffering.* Downers Grove, IL: IVP Academic, 2008.

Hegel, Georg. *Hegel's Philosophy of Right.* Translated by T. M. Knox. New York: Oxford University Press, 1967.

Heil, John. "Relations." In *The Routledge Companion to Metaphysics*, edited by Robin Le Poidevin, Peter Simons, Andrew McGonigal and Ross P. Cameron, pp. 310-21. New York: Routledge, 2009.

Helm, Paul. *The Eternal God: A Study of God Without Time*. Oxford: Clarendon, 1988.

———. "God and Spacelessness." In *Contemporary Philosophy of Religion*, edited by Steven M. Cahn and David Shatz, pp. 99-110. Oxford: Oxford University Press, 1982.

———. *The Providence of God*. Downers Grove, IL: InterVarsity Press, 1993.

Hempel, Carl. *Aspects of Scientific Explanation and Other Essays in the Philosophy of Science*. New York: Free Press, 1965.

———. *Philosophy of Science*. Englewood Cliffs, NJ: Prentice-Hall, 1966.

Heppe, Heinrich. *Reformed Dogmatics: Set Out and Illustrated from the Sources*. Rev. and ed. E. Bizer. Translated by G. T. Thomson. 1950. Reprint, London: Wakeman Great Reprints, n.d.

Highfield, Ron. "Divine Self-Limitation in the Theology of Jürgen Moltmann: A Critical Appraisal." *Christian Scholar's Review* 32 (2002): 47-71.

———. "Does the World Limit God: Assessing the Case for Open Theism." *Stone-Campbell Journal* 5 (2002): 69-92.

———. "The Function of Divine Self-Limitation in Open Theism: Great Wall or Picket Fence?" *Journal of the Evangelical Theological Society* 45 (2002): 279-99.

———. "Galileo, Scientific Creationism, and Biblical Hermeneutics." *Restoration Quarterly* 36 (1994): 279-90.

———. *Great Is the Lord: Theology for the Praise of God*. Grand Rapids: Eerdmans, 2008.

———. "The Problem with the 'Problem of Evil': A Response to Gregory Boyd's Open Theist Solution." *Restoration Quarterly* 45 (2003): 165-80.

———. "Response to William Lane Craig." in *Four Views on Divine Providence*, edited by Dennis W. Jowers, pp. 114-22. Grand Rapids: Zondervan, 2011.

Holmes, Steven R. "Something Much Too Plain to Say: Towards a Defense of the Doctrine of Divine Simplicity." In *Listening to the Past: The Place of Tradition in Theology*, pp. 50-67. Grand Rapids: Baker Academic, 2002.

Hughes, Christopher. *On a Complex Theory of a Simple God: An Investigation in Aquinas' Philosophical Theology*. Ithaca, NY: Cornell University Press, 1989.

Hughes, John. "Creatio ex Nihilo and the Divine Ideas in Aquinas: How Fair Is Bulgakov's Critique?" *Modern Theology* 29 (2013): 124-37.

Hunt, David. "Divine Providence and Simple Foreknowledge." *Faith and Philosophy* 10 (1993): 394-416.

———. "The Simple Foreknowledge View." In *Divine Foreknowledge: Four Views*, edited by James K. Beilby and Paul R. Eddy, pp. 65-103. Downers Grove, IL: InterVarsity Press, 2001.

Hurtado, Larry. *Lord Jesus Christ: Devotion to Jesus in Earliest Christianity.* Grand Rapids: Eerdmans, 2003.

Ireneaus of Lyons. *Demonstration of the Apostolic Preaching.* Translated by Armitage Robinson. London: SPCK, 1920.

Jensen, Robert. *Systematic Theology.* 2 vols. Oxford: Oxford University Press, 1997–1999.

Johansen, Thomas Kjeller. *Plato's Natural Philosophy: A Study of the Timaeus-Critias.* Cambridge: Cambridge University Press, 2004.

Kant, Immanuel. *Critique of Pure Reason.* Translated by Norman Kemp Smith. New York: Palgrave, 1933.

Kennedy, Darren M. *Providence and Personalism: Karl Barth in Conversation with Austin Farrer, John Macmurray and Vincent Brümmer.* New York: Peter Lang, 2011.

Kierkegaard, Søren. *Christian Discourses.* Translated by Walter Lowrie. Princeton, NJ: Princeton University Press, 1971.

———. *Works of Love.* Translated by Howard Hong and Edna Hong. New York: HarperCollins, 2009.

Kittel, Gerhard, and Gerhard Friedrich, eds. *Theological Dictionary of the New Testament.* Translated by Geoffrey W. Bromiley. 10 vols. Grand Rapids: Eerdmans, 1964–1976.

Knuuttila, Simo. "Time and Creation in Augustine." In *The Cambridge Companion to Augustine*, edited by Eleonore Stump and Norman Kretzmann, pp. 103-15. Cambridge: Cambridge University Press, 2011.

Kondoleon, Theodore J. "The Immutability of God: Some Recent Challenges." *The New Scholasticism* 58 (1984): 293-315.

Kosso, Peter. *Reading the Book of Nature: An Introduction to the Philosophy of Science.* Cambridge: Cambridge University Press, 1992.

Kretzmann, Norman. *The Metaphysics of Creation: Aquinas's Natural Theology in Summa Contra Gentiles II.* New York: Oxford University Press, 1998.

Larrimore, Mark. "Evil." In *New Dictionary of the History of Ideas*, edited by Maryanne Cline Horowitz, pp. 2:744-50. Detroit: Charles Scribner's Sons, 2005.

———. *The Problem of Evil: A Reader.* Malden, MA: Blackwell, 2001.

Lindberg, David C. *The Beginnings of Western Science: The European Scientific Tradition in Philosophical, Religious, and Institutional Context, 600 B.C. to A.D. 1450.* Chicago: University of Chicago Press, 1992.

Luther's Works. American Edition. Edited by Jaroslav Pelikan and Helmut Lehmann. 55 vols. Philadelphia: Fortress; St. Louis: Concordia, 1955–1986.

MacCallum, Gerald C., Jr. "Negative and Positive Freedom." *Philosophical Review* 76 (1967): 312-34.

Mackie, John L. "Evil and Omnipotence." *Mind* 64 (1955): 200-212.

Macquarrie, John. *In Search of Deity: An Essay in Dialectical Theism.* New York: Crossroad, 1987.

———. *Twentieth Century Religious Thought: The Frontiers of Philosophy and Theology 1900–1980.* Rev. ed. New York: Charles Scribner's Sons, 1981.

Marcel, Gabriel. *The Mystery of Being: Reflection and Mystery.* London: Harvill, 1950.

Marcus Aurelius. *Marcus Aurelius and His Times.* Introduction by Irwin Edman. Roslyn, NY: Black, 1945.

Marquart, Odo. "Unburdenings: Theodicy Motives in Modern Philosophy." In *In Defense of the Accidental: Philosophical Studies,* pp. 8-28. Translated by Robert M. Wallace. New York: Oxford University Press, 1991.

May, Gerhard. *Creatio ex Nihilo: The Doctrine of "Creation Out of Nothing" in Early Christian Thought.* Translated by A. S. Worrall. London: T & T Clark, 2004.

McDonough, Sean M. *Christ as Creator: The Origins of a New Testament Doctrine.* New York: Oxford University Press, 2009.

McMullin, Ernan. "Darwin and the Other Christian Tradition." *Zygon* 46 (2011): 291-316.

———. "How Should Cosmology Relate to Theology?" In *The Sciences and Theology in the Twentieth Century,* edited by Arthur Peacocke, pp. 17-57. Notre Dame, IN: University of Notre Dame Press, 1981.

———. *The Inference That Makes Science.* Milwaukee: Marquette University Press, 1992.

———. "Natural Science and Belief in a Creator." In *Physics, Philosophy and Theology: A Common Quest for Understanding,* edited by Robert John Russell, William R. Stoeger and George V. Coyne, pp. 49-79. Vatican: Vatican Observatory, 1988.

Meadors, Gary, ed. *Four Views on Moving Beyond the Bible to Theology.* Grand Rapids: Zondervan, 2009.

Molina, Luis de. *On Divine Foreknowledge (Part IV of the Concordia)*. Translated by Alfred J. Freddoso. Ithaca, NY: Cornell University Press, 1988.

Moltmann, Jürgen. *The Crucified God: The Cross of Christ as the Foundation and Criticism of Christian Theology*. Translated by R. A. Wilson and John Bowden. New York: Harper & Row, 1974.

———. *God in Creation: A New Theology of Creation and the Spirit of God*. Translated by Margaret Kohl. Minneapolis: Fortress, 1993.

———. *The Trinity and the Kingdom*. Translated by Margaret Kohl. Minneapolis: Fortress, 1993.

Moo, Douglas J. *The Epistle to the Romans*. Grand Rapids: Eerdmans, 1996.

Moore, G. E. "External and Internal Relations." *Proceedings of the Aristotelian Society* 20 (1919–1920): 40-62.

Muller, Richard A. *Post-Reformation Reformed Dogmatics*. 4 vols. Grand Rapids: Baker Academic, 2003.

Nightingale, Andrea. *Once out of Nature: Augustine on Time and the Body*. Chicago: University of Chicago Press, 2011.

O'Hanlon, Gerry, SJ. "Does God Change?—H. U. von Balthasar on the Immutability of God." *Irish Theological Quarterly* 53 (1987): 161-83.

O'Keefe, John J., and R. R. Reno. *Sanctified Vision: An Introduction to Early Christian Interpretation of the Bible*. Baltimore: Johns Hopkins University Press, 2005.

O'Meara, Dominic J. "The Hierarchical Ordering of Reality in Plotinus." *The Cambridge Companion to Plotinus*, edited by Lloyd P. Gerson, pp. 66-81. Cambridge: Cambridge University Press, 1996.

Ogden, Schubert. *The Reality of God*. New York: Harper & Row, 1963.

Olbricht, Thomas H. *He Loves Forever: The Message of the Old Testament*. Joplin, MO: College Press, 2000.

Osborn, Eric. "Origen: The Twentieth Century Quarrel and Its Recovery." In *Origeniana Ouinta*, edited by R. Daly, pp. 26-39. Leuven: Leuven University Press, 1992.

———. *Irenaeus of Lyon*. Cambridge: Cambridge University Press, 2001.

Owens, Joseph. "Matter and Predication in Aristotle." In *The Concept of Matter in Greek and Medieval Philosophy*, edited by Ernan McMullin, pp. 70-93. Notre Dame, IN: University of Notre Dame Press, 1963.

Padgett, Alan. "Eternity as Relative Timelessness." In *God and Time: Four Views*, edited by Gregory E. Gannsle, pp. 92-110. Downers Grove, IL: InterVarsity Press, 2001.

————. *God, Eternity and the Nature of Time.* New York: St. Martin's, 1992.

Pailin, David A. "The Utterly Absolute and the Totally Related: Change in God." *New Blackfriars* 68 (1987): 243-55.

Pambrun, James R. "Creatio ex Nihilo and Dual Causality." In *Creation and the God of Abraham*, edited by David Burrell, Carlo Cogliati, Janet Soskice and William Stoeger, pp. 192-220. Cambridge: Cambridge University Press, 2010.

Pannenberg, Wolfhart. *Human Nature, Election, and History.* Philadelphia: Westminster, 1977.

————. *Systematic Theology.* 3 vols. Translated by Geoffrey W. Bromiley. Grand Rapids: Eerdmans, 1991–1998.

Pascal, Blaise. *Pensées.* Translated by A. J. Krailscheimer. London: Penguin, 1966.

Pelikan, Jaroslav, and Valerie Hotchkiss. *Creeds and Confessions of Faith in the Christian Tradition.* 3 vols. New Haven, CT: Yale University Press, 2003.

Pennington, Jonathan T., and Sean M. McDonough, eds. *Cosmology and New Testament Theology.* New York: T & T Clark, 2008.

Pictet, Benedict. *Christian Theology.* Translated by Frederick Reyroux. London: Seeley and Burnside, 1834.

Pinnock, Clark H. "God Limits His Knowledge." In *Predestination and Free will: Four Views of Divine Sovereignty and Human Freedom*, edited by David Basinger and Randall Basinger, pp. 141-62. Downers Grove, IL: InterVarsity Press, 1986.

————. *Most Moved Mover: A Theology of God's Openness.* Grand Rapids: Baker Academic, 2001.

Plantinga, Alvin. *Does God Have a Nature?* Milwaukee: Marquette University Press, 1980.

————. *The Nature of Necessity.* Oxford: Oxford University Press, 1974.

Plato. *The Complete Works of Plato.* Edited by John M. Cooper. Indianapolis: Hackett, 1997.

————. *Timaeus.* Translated by Donald J. Zeyl. Indianapolis: Hackett, 2000.

Plotinus. *The Enneads.* Translated by Stephen Mackenna. Abridged by John Dillon. London: Penguin, 1991.

Polkinghorne, John. "Divine Action in a World of Chaos." *Faith and Philosophy* 14 (1997): 41-61.

————. *The Faith of a Physicist: Reflections of a Bottom Up Thinker.* Minneapolis: Fortress, 1996.

————. "Kenotic Creation and Divine Action." In *The Work of Love: Creation as Kenosis*, edited by John Polkinghorne, pp. 90-106. Grand Rapids: Eerdmans, 2001.

————. "The Metaphysics of Divine Action." In *Chaos and Complexity: Scientific*

Perspectives on Divine Action, edited by Robert John Russell, Nancey Murphy and Arthur R. Peacocke, pp. 147-56. Vatican City and Berkeley, CA: Vatican Observatory Publications and The Center for Theology and the Natural Sciences, 1997.

———. *Science and Providence: God's Interaction with the World.* Boston: Shambhala, 1989.

Pollard, T. E. "The Impassibility of God." *Scottish Journal of Theology* 8 (1955): 353-64.

Pope, Alexander. "An Essay on Man." In *An Essay on Man: Moral Essays and Satires.* London: Cassell, 1891. www.gutenberg.org/files/2428/2428-h/2428-h.htm.

Pope, Marvin. *Job: A New Translation with Introduction and Commentary.* Anchor Bible 15. Garden City, NY: Doubleday, 1973.

Pratt, Douglas. *Relational Deity: Hartshorne and Macquarrie on God.* Lanham, MD: University Press of America, 2002.

Prenter, Regin. *Creation and Redemption.* Translated by Theodor I. Jensen. Philadelphia: Fortress, 1967.

Preus, Robert D. *The Theology of Post-Reformation Lutheranism.* 2 vols. Saint Louis: Concordia, 1970–1972.

Rahner, Karl, ed. *Encyclopedia of Theology: The Concise Sacramentum Mundi.* New York: Crossroad, 1975.

———. *The Trinity.* Translated by Joseph Donceel. London: Burns & Oates, 1970.

Ricoeur, Paul. "Evil: A Challenge to Philosophy and Theology." *Journal of the American Academy of Religion* 52 (1985): 635-48.

———. *Fallible Man: Philosophy of the Will.* Translated by Charles Kelbley. Chicago: Regnery, 1965.

Ridgley, Thomas. *Body of Divinity.* Edited by Rev. John M. Wilson. New York: Robert Carter & Brothers, 1855.

Robinette, Brian. "The Difference Nothing Makes: *Creatio ex Nihilo*, Resurrection, and Divine Gratuity." *Theological Studies* 72 (2011): 525-57.

Robson, Mark Ian Thomas. *Ontology and Providence in Creation: Taking ex Nihilo Seriously.* London: Continuum, 2008.

Rogers, Katherin A. "Eternity Has No Duration." *Religious Studies* 30 (1994): 1-16.

———. *Perfect Being Theology.* Edinburgh: Edinburgh University Press, 2002.

———. "The Traditional Doctrine of Divine Simplicity." *Religious Studies* 32 (1996): 165-86.

Rorty, Richard. "Relations, Internal and External." In *The Encyclopedia of Philosophy*, edited by Paul Edwards, 7:125-33. New York: Macmillan, 1967.

Roth, John K. "A Theodicy of Protest." In *Encountering Evil: Live Options in*

Theology, edited by Stephen T. Davis, pp. 1-20. Louisville, KY: Westminster John Knox, 2001.

Rowe, William. "The Problem of Evil and Some Varieties of Atheism." In *Contemporary Perspectives on Religious Epistemology*, edited by R. Douglas Geivett and Brendan Sweetman, pp. 33-42. New York: Oxford University Press, 1992.

Rubenstein, Richard. *After Auschwitz*. Indianapolis: Bobbs-Merrill, 1966.

Salmon, Wesley. "Statistical Explanation." In *Statistical Explanation and Statistical Relevance*, edited by Wesley Salmon, pp. 29-87. Pittsburgh: University of Pittsburgh Press, 1971.

————. *Explanation and the Causal Structure of the World*. Princeton, NJ: Princeton University Press, 1984.

Sanders, John. *The God Who Risks: A Theology of Providence*. 2nd ed. Downers Grove: IVP Academic, 2007.

————. "Why Simple Foreknowledge Offers No More Providential Control Than the Openness of God." *Faith and Philosophy* 14 (1997): 26-40.

Sarna, Nahum M. *Understanding Genesis: The Heritage of Biblical Israel*. New York: Schocken, 1970.

Schaeffer, Jonathan. "The Internal Relatedness of All Things." *Mind* 119 (2010): 341-76.

Scheffczyk, Leo. *Creation and Providence*. Translated by Richard Strachan. New York: Herder & Herder, 1970.

Schmid, Heinrich. *The Doctrinal Theology of the Evangelical Lutheran Church*. Edited and translated by C. A. Hay and H. E. Jacobs. 3rd ed. Minneapolis: Augsburg, 1961.

Schwarz, Hans. *Creation*. Grand Rapids: Eerdmans, 2002.

Scott, R. B. Y. *Proverbs, Ecclesiastes*. Anchor Bible 18. Garden City, NY: Doubleday, 1965.

Sedley, David. *Creationism and Its Critics in Antiquity*. Berkeley: University of California Press, 2007.

Sennett, James. "The Inscrutable Evil Defense Against the Inductive Argument from Evil." *Faith and Philosophy* 10 (1993): 220-29.

Sia, Santiago. "The Doctrine of God's Immutability: Introducing the Modern Debate." *New Blackfriars* 68 (1987): 220-31.

Simons, Eberhard. "Dualism." In *Encyclopedia of Theology: The Concise Sacramentum Mundi*, edited by Karl Rahner, pp. 370-74. New York: Crossroad, 1975.

Smedes, Taede A. *Chaos, Complexity, and God: Divine Action and Scientism*. Leuven: Peeters, 2004.

Smulders, Pieter. "Creation." In *Encyclopedia of Theology: The Concise Sacra-mentum Mundi*, edited by Karl Rahner, pp. 313-19. New York: Crossroad, 1975.

Sokolowski, Robert. "Creation and Christian Understanding." In *God and Creation: An Ecumenical Symposium*, edited by David B. Burrell and Bernard McGinn, pp. 179-92. Notre Dame, IN: University of Notre Dame Press, 1990.

———. *The God of Faith and Reason*. Washington, DC: Catholic University of America Press, 1995.

Soskice, Janet. "Creatio ex nihilo: Jewish and Christian Foundations." In *Creation and the God of Abraham*, edited by David Burrell, Carlo Cogliati, Janet Soskice and William Stoeger, pp. 24-39. Cambridge: Cambridge University Press, 2010.

Spiegel, James S. *The Benefits of Providence: A New Look at Divine Sovereignty*. Wheaton, IL: Crossway, 2005.

Staniloae, Dumitru. *The Experience of God*. 2 vols. Translated and edited by Ioan Ionita and Robert Barringer. Brookline, MA: Holy Cross Orthodox Press, 1994–2000.

Stead, Christopher. "Divine Simplicity as a Problem for Orthodoxy." In *The Making of Orthodoxy: Essays in Honour of Henry Chadwick*, edited by Rowan Williams, pp. 255-69. Cambridge: Cambridge University Press, 1989.

Stoeger, William R. "The Big Bang, Quantum Cosmology and Creatio ex Nihilo." In *Creation and the God of Abraham*, edited by David Burrell, Carlo Cogliati, Janet Soskice and William Stoeger, pp. 152-75. Cambridge: Cambridge University Press, 2010.

———. "Contemporary Cosmology and Its Implications for the Science-Religion Dialogue." In *Physics, Philosophy and Theology: A Common Quest for Understanding*, edited by Robert John Russell, William R. Stoeger and George V. Coyne, pp. 219-47. Vatican City: Vatican Observatory, 1988.

———. "Key Developments in Physics Challenging Philosophy and Theology." In *Religion and Science: History, Method, Dialogue*, edited by W. M. Richardson and W. J. Wildman, pp. 183-200. London: Routledge, 1996.

Stump, Eleonore. "Providence and Evil." In *Christian Philosophy*, edited by Thomas Flint, pp. 51-91. Notre Dame, IN: University of Notre Dame Press, 1990.

Stump, Eleonore, and Norman Kretzmann. "Being and Goodness." In *Divine and Human Action: Essays in the Metaphysics of Theism*, edited by Thomas V. Morris, pp. 281-312. Ithaca, NY: Cornell University Press, 1988.

———. "Eternity." *Journal of Philosophy* 79 (1981): 429-58.

Surin, Kenneth. *Theology and the Problem of Evil.* Oxford: Blackwell, 1986.

Swinburne, Richard. *The Providence of God.* 2nd ed. Oxford: Oxford University Press, 1993.

Tanner, Kathryn. "*Creatio ex Nihilo* as Mixed Metaphor." *Modern Theology* 29 (2013): 138-55.

———. *God and Creation in Christian Theology.* Minneapolis: Fortress, 2005.

Thomas Aquinas. *Basic Writings of Saint Thomas Aquinas.* 2 vols. Edited by Anton C. Pegis. New York: Random House, 1945.

———. *On the Eternity of the World.* Translated by Robert T. Miller. *Internet Medieval Sourcebook.* 1997. www.fordham.edu/halsall/basis/aquinas-eternity.html.

———. *Selected Philosophical Writings.* Edited and translated by Timothy Mc-Dermott. Oxford World's Classics. New York: Oxford University Press, 1993.

———. *Summa Contra Gentiles, Book Two: Creation.* Translated by James F. Anderson. Notre Dame, IN: University of Notre Dame Press, 1975.

Tiessen, Terrance. *Providence and Prayer: How Does God Work in the World?* Downers Grove, IL: InterVarsity Press, 2000.

Torrance, Thomas F. *The Christian Doctrine of God: One Being, Three Persons.* Edinburgh: T & T Clark, 1996.

———. "God and the Contingent World." *Zygon* 14 (1979): 329-48.

Tracy, Thomas F. "God and Creatures Acting: The Idea of Double Agency." In *Creation and the God of Abraham*, edited by David Burrell, Carlo Cogliati, Janet Soskice and William Stoeger, pp. 221-37. Cambridge: Cambridge University Press, 2010.

———. "Narrative Theology and the Acts of God." In *Divine Action: Studies Inspired by the Philosophical Theology of Austin Farrer*, edited by Brian Hebblethwaite and Edward Henderson, pp. 173-210. Edinburgh: T & T Clark, 1990.

Tur-Sinai, N. H. *The Book of Job: A New Commentary.* Jerusalem: Kiryath Sepher, 1967.

Veldhuijsen, P. Van. "The Question on the Possibility of an Eternally Created World: Bonaventura and Thomas Aquinas." In *The Eternity of the World in the Thought of Thomas Aquinas and His Contemporaries*, edited by J. B. M. Wissink, pp. 20-38. Leiden: Brill, 1990.

Voltaire, *Candide.* New York: Boni and Liveright, 1918.

von Rad, Gerhard. *Old Testament Theology.* 2 vols. Translated by D. M. G. Stalker. New York: Harper & Row, 1962.

———. *Genesis: A Commentary.* Translated by John H. Marks. Old Testament Library. London: SCM, 1961.

Walton, John H. *The Lost World of Genesis One: Ancient Cosmology and the Origins Debate.* Downers Grove, IL: IVP Academic, 2009.

Weber, Otto. *Foundations of Dogmatics.* Translated by Darrell L. Guder. 2 vols. Grand Rapids: Eerdmans, 1981.

Webster, John. "Love Is Also a Lover of Life: Creatio ex Nihilo and Creaturely Goodness." *Modern Theology* 29 (2013): 156-71.

Weinandy, Thomas. "Aquinas: God IS Man: The Marvel of the Incarnation." In *Aquinas on Doctrine: A Critical Introduction*, edited by Thomas Weinandy, Daniel Keating and John Yocum, pp. 67-89. London: T & T Clark, 2004.

———. *Does God Change? The Word's Becoming in the Incarnation.* Still River, MA: St. Bede's, 1985.

Welker, Michael. *Creation and Reality.* Translated by John E. Hoffmeyer. Minneapolis: Fortress, 1999.

Westermann, Claus. *Creation.* Philadelphia: Fortress, 1974.

———. *Isaiah 40–66: A Commentary.* Translated by David M. G. Stalker. Philadelphia: Westminster, 1969.

Wheelwright, Philip, ed. *The Presocratics.* New York: Odyssey, 1966.

Whitehead, Alfred North. *Process and Reality: An Essay in Cosmology.* New York: Macmillan, 1929.

Wilson, Jonathan R. *God's Good World: Reclaiming the Doctrine of Creation.* Grand Rapids: Baker Academic, 2013.

Wittgenstein, Ludwig. *Tractatus Logico-Philosophicus.* Translated by D. F. Pears and B. F. McGuinness. London: Routledge & Kegan Paul, 1961.

Wolf, Susan. *Freedom Within Reason.* Oxford: Oxford University Press, 1990.

Wolterstorff, Nicholas. "God Everlasting." In *Contemporary Philosophy of Religion*, edited by S. Cahn and D. Shatz, pp. 77-98. Oxford: Oxford University Press, 1982.

———. "Unqualified Divine Temporality." In *Four Views: God and Time*, edited by Gregory E. Ganssle, pp. 186-213. Downers Grove, IL: InterVarsity Press, 2001.

Wood, Charles M. *The Question of Providence.* Louisville, KY: Westminster John Knox, 2008.

Woodward, James. "Scientific Explanation." In *Stanford Encyclopedia of Philosophy* (Winter 2014 ed.), edited by Edward N. Zalta, http://plato.stanford .edu/entries/scientific-explanation/.

Wright, John H. *Divine Providence in the Bible: Meeting the Living and True God.* 2 vols. Mahwah, NJ: Paulist Press, 2009.

Young, Davis A. "The Contemporary Relevance of Augustine's View of Creation." *Perspectives on Science and Christian Faith* 40 (1988): 42-45.

Author Index

Adler, Mortimer, 275-76, 279
Alston, William, 81
Anselm of Canterbury, 62, 240, 265, 284, 305
Antiochus of Ascalon, 87
Aquinas, Thomas, 17, 19, 60, 67, 90, 103, 107-8, 155, 169, 177-80, 240-42, 247, 284, 286-87, 289, 290, 294-97
Aristotle, 58, 64, 80, 87, 96-97, 100, 131, 140-41, 145-47, 150-51, 177, 274, 316
Arnold, Matthew, 143
Athanasius, 78
Augustine of Hippo, 17, 20, 60, 65, 78, 111, 125, 135-37, 168-69, 175-77, 180, 208, 235, 240, 280, 282, 304-5, 311-12, 324-25
Barbour, Ian, 17, 152-54
Barth, Karl, 16, 19, 75, 90, 110, 120-22, 128, 129, 183, 240, 247-56, 312-13, 318, 319
Basil of Caesarea, 20, 305
Bauckham, Richard, 44, 277
Bellarmine, Robert, 17, 169-71
Bernard of Clairvaux, 19, 240, 282-84, 286, 289, 290-91, 293, 297
Betz, Hans Dieter, 278
Blowers, Paul, 116
Boethius, 63-64, 179, 182, 266
Boyd, Gregory, 19, 258-66, 268-72, 285
Bradley, F. H., 101
Braunius, Johannes, 243, 245
Broadie, Sarah, 86, 87
Brom, Luco van den, 127
Brown, William, 30
Bucanus, Gulielmus, 247
Calvin, John, 169, 240, 243
Camus, Albert, 348-50, 352, 355, 356, 357
Childs, Brevard, 31
Chrysostom, John, 312
Clement of Alexandria, 118

Clifford, Richard J., 31
Collingwood, R. G., 96
Copan, Paul, 137, 156-57
Copleston, Frederick, 136, 163
Cox, Ronald, 51
Craig, William Lane, 19, 137-39, 156-57, 230-39, 285
Crain, Stephen Dale, 92
Davies, Brian, 61
Dawkins, Richard, 17, 160-61, 164
Dembski, William A., 17, 161-64
Dillon, John, 88
Dionysius the Areopagite, 61-62, 67
Dodds, Michael, 81, 92, 97-99
Dostoevsky, Fyodor, 20, 339, 343, 346, 348, 365
Edwards, Jonathan, 19, 240, 295-96
Eudorus of Alexandria, 87
Farrer, Austin, 89-92
Fergusson, David, 73
Fichte, J. G., 105, 163-64
Fretheim, Terence, 36, 37, 103-4
Galileo Galilei, 17, 97, 147-48, 170, 171
Gould, Stephen Jay, 17, 159-60
Gregory of Nyssa, 60
Griffin, David, 102-3
Gunton, Colin, 120, 136
Hartshorne, Charles, 68-69, 99, 102
Hasker, William, 20, 103, 228, 329-30
Hegel, Georg Wilhelm Friedrich, 101, 288-89, 325
Heidegger, Johannes, 243
Helm, Paul, 19, 295-96
Hempel, Carl, 149
Heppe, Heinrich, 243
Heraclitus, 84, 85, 114, 145
Hunt, David, 228
Irenaeus of Lyons, 115-18, 120,

128, 132-34, 201
Jenson, Robert, 16, 70, 126, 128, 130, 200, 201, 202
Jeremiah the prophet, 26, 34, 189-90
John the Evangelist, 43-47, 109, 110, 129, 277, 278, 289
John of Damascus, 60
Kant, Immanuel, 167, 325
Kierkegaard, Søren, 362
Kretzmann, Norman, 180
Laplace, Pierre, 91-92
Larrimore, Mark, 323-24, 325
Leibniz, Wilhelm Gottfried, 325, 333-35
Lemaître, Georges, 155-56
Mackie, J. L., 20, 325-26
Marcel, Gabriel, 74
Marcus Aurelius, 322
Mastricht, Petrus Van, 245
Matthew the evangelist, 21, 361-62
May, Gerhard, 131-34
McDonough, Sean, 49, 52-54
McMullin, Ernan, 150, 154
Molina, Luis de, 138, 228-29
Moltmann, Jürgen, 16, 20, 122-24, 128, 352-59
Moore, G. E., 101, 102
Musculus, Wolfgang, 72
Newton, Isaac, 97
Origen, 16, 115, 118-20, 125, 128
Pannenberg, Wolfhart, 16, 109, 124-25, 128, 278
Parmenides, 84, 131, 145
Pascal, Blaise, 308
Paul the apostle, 14, 21, 40, 41-43, 48-53, 109, 111-12, 129, 130, 134, 170, 192-93, 200, 210, 214-15, 219, 223, 277-78, 289, 291, 310-13, 363-65
Peirce, Charles, 150
Peter the apostle, 14-15, 40, 41, 192, 278
Philo of Alexandria, 113
Pictet, Benedict, 62, 269

Pinnock, Clark, 69
Plantinga, Alvin, 20, 326-28
Plato, 16, 64, 85-87, 101, 134, 140, 145, 303
Plotinus, 58, 88-89, 101
Polanus, Amandus, 243, 246
Polkinghorne, John, 16, 89-94
Pope, Alexander, 333-34
Prenter, Regin, 128
Robson, Mark Ian Thomas, 139
Roth, John K., 20, 350-52
Rowe, William, 20, 328, 330
Rozanov, Vasily, 348
Rubenstein, Richard, 352
Russell, Bertrand, 101

Sanders, John, 103
Scheffczyk, Leo, 115
Scott, R. Y. B., 33
Sennett, James, 328
Sextus Empiricus, 323
Smedes, Taede A., 93
Soskice, Janet, 157
Spiegel, James S., 207, 286, 365
Stoeger, William, 154, 156, 157
Suarez, Francisco, 169
Tanner, Kathryn, 59, 81-82, 181, 283, 284
Tertullian, 132, 324
Thales, 145
Timaeus, 85-86
Torrance, Thomas, 70, 71, 110

Ursinius, Zacharias, 245
von Rad, Gerhard, 26, 30, 35
Voltaire, 20, 325, 334-36
Walton, John, 39
Weinandy, Thomas, 107
Westermann, Claus, 34, 35
Whitehead, Alfred North, 101, 103
Wiesel, Elie, 352
Wittgenstein, Ludwig, 167
Wolf, Susan, 286
Wood, Charles M., 210, 212, 213, 215
Wright, John H., 188, 191, 196-97, 198

Subject Index

Abram (Abraham), 170,
187-88, 197, 199, 310
Adam, 43, 118, 119, 195, 199,
280, 281, 308, 310-14
angels, 19, 41, 53, 54, 117, 132,
176, 191, 199, 224, 236-37, 302
delegation to, 18, 210, 213-14
fallen, 304, 309
anxiety, 13, 18, 184, 192, 223,
251, 267, 320, 360, 362, 365
Apologists, 115
Apostolic Fathers, 115
atheism, 59, 97, 147, 164-66,
323-25, 327-28, 334, 348, 353,
355
protest, 20, 349, 350, 352,
353, 355-59, 364
Brothers Karamazov, The
(Dostoevsky), 20, 339-50
Candide (Voltaire), 20, 335-39
causality, 16, 74, 76, 92, 93,
95-100, 127, 146, 228, 233,
241-42, 249-50, 294-95
Christian pantheism, 123
Christian theism, 123
concurrence, 215-16, 222,
243-44, 248
contingency, 18, 145, 161, 210, 214,
218-23, 233, 241, 244-45, 257
corruption, 20, 277, 280, 290,
301, 302, 306-9, 314-16,
319-21, 323
cosmogony, 144
cosmology, 17, 51, 53, 54, 98,
115, 144, 152, 170-71
big bang, 17, 155-57, 159, 162
Copernican, 170-71
empirical-mechanical,
144, 147
mythological, 144-45
rational-esthetic, 145-46
Council of Trent, 170, 313
covenant, 14, 56, 103, 187-88,
190, 197-98, 248, 309
creatio ex nihilo, 80, 131, 142,
157, 220, 233

biblical meaning of,
134-35
development of, 131-34
creation
and causation, 95-99
and Christ, 44, 115, 120-22,
127-30
in Colossians, 51-53
in Corinthians, 47-51
as divine action, 75-81
in Genesis, 26-30, 143-44
God's motivation for,
108-12
in Hebrews, 53-55
in Job, 30-31
in John, 45-47
noetic basis of, 120, 121, 125
from nothing (*see creatio
ex nihilo*)
ontic basis of, 120, 121, 125
in Proverbs, 32-33
in Psalms, 31, 173, 204-5
and relationality, 99-104
the Trinity and, 69-72, 122,
126, 181
Creator-creature relation, 17,
36-38, 41, 42, 74, 82, 104, 106,
133, 142, 151-52, 210-12, 215,
220, 222, 224, 296, 304
deductive-nomological (DN)
model, 149-50
determinism, 15, 91-92, 213,
229, 232, 238, 282
psychological, 242, 285,
295-97
directing, 216-18, 220-22, 227,
294
divine perfection, 58, 61, 66,
86-89, 136, 146, 207, 208,
265-66, 270
divine plan, 18, 193, 195-96,
209, 222-24, 227, 243-45, 271
divine providence, 17-19, 82,
91, 93, 294, 296, 311, 322-23,
333, 363-64
biblical teaching on,

194-208, 209
doctrine of, 210-25
foreknowledge model of,
19, 137-38, 227-39, 240
in the New Testament,
191-93
in the Old Testament,
187-91
omnipotence model of, 19,
240-56, 257, 274, 294
DNA, 158, 161
dramatis dei personae, 126
dualism, 131, 164, 302-5
creational, 303
metaphysical, 19-20,
303-4, 323
moral, 303-4, 323
radical, 303, 324
empirical science, 92, 148, 159,
166, 168
Epicureanism, 323-24
error. *See* fallibility
eternity, in the doctrine of
God, 16, 59, 62-64, 69,
265-66
Eve, 118, 308, 310
evil, 15, 18, 19-20, 252-54, 258-60,
264-65, 301-2, 322, 332
ancient problem of, 322-25
definition of, 316-19
dualism of good and, 104,
119, 183, 281, 302-5, 316,
337, 341, 343
evidential problem of,
328-31
logical problem of, 325-28
modern problem of, 325
sin and (*see* sin: and evil)
evolutionary biology, 17, 143,
152, 157-60, 210
fall, the, 20, 30, 62, 108, 119,
128, 187, 280-82, 284, 291-92,
308-11, 313, 315, 325
fallibility, 20, 285, 292-94,
306-9, 311, 314
fatalism, 213-14

fate, 18, 43, 48, 210, 213-15, 235, 245, 313, 322, 334, 362

feasible worlds, 224, 230-31, 234-36, 239

First Principles (Origen), 118, 119-20

flood, 29, 30, 187, 199, 310, 344

foreknowledge. *See* divine providence: foreknowledge model of

free will, 15, 19, 117, 246, 258, 283, 285, 286, 288, 290, 291, 313

freedom
  biblical idea of, 275-76
  eschatological, 276-80, 293, 297, 298
  as gift from God, 297-98
  incompatibilist libertarian, 238, 284-89
  and moral responsibility, 276, 280-89
  as noncompulsion, 282
  and providence, 294-98
  and repentance and forgiveness, 289-93

future, 13, 19, 64-64, 176, 177, 183-84, 196, 221, 227, 261-64, 267-70, 278, 290, 360, 362

Gnostics, 58, 65-66, 104, 115, 116, 118, 119, 128, 132, 165, 201-3, 252, 303-4

God-creature relation. *See* Creator-creature relation

governance, 215-16, 218-19, 244

Hellenistic Judaism, 48, 50, 51, 133

human beings, image and likeness to God, 29, 38-39, 43, 114, 117, 132, 152, 324, 344

immutability, in the doctrine of God, 16, 59, 64-66, 69, 265-66

infeasible worlds, 235-39

intelligent design theory, 161-62, 164

Israel, 14, 25, 29, 31, 33, 35, 46, 49, 56, 57, 75, 82, 122, 188-89, 192-93, 195, 198-99, 265, 361

knowledge
  free, 229, 232
  middle, 19, 138-39, 228-37, 271, 274

natural, 229-30, 232, 235

literal vs. figurative, 202-6, 208

Logos, 45, 51, 114-15, 123, 128

Manichaeism, 65, 128, 168, 202, 252, 303-4, 324, 337

Marcionism, 65, 201, 324

matter, 42, 60, 80, 85, 87, 88, 98, 117, 132, 134, 135, 140-42, 145, 164, 165, 217, 303, 335
  unformed, 114, 133, 135, 138, 175, 303

mediation, 16, 61, 63, 71, 76, 79, 84-85, 87-89, 96, 100, 104-6, 113-15, 125, 126, 150, 175, 234

metaphysics, 19-20, 60, 71, 92, 104, 118, 142, 144-45, 148, 151, 153, 154, 157, 164-68, 203, 252, 302-4

Middle Platonism, 50, 87-88, 113, 114

miracles, 18, 44, 189, 191, 194, 209

Molinism, 138, 139, 228-39

Moses, 53, 112, 188, 198, 199, 277

necessity, 18, 145, 163, 210, 214, 218-20, 223, 241, 244-45, 257, 282-84, 288-93

Neo-Platonism, 87, 88

nonoverlapping magisteria (NOMA), 159-61

nothingness, 81, 184, 216, 249, 252-56, 308, 315, 320

omnipotence, in the doctrine of God, 16, 59, 66-69, 265-66, 269

open theism, 15, 16, 19, 69, 103, 108, 232, 257-58, 326, 364
  and confidence in God's care, 270-73
  and divine perfection, 265-68
  Gregory Boyd on, 257, 261-62, 264-65
  and omniscience, 268-70

ordering, 44, 216-18, 220-22, 227, 244, 246, 294

paradise, 309-10, 343, 345

physics, 16, 83, 87, 98, 141, 146, 148-50, 156, 157

protest atheism. *See* atheism: protest

Protestantism, 60, 63, 169, 170, 243, 313, 314
  Orthodoxy, 90, 96
  Reformed, 19, 72

Pyrrhonism, 323-24

recapitulation, 117

Reformed Orthodox, 118, 211, 242-47, 314

Roman Catholicism, 60, 63, 132, 169, 170, 179, 180, 202, 241, 313, 314

salvation, 115-20, 128-30, 252, 272, 278
  creation and, 33-35, 108-11, 353-55
  economy of, 41, 57, 105, 120, 122, 124, 125, 127-29, 181-82, 201
  in the New Testament, 41-43, 45, 48, 50-51, 53, 55
  the Trinity and, 71-72

simplicity, in the doctrine of God, 16, 59-62, 63, 69, 136, 265-66, 269

sin, 306-8
  and evil, 15, 20, 195, 252-54, 305-6, 308, 316-22
  original, 20, 292, 308-14, 325, 335 (*see also* fall)

Spirit, 70-72, 122-24, 126-29, 179, 182-83, 215, 251, 277-78, 354, 363-65

spiritual warfare, 259, 264

Stoicism, 85, 87, 113-14, 176, 213, 322-24

sustenance, 215-16, 222, 315

theodicy, 318, 325, 350-52

Theodicy (Leibniz), 333

theology of nature, 152-53

time, 173-74
  Aquinas on, 177-81
  Augustine on, 175-77
  and Christ, 183-84

transcendence, 15, 58-59, 61, 71, 87, 88, 99, 107, 116, 172-74, 284

Trinity, 62, 63, 124, 182, 215, 353-54
  See also creation: the Trinity and

Word, 44, 45-47, 56, 75, 79, 94, 105, 117, 121-22, 128, 132, 179, 183, 216, 233, 251

Scripture Index

Old Testament

Genesis
1, *26, 42, 46, 54, 118,*
  169, 281
1–2, *26, 284, 309*
1–2:4, *26*
1–4, *119*
1–11, *310*
1:1, *16, 36, 46, 56, 57,*
  143, 173, 175, 187
1:1–2:3, *28*
1:2, *47, 56*
1:11, *169*
1:20, *169*
1:24, *169*
1:26, *118*
1:27, *43, 114*
1:28, *281*
1:31, *37*
2, *26, 281*
2:4, *26, 29, 30*
2:17, *281*
2:21-25, *309*
2:24, *43*
3, *187, 310*
3:1-7, *309*
3:1-24, *30*
3:8-9, *199*
6, *310*
6–9, *187*
6:1–8:22, *30*
6:5, *310*
6:6, *199*
11, *187*
11:5, *199*
12, *197*
12:2, *188*
15, *188, 197*
15:4, *197*
15:17, *197*
17, *197*
17:1-2, *197*
17:9-14, *198*
18:22-33, *199*
21:1-7, *188*
22:12, *199*

45, *199*
45:5-8, *188*

Exodus
2:1-10, *188*
20:1-17, *281*
20:2-3, *361*
20:5, *50*
24:3-4, *198*
32, *199*
34:10-28, *198*

Numbers
32:9-11, *199*

Deuteronomy
6:4, *49, 50*
6:4-5, *361*
6:15, *199*
7:9, *14*
28–29, *198*

Joshua
1–12, *189*

1 Samuel
15:35, *199*

2 Kings
20:5, *199*

1 Chronicles
16:30, *170*

Job
9:19, *190*
38, *30*
38–41, *26, 30, 333*
38:2-8, *30*
39:13-18, *35*
39:17, *35*
41, *30*
42:2, *190*

Psalms
8, *26, 190*

19, *26*
19:1, *191*
22:27-28, *191*
24, *26, 190*
24:1, *42*
33, *26*
33:9, *38*
33:10-11, *191*
44, *31*
46:1-2, *173*
47:3, *191*
74, *26*
74:13-16, *31*
77, *31*
82:8, *191*
89, *26, 31*
90, *26*
90:2-4, *173*
93, *26*
95:3-5, *191*
102, *26*
104, *26*
135:5-6, *191, 199*
136, *190*
139, *205*
148, *26, 190*

Proverbs
1:31-32, *32*
6:6-8, *35*
6:7-8, *35*
8, *114*
8:22-31, *26, 32*
8:35-36, *32*
16:33, *190*
19:21, *190*
21:1, *190*

Ecclesiastes
1:5, *170*
9:1-2, *190*

Isaiah
14:26-27, *189*
29:16, *41*
40, *33*

40–45, *26*
40–66, *33, 34*
40:9, *33*
40:10, *33*
40:12, *34*
40:26, *34*
41–48, *205*
46:10-11, *189, 199*

Jeremiah
1:5, *190*
3:7, *199*
10, *34*
10:11, *198*
10:12-16, *26, 34*
22:9, *198*
23:23-24, *205*
31:31-34, *190*
32:35, *199*

Ezekiel
16:59, *198*

Amos
5:4-6, *189*

Apocrypha

2 Maccabees
7:28, *133*

**Wisdom of
Solomon**
11:17, *133*

New Testament

Matthew
1–2, *191*
6:11, *42*
6:24, *361*
6:25, *13, 361*
6:25-34, *21, 192, 361*
6:31-32, *362*
6:34, *13*
10:30-31, *192*
11:25, *41*

12:32, *290*
16:16, *44*
19:4-6, *43*
25:31-46, *44*

MARK
1:1, *44*
1:3, *192*
3:9, *290*
7:14-23, *42, 43*
10:6, *40*
13, *40*
13:19, *40*
14:24, *198*

LUKE
11:14-26, *191*
12:7, *13, 14*
12:10, *290*
12:32, *14*
15, *43*
23:34, *291*
24:44-49, *192*

JOHN
1–12, *47*
1:1, *44, 46*
1:1-2, *46*
1:1-5, *16, 44, 45, 113,*
  114
1:3, *42, 46*
1:4, *47*
1:10, *16, 44*
1:10-14, *45, 113, 114*
1:13, *278*
1:14, *44, 45, 47*
1:18, *44, 46*
3:3-8, *278, 292*
3:16, *43, 109, 293*
4:24, *205*
5:22, *44*
8:34, *291*
8:34-36, *277*
9, *192*
9:3, *192*
16:27, *109*
17, *110*
17:5, *40, 46, 174*
17:6, *44*
17:11, *46*
17:21-24, *46*
17:24, *40, 110, 129, 174*

ACTS
2:23-24, *192*
2:36, *44*
4:12, *44*
5:31, *44*
8:4, *192*
8:9-40, *192*
9, *192*
9:15-16, *192*
10–11, *192*
17:24-25, *41*
17:24-28, *205*
17:28, *41*
22, *192*
24:23, *277*
26, *192*

ROMANS
1, *162*
1–5, *277*
1:16, *293*
1:20, *40*
2:7, *293*
3:16, *291*
3:20, *319*
3:23, *289*
4:17, *38, 130, 134*
4:25, *293*
5, *43, 363*
5:2, *363*
5:3-4, *363*
5:8, *43, 109*
5:12, *311, 312*
5:12-21, *43, 310*
5:18, *293*
5:19, *312*
5:20-21, *112*
6, *277*
6–7, *278*
6:4, *293*
6:5, *293*
6:6, *291*
6:6-7, *278*
6:7-18, *277*
6:20, *277*
6:22, *277*
7, *277*
7:7, *319*
7:8, *313*
7:11, *291*
7:13, *112*
7:14, *291*

7:24, *277*
8, *277, 294, 363*
8:1-4, *277*
8:2, *277*
8:9, *43*
8:12-27, *215*
8:15, *293*
8:17, *193, 364*
8:18, *21, 363, 364, 365*
8:19-21, *278*
8:20-21, *112*
8:21, *283, 293*
8:22-25, *278*
8:26, *124*
8:28, *18, 195, 199, 207,*
  209, 210, 215, 219,
  321
8:28-30, *21, 193, 363*
8:28-39, *214, 223*
8:31, *184, 364*
8:32, *214*
8:37, *363, 364*
8:39, *109, 215*
9–11, *193*
9:20, *41*
10:9, *44*
11:11-25, *193*
11:31-32, *193*
11:36, *41, 42, 53, 193,*
  199, 207
14:14, *42, 43*

1 CORINTHIANS
1:3, *50*
1:8-9, *14*
1:24, *44, 51*
2:2, *51*
3:1-23, *48*
4:1-21, *48*
4:10, *51*
5:1-13, *48*
6:11, *293*
6:12-20, *48*
7:1-40, *48*
8, *49*
8–10, *48, 50*
8:1, *51*
8:1-10:33, *48*
8:4-6, *16, 44, 45, 47,*
  49, 113, 114
8:6, *41, 42, 47, 50, 51, 53*
9, *278*

10, *49*
10:1-13, *199*
10:1-22, *49*
10:11-13, *193*
10:14-22, *50*
10:20, *49*
10:22, *50*
10:25-26, *42*
11:1-34, *48*
11:25, *198*
12:1–14:40, *48*
15, *42, 43*
15:1-20, *42*
15:1-58, *48*
15:8, *192*
15:28, *183, 302*
15:45-49, *43*
15:49, *43*

2 CORINTHIANS
1:9, *193*
1:20, *14*
4, *43*
4:4, *44, 52, 291*
4:6, *41*
5:10, *44*
5:14-15, *43, 129, 130*
5:16-18, *278*
5:17, *293*
5:18-20, *293*

GALATIANS
1:3, *50*
1:15-16, *192*
2:4, *277*
2:20, *44, 109*
3:24, *112, 319*
5:1, *277*
6:15, *293*

EPHESIANS
1, *193*
1:3-10, *109, 111*
1:4, *40, 43, 174*
1:5, *293*
1:9-10, *128*
1:10, *193*
1:11, *193, 199, 207,*
  222
1:23, *205*
2:1, *291*
2:4, *109*

3:9, 42
3:11, 222
3:20-21, 207
4:18, 291
4:22-23, 279
5:1-2, 115
5:2, 109

PHILIPPIANS
1:2, 50
2:1-11, 115
2:7, 129
2:11, 44
2:12-13, 193
3:11, 293
3:20, 44

COLOSSIANS
1, 52
1:14, 52
1:15, 43, 44, 52
1:15-16, 52
1:15-17, 45, 52, 113, 114
1:15-20, 51, 52, 110

1:16, 16, 42, 44, 47, 53, 54, 112
1:17, 47
1:17-20, 53
1:18-20, 52
2:2-3, 44
3:9-10, 279
3:10, 43, 279

1 TIMOTHY
1:13, 291
1:15, 302
4:3-5, 42
4:4, 43

2 TIMOTHY
1:10, 44, 302
2:10, 293

TITUS
3:3, 291
3:4, 109
3:5, 293

HEBREWS
1:1-3, 110
1:1-4, 45, 53, 55, 113, 114
1:2, 16, 44, 54
1:2-3, 54
1:3, 42, 44, 47, 52, 54
1:10, 16, 44, 45, 54, 113, 114
2:10, 41, 42, 129
6:4, 293
7:22, 53
10:29, 293
11:3, 41, 54, 134

JAMES
1:16-17, 42
3:9, 43

1 PETER
1:3, 278
1:11, 44
1:20, 40, 174
4:19, 14, 41

2 PETER
2:10, 291
3:8, 173

1 JOHN
1:1, 44
1:8, 289
2:11, 291
3:1, 43, 109
3:16, 43
4:4, 44
4:9, 288
4:9-11, 109
4:9-12, 43
4:16, 109
5:20, 44

REVELATION
4:11, 42
10, 41
10:6, 41
19:13, 44